T0178378

Universitext

Universitext

Universitext is a series of textbooks that presents material from a wide variety of mathematical disciplines at master's level and beyond. The books, often well class-tested by their author, may have an informal, personal even experimental approach to their subject matter. Some of the most successful and established books in the series have evolved through several editions, always following the evolution of teaching curricula, to very polished texts.

Thus as research topics trickle down into graduate-level teaching, first textbooks written for new, cutting-edge courses may make their way into *Universitext*.

More information about this series at http://www.springer.com/series/223

Vilmos Komornik

Lectures on Functional Analysis and the Lebesgue Integral

 Springer

Vilmos Komornik
University of Strasbourg
Strasbourg, France

Translation from the French language edition:
Précis d'analyse réelle - Analyse
fonctionnelle, intégrale de Lebesgue, espaces
fonctionnels, vol - 2 by Vilmos Komornik
Copyright © 2002 Edition Marketing S.A.
www.editions-ellipses.fr/
All Rights Reserved

ISSN 0172-5939   ISSN 2191-6675   (electronic)
Universitext
ISBN 978-1-4471-6810-2   ISBN 978-1-4471-6811-9   (eBook)
DOI 10.1007/978-1-4471-6811-9

Library of Congress Control Number: 2016941752

Mathematics Subject Classification: 46-01, 46E10, 46E15, 46E20, 28-01, 28A05, 28A20, 28A25, 28A35, 41A10, 41A36

Printed on acid-free paper

This Springer imprint is published by Springer Nature
The registered company is Springer-Verlag London Ltd.

Preface

This book is based on lectures given by the author at the University of Strasbourg.

Functional analysis is presented first, in a nontraditional way: we try to generalize some elementary theorems of plane geometry to spaces of arbitrary dimension. This approach leads us to the basic notions and theorems in a natural way. The results are illustrated in the small ℓ^p spaces.

The *Lebesgue integral* is treated next by following F. Riesz. Starting with two innocent-looking lemmas on step functions, the whole theory is developed in a surprisingly short and clear manner. His constructive definition of measurable functions quickly leads to optimal versions of the classical theorems of Fubini–Tonelli and Radon–Nikodým.

These two parts are essentially independent of each other, and only basic topological results are used. In the last part, they are combined to study various *function spaces* of continuous and integrable functions.

We indicate the original sources of most notions and results. Some other novelties are mentioned on page 375. The material marked by the symbol $*$ may be skipped during the first reading.

Each chapter ends with a list of exercises. However, the most important exercises are incorporated in the text as examples and remarks, and the reader is expected to fill in the missing details.

We list on p. xi some interesting papers of the general mathematical culture.

We have put a great deal of effort into selecting the material, formulating aesthetic and general statements, seeking short and elegant proofs, and illustrating the results with simple but pertinent examples. Our work was strongly influenced by the beautiful lectures of Á. Császár and L. Czách at the Eötvös Loránd University, Budapest, in the 1970s, and more generally by the Hungarian mathematical tradition created by Leopold Fejér, Frédéric Riesz, Paul Turán, Paul Erdős, and others.

We also thank C. Baud, B, Beeton, Á. Besenyei, T. Delzant, C. Disdier, O. Gebuhrer, V. Kharlamov, P. Loreti, C.-M. Marle, P. Martinez, P.P. Pálfy, P. Pilibossian, J. Saint Jean Paulin, Z. Sebestyén, A. Simonovits, Mrs B. Szénássy, J. Vancostenoble, and the editors of Springer for their precious help.

This book is dedicated to the memory of my father.

Strasbourg, France   Vilmos Komornik
May 23, 2016

Contents

Some Papers of General Interest

1. G.D. Birkhoff, What is the ergodic theorem? Am. Math. Mon. **49**, 222–226 (1942)
2. J.A. Clarkson, P. Erdős, Approximation by polynomials. Duke Math. J. **10**, 5–11 (1943)
3. R. Courant, Reminiscences from Hilbert's Göttingen. Math. Intell. **3**, 154–164 (1980/81)
4. J.L. Doob, What is martingale? Am. Math. Mon. **78**, 451–463 (1971)
5. L.E. Dubins, E.H. Spanier, How to cut a cake fairly. Am. Math. Mon. **68**, 1–4 (1961)
6. P. Erdős, Beweis eines Satzes von Tschebyschef. Acta Sci. Math. (Szeged) **5**, 194–198 (1930–32)
7. P. Erdős, Über die Reihe $\sum 1/p$. Mathematica, Zutphen B. **7**, 1–2 (1938)
8. L. Fejér, On some characterization of some remarkable systems of points of interpolation by means of conjugate points. Am. Math. Mon. **41**, 1–14 (1934); see in *Gesammelte Arbeiten von Leopold Fejér I-II* (Akadémiai Kiadó, Budapest, 1970), II, pp. 527–539
9. W. Feller, The problem of n liars and Markov chains. Am. Math. Mon. **58**, 606–608 (1951)
10. P.R. Halmos, The foundations of probability. Am. Math. Mon. **51**, 493–510 (1944)
11. P.R. Halmos, The legend of John von Neumann. Am. Math. Mon. **80**, 382–394 (1973)
12. P.R. Halmos, The heart of mathematics. Am. Math. Mon. **87**, 519–524 (1980)
13. R.W. Hamming, An elementary discussion of the transcendental nature of the elementary transcendental functions. Am. Math. Mon. **77**, 294–297 (1970)
14. G.H. Hardy, An introduction to the theory of numbers. Bull. Am. Math. Soc. **35**, 778–818 (1929)
15. G.H. Hardy, The Indian mathematician Ramanujan. Am. Math. Mon. **44**, 137–155 (1937)
16. D. Hilbert, Mathematische probleme. Göttinger Nachrichten, 253–297 (1900), and Arch. Math. Phys. **1**(3), 44–63, 213–237 (1901). English translation: Mathematical problems. Bull. Am. Math. Soc. **8**, 437–479 (1902)
17. H. Hochstadt, Eduard Helly, father of the Hahn–Banach theorem. Math. Intell. **2**(3), 123–125 (1979)
18. J. Horváth, An introduction to distributions. Am. Math. Mon. **77**, 227–240 (1970)
19. D.K. Kazarinoff, A simple derivation of the Leibnitz–Gregory series for $\pi/4$. Am. Math. Mon. **62**, 726–727 (1955)
20. K.M. Kendig, Algebra, geometry, and algebraic geometry: some interconnections. Am. Math. Mon. **90**(3), 161–174 (1983)
21. J. Milnor, Analytic proofs of the "hairy ball theorem" and the Brouwer fixed-point theorem. Am. Math. Mon. **85**, 521–524 (1978)
22. J. von Neumann, Zur Theorie der Gesellschaftsspiele. Math. Ann. **100**, 295–320 (1928); [25] VI, 1–26. English translation: On the theory of game of strategy, in *Contributions to the Theory of Games*, vol. IV (AM-40), ed. by A.W. Tucker, R.D. Luce (Princeton University Press, Princeton, 1959), pp. 13–42.

23. J. von Neumann, The mathematician, in *The Works of the Mind*, ed. by R.B. Heywood (University of Chicago Press, Chicago, 1947), pp. 180–196; [25] I, 1–9

24. J. von Neumann, The role of mathematics in the sciences and in society, in *Address at the 4th Conf. of Assoc. of Princeton Graduate Alumni* (1954); [25] VI, 477–490

25. J. von Neumann, *Collected Works I-VI* (Pergamon Press, Oxford, 1972–1979)

26. D.J. Newman, Simple analytic proof of the prime number theorem. Am. Math. Mon. **87**, 693–696 (1980)

27. B. Riemann, *Ueber die Anzahl der Primzahlen unter einer gegebenen Grösse*, Monatsberichte der Berliner Akademie (1859); in *Gesammelte mathematische Werke* (Teubner, Leipzig, 1876), pp. 135–144; H.M. Edwards, English translation: On the number of primes less than a given magnitude, in *Riemann's Zeta Function* (Academic, New York, 1974), pp. 299–305

28. F. Riesz, Sur les valeurs moyennes des fonctions. J. Lond. Math. Soc. **5**, 120–121 (1930); [31] I, 230–231

29. F. Riesz, L'évolution de la notion d'intégrale depuis Lebesgue. Ann. Inst. Fourier **1**, 29–42 (1949); [31] I, 327–340

30. F. Riesz, Les ensembles de mesure nulle et leur rôle dans l'analyse. Az I. Magyar Mat. Kongr. Közl., *Proceedings of the First Hungarian Mathematical Congress*, pp. 214–224 (1952); [31] I, 363–372

31. F. Riesz, *Oeuvres Complètes, I-II* (Akadémiai Kiadó, Budapest, 1960)

32. C.A. Rogers, A less strange version of Milnor's proof of Brouwer's fixed-point theorem. Am. Math. Mon. **87**, 525–527 (1980)

33. S. Russ, Bolzano's analytic programme. Math. Intell. **14**(3), 45–53 (1992)

34. A. Seidenberg, A simple proof of a theorem of Erdös and Szekeres. J. Lon. Math. Soc. **34**, 352 (1959)

35. S. Smale, What is global analysis? Am. Math. Mon. **76**, 4–9 (1969)

36. K. Stromberg, The Banach–Tarski paradox. Am. Math. Mon. **86**, 151–161 (1979)

37. G. Szegö, Über eine Eigenschaft der Exponentialreihe. Sitzungsber. Berl. Math. Ges. **23**, 50–64 (1924); see in *The Collected Papers of Gábor Szegö I-III* (Birkhäuser, Basel, 1982)

38. F. Tréves, Applications of distributions to PDE theory. Am. Math. Mon. **77**, 241–248 (1970)

39. E.M. Wright, A prime-representing function. Am. Math. Mon. **58**, 616–618 (1951)

40. F.B. Wright, The recurrence theorem. Am. Math. Mon. **68**, 247–248 (1961)

41. D. Zagier, A one-sentence proof that every prime $p \equiv 1 \pmod 4$ is a sum of two squares. Am. Math. Mon. **97**, 144 (1990)

42. L. Zalcman, Real proofs of complex theorems (and vice versa). Am. Math. Mon. **81**, 115–137 (1974)

Topological Prerequisites

We briefly recall some basic notions and results that we will use in this book. The proofs may be found in most textbooks on topology, e.g., in Kelley 1965.

Topological Spaces

By a *topological space* we mean a nonempty set X endowed with a *topology* on X, i.e., a family $\mathcal{T}$ of subsets of X that contains $\varnothing$ and X and is stable under finite intersections and arbitrary unions. For example, the *discrete topology* contains all subsets of X, while the *anti-discrete topology* contains only $\varnothing$ and X.

The elements of the topology are called the *open sets* and their complements the *closed sets* of the topological space.

Given a set A in a topological space X, there exists a largest open set contained in A and a largest open set contained in $X \setminus A$. They are called the *interior* and *exterior* of A and denoted by $\operatorname{int} A$ and $\operatorname{ext} A$. The remaining set $X \setminus (\operatorname{int} A \cup \operatorname{ext} A)$ is called the *boundary* of A and denoted by ∂A. The three sets $\operatorname{int} A$, $\operatorname{ext} A$, and ∂A form a *partition* of X: they are pairwise disjoint, and their union is equal to X.

If $a \in \operatorname{int} A$, then we also say that A is a *neighborhood* of a.

The sets ∂A and $\overline{A} := \operatorname{int} A \cup \partial A = X \setminus \operatorname{ext} A$ are closed; the latter is the smallest closed set containing A and is called the *closure* of A. A set $D \subset A$ is said to be *dense* in A if $A \subset \overline{D}$. A topological space X is called *separable* if it contains a countable dense set.

A set K in a topological space X is called *compact* if every open cover of A has a finite subcover. For example, the finite subsets are compact.

Theorem 1 (Cantor's Intersection Theorem) *If (K_n) is a decreasing sequence of nonempty compact sets, then $\cap K_n$ is nonempty.*

Let X and Y be two topological spaces. We say that a function $f : X \to Y$ is *continuous* at $a \in X$ if for every neighborhood V of $f(a)$ in Y there exists a neighborhood U of a in X such that $f(U) \subset V$. Furthermore, we say that f is *continuous* if it is continuous at each point $a \in X$.

Theorem 2 (Hausdorff) *Let X and Y be two topological spaces and $f : X \to Y$.*

(a) *f is continuous $\iff$ the preimage $f^{-1}(V)$ of every open set $V \subset Y$ is open in X, or equivalently, if the preimage $f^{-1}(F)$ of every closed set $F \subset Y$ is closed in X.*
(b) *If $K \subset X$ is compact and f is continuous, then $f(K) \subset Y$ is compact, i.e., the continuous image of a compact set is compact.*

The last result implies another important theorem:

Theorem 3 (Weierstrass) *Let X be a compact topological space and $f : X \to \mathbb{R}$ a continuous function. Then f is bounded; moreover, it has maximal and minimal values.*

If Z is a nonempty subset of a topological space X, then there exists a smallest topology on Z such that the *embedding*[1] of Z into X is continuous. This is called the *subspace topology* of Z. A nonempty set in a topological space X is compact $\iff$ the corresponding subspace topology is compact. A closed subspace of a compact space is also compact.

A topological space X is called *separated* or a *Hausdorff space* if any two distinct points of X belong to two disjoint open sets. Hausdorff spaces have many open and closed sets; in particular, the compact sets of Hausdorff spaces are always closed.

A topological space X is called *connected* if $\varnothing$ and X are the only sets that are simultaneously open and closed. A nonempty subset of a topological space X is called connected if it is connected as a subspace. The empty set is also considered to be connected.

Theorem 4

(a) *The closure of a connected set is also connected.*
(b) *If a family of connected sets C_i has a nonempty intersection, then $\cup C_i$ is also connected.*
(c) *(Bolzano) The continuous image of a connected set is connected.*

If X is the direct product of an arbitrary nonempty family of topological spaces X_i, then there exists a smallest topology on X such that all *projections* $X \to X_i$ are continuous. This is called the *(Tychonoff)* product of the spaces X_i.

Theorem 5

(a) *(Tychonoff) The product of compact spaces is compact.*
(b) *The product of connected spaces is connected.*
(c) *The product of separated spaces is separated.*

---

[1] The embedding of Z into X is the function $Z \ni z \mapsto z \in X$.

Many topological properties may be conveniently characterized by a generalization of convergent sequences. By a *net* in a set X we mean a function $x : I \to X$ where I is endowed with a partial ordering $\geq$, i.e., a reflexive and transitive binary relation having the following extra property: for any $i, j \in I$ there exists a $k \in I$ satisfying $k \geq i$ and $k \geq j$. We often write x_i instead of $x(i)$ and (x_i) instead of x.

We say that a net (x_i) *converges* to a point a in a *topological space* X if for each open set $U \subset X$ containing a, the net (x_i) *eventually belongs to* U, i.e., there exists a $j \in I$ such that $x_i \in U$ for all $i \geq j$. Then we write $x_i \to a$ or $\lim x_i = a$, and a is called a *limit* of (x_i).

Proposition 6 *Let X and Y be topological spaces and $a \in A \subset X$.*

(a) $a \in \overline{A} \iff$ *there exists a net in A converging to a.*
(b) *A is closed $\iff$ no net in A converges to any point of $X \setminus A$.*
(c) *A function $f : X \to Y$ is continuous at $a \iff \lim f(x_i) = f(a)$ in Y for every converging net $\lim x_i = a$ in X.*
(d) *X is a Hausdorff space $\iff$ no net has more than one limit.*

In order to characterize compactness, we introduce accumulation points and subnets. By a *subnet* of a net $x : I \to X$, we mean a net $x \circ f : J \to X$ where $f : J \to I$ is a function having the following property: for every $i \in I$ there exists a $j \in J$ such that $k \geq j \implies f(k) \geq i$.

We say that a is an *accumulation point* of a net (x_i) in a topological space X if for each open set $U \subset X$ containing a, the net (x_i) *often belongs to* U, i.e., for every $i \in I$ there exists a $j \geq i$ such that $x_j \in U$.

Proposition 7 *Let X be a topological space and let $a \in A \subset X$.*

(a) *a is an accumulation point of a net $(x_i) \iff$ there exists a subnet converging to x.*
(b) *A is compact $\iff$ each net in A has at least one accumulation point in A.*
(c) *Equivalently, A is compact $\iff$ each net in A has a subnet converging to some point of A.*

Metric Spaces

By a *metric* on a nonempty set X, we mean a *nonnegative* and *symmetric* function $d : X \times X \to \mathbb{R}$ satisfying the relation $d(x, y) = 0 \iff x = y$, and the *triangle inequality*

$$d(x, y) \leq d(x, z) + d(z, y)$$

for all $x, y, z \in X$.

By a *metric space* we mean a nonempty set X endowed with a *metric*.

For example, the usual distance $d(x,y) := |x-y|$ between real numbers is a metric on $\mathbb{R}$, and the Euclidean distance between the points of $\mathbb{R}^n$ is a metric on $\mathbb{R}^n$. The *discrete metric* on an arbitrary nonempty set X is defined by $d(x,x) = 0$ for all $x \in X$, and $d(x,y) = 1$ whenever $x \neq y$.

Every metric space has a natural topology as follows. By a *ball* of radius $r > 0$ centered at $a \in X$, we mean the set $B_r(a) := \{x \in X : d(x,a) < r\}$. A set $U \subset X$ is called open if for each $a \in U$ there exists an $r > 0$ such that $B_r(a) \subset U$. Then the balls are open. In this way every metric space is a Hausdorff space.

We define the *diameter* of a set A in a metric space by the formula $\operatorname{diam} A := \sup\{d(x,y) : x,y \in A\}$. A set A is called *bounded* if $\operatorname{diam} A < \infty$.

If K is a nonempty set and X is a metric space, then the *bounded* functions $f : K \to X$ form a metric space $\mathcal{B}(K,X)$ with respect to the metric

$$d_\infty(f,g) := \sup_{t \in K} d(f(t),g(t)).$$

The boundedness of f means that its range (or image) is a bounded set in X.

In metric spaces the convergence $x_i \to a$ is equivalent to $d(x_i,a) \to 0$. The nets and subnets may be replaced by sequences (nets defined on $I = \mathbb{N}$) and subsequences (subnets $x \circ f$ with an increasing function $f : \mathbb{N} \to \mathbb{N}$):

Proposition 8 *Let X and Y be metric spaces and $a \in A \subset X$.*

(a) *$a \in \overline{A} \iff$ there exists a sequence in A converging to a.*
(b) *A is closed $\iff$ no sequence in A converges to any point of $X \setminus A$.*
(c) *A function $f : X \to Y$ is continuous at $a \iff \lim f(x_i) = f(a)$ in Y for every converging sequence $\lim x_i = a$ in X.*
(d) *a is an accumulation point of a sequence $\iff$ there exists a subsequence converging to x.*
(e) *A is compact $\iff$ each sequence in A has at least one accumulation point in A.*
(f) *Equivalently, A is compact $\iff$ each sequence in A has a subsequence converging to some point of A.*

We will often use the following properties of compact sets:

Proposition 9 *Consider two nonempty compact sets K,L in a metric space.*

(a) *The diameter of K is attained: there exist $a,b \in K$ such that $\operatorname{diam} K = d(a,b)$.*
(b) *The distance between K and L is attained: there exist $a \in K$ and $b \in L$ such that $d(a,b) \leq d(x,y)$ for all $x \in K$ and $y \in L$.*

An important property of compact metric spaces is the following:

Theorem 10 (Heine) *Let $(X,d), (X',d')$ be two metric spaces and $f : X \to X'$ a continuous function. If X is compact, then f is uniformly continuous, i.e., for each $\varepsilon > 0$ there exists a $\delta > 0$ such that*

$$x,y \in X \quad and \quad d(x,y) < \delta \implies d'(f(x),f(y)) < \varepsilon.$$

Next we study the metric spaces for which the Cauchy criterion may be generalized. A sequence in a metric space is called a *Cauchy sequence* if $\operatorname{diam}\{x_k : k \geq n\} \to 0$ as $n \to \infty$. Every convergent sequence is a Cauchy sequence. A metric space is called *complete* if, conversely, every Cauchy sequence is convergent.

For example, the discrete metric spaces are complete, and the spaces $\mathbb{R}^n$ are complete with respect to the Euclidean metrics. If X is a complete metric space, then the metric spaces $\mathcal{B}(K, X)$ are complete.

Cantor's intersection theorem has a useful variant:

Theorem 11 (Cantor's Intersection Theorem) *Let (F_n) be a decreasing sequence of nonempty closed sets in a complete metric space. If* $\operatorname{diam} F_n \to 0$, *then* $\cap F_n$ *is nonempty.*

Next we consider a strengthening of uniform continuity. Let (X, d) and (X', d') be two metric spaces. A function $f : X \to X'$ is *Lipschitz continuous* if there exists a constant L such that $d'(f(x), f(y)) \leq L d(x, y)$ for all $x, y \in X$. If, moreover, $L < 1$, then f is called a *contraction*.

Theorem 12 (Banach–Cacciopoli) *In a complete metric space X, every contraction $f : X \to X$ has a unique fixed point, i.e., a point $a \in X$ satisfying $f(a) = a$.*

The following extension theorem is often applied in classical analysis, for example, to define integrals of continuous functions.

Theorem 13 *Let X, X' be two metric spaces, $A \subset X$ and $f : A \to X'$ a uniformly continuous function. If X' is complete, then f may be extended in a unique way to a uniformly continuous function $F : \overline{A} \to X'$.*

If, moreover, f is Lipschitz continuous, then F is Lipschitz continuous with the same constant L.

Every metric space may be completed. More precisely:

Theorem 14 *For every metric space X, there exists a complete metric space X' and an isometry $f : X \to X'$ such that $f(X)$ is dense in X'.*

The *isometry* means that f preserves the distances. This *completion* is essentially unique.

A nonempty subset of a metric space may be considered as a *metric subspace* with respect to the restriction of the metric to this set. A set in a metric space is called *complete* if it is empty or if the corresponding metric subspace is complete. A complete set is always closed, and a closed subspace of a complete metric space is also complete.

For example, if K is a topological space and X is a metric space, then the continuous functions in $\mathcal{B}(K, X)$ form a closed subspace $C_b(K, X)$. If X is complete, then $C_b(K, X)$ is also complete.

We end this section with another characterization of compactness.

A set A in a metric space is called *totally (or completely) bounded* if for each fixed $\varepsilon > 0$ it has a finite cover by sets of diameter $< \varepsilon$ or, equivalently, if for each fixed $r > 0$ it has a finite cover by balls of radius r.

Theorem 15

(a) *A set A in a metric space is compact $\Longleftrightarrow$ it is complete and totally bounded.*
(b) *A set A in a complete metric space is compact $\Longleftrightarrow$ it is closed and totally bounded.*

Normed Spaces

By a *seminorm* on a vector space X, we mean a *nonnegative, positively homogeneous* function $p : X \to \mathbb{R}$ satisfying $p(0) = 0$ and the *triangle inequality $p(x + y) \leq p(x) + p(y)$* for all $x, y \in X$. If we have also $p(x) > 0$ for all $x \neq 0$, then p is called a *norm*, and we often write $\|x\|$ instead of $p(x)$. A *normed space* is a vector space X endowed with a norm.

Every normed space is also a metric (and hence a topological) space with respect to the metric $d(x, y) := \|x - y\|$.

For example, $\mathbb{R}^n$ is a normed space with respect to each of the norms

$$\|x\|_p := (|x_1|^p + \cdots + |x_n|^p)^{1/p} \quad (1 \leq p < \infty)$$

and

$$\|x\|_\infty := \max \{|x_1|, \ldots, |x_n|\}.$$

If I is a non-degenerate compact interval in $\mathbb{R}$, then the vector space $C(I, \mathbb{R})$ of continuous functions $f : I \to \mathbb{R}$ is a normed space with respect to each of the norms

$$\|f\|_p := \left(\int_I |f|^p \right)^{1/p} \quad (1 \leq p < \infty) \quad \text{and} \quad \|f\|_\infty := \sup |f|.$$

If X is a normed space, then $\mathcal{B}(K, X)$ is a normed space for every nonempty set K, and $C_b(K, X)$ is a normed space for every topological space X.

If X, Y are normed spaces, then the continuous linear maps $A : X \to Y$ form a normed space $L(X, Y)$ with respect to the norm

$$\|L\| := \sup \{\|Ax\|_Y : x \in X, \|x\|_X \leq 1\}.$$

More generally, for each positive integer k the continuous k-linear maps $A : X^k \to Y$ form a normed space $L^k(X^k, Y)$ with respect to the norm

$$\|L\| := \sup\{\|A(x_1,\ldots,x_k)\|_Y \ : \ x_i \in X \quad \text{and} \quad \|x_i\|_X \le 1, \quad i = 1,\ldots,k\}.$$

Let X, Y be normed spaces, $U \subset X$ a nonempty open set, and k a positive integer, and consider the set $C_b^k(U,Y)$ of C^k functions $f \ : \ U \to Y$ for which f and its derivatives $f^{(j)} \ : \ U \to L^j(X^j, Y)$ are bounded for $j = 1,\ldots,k$. Then $C_b^k(U,Y)$ is a normed space with respect to the norm

$$\|f\| := \|f\|_\infty + \|f'\|_\infty + \cdots + \|f^{(k)}\|_\infty.$$

By a *scalar product* on a vector space X, we mean a nonnegative, symmetric bilinear functional $(\cdot,\cdot) \ : \ X \times X \to \mathbb{R}$ satisfying $(x,x) > 0$ whenever $x \ne 0$. By a *Euclidean space*, we mean a vector space endowed with a scalar product.

Every Euclidean space is also a normed space with respect to the norm $\|x\| := \sqrt{(x,x)}$. Moreover, this norm satisfies the *parallelogram identity*

$$\|x+y\|^2 + \|x-y\|^2 = 2\|x\|^2 + 2\|y\|^2$$

and the *Cauchy–Schwarz inequality*

$$|(x,y)| \le \|x\| \cdot \|y\|$$

for all $x, y \in X$.

The balls of normed spaces are *convex*, i.e., if $x, y \in B_r(a)$, then the whole *segment* $[x,y] := \{tx + (1-t)y \ : \ 0 \le t \le 1\}$ lies in $B_r(a)$.

The connected open sets have a simple geometric characterization in normed spaces. By a *broken line* in a vector space, we mean a finite union of segments $L := \cup_{i=1}^k [x_{i-1}, x_i]$. We say that it *connects* x_0 and x_k, and we say that it *lies* in a set U if $L \subset U$.

Proposition 16 *An open set U in a normed space X is connected $\Longleftrightarrow$ any two points $a, b \in U$ may be connected by a broken line lying in U.*

The theory of finite-dimensional normed spaces is considerably simplified by the following results:

Theorem 17 (Tychonoff)

(a) *On a finite-dimensional vector space X, all norms are equivalent, i.e., for any two norms $\|\cdot\|$ and $\|\cdot\|'$ there exist two positive constants c_1, c_2 such that*

$$c_1\|x\| \le \|x\|' \le c_2\|x\|$$

for all $x \in X$.

(b) *Consequently, if X is a finite-dimensional normed space, then*

- *X is complete.*
- *Every bounded set in X is totally bounded.*

- *A set in X is compact $\iff$ it is bounded and closed.*
- *X is separable.*
- *Every bounded sequence in X has a convergent subsequence.*

(c) *Every linear map $A : X \to Y$, where X, Y are normed spaces and X is finite-dimensional, is continuous.*

We emphasize that the Bolzano–Weierstrass theorem remains valid in every *finite dimensional* normed space.

Part I
Functional Analysis

Geometrical and physical problems led to the birth of functional analysis at the end of the nineteenth century. Following the works of Dini, Ascoli, Peano, Arzelà, Volterra, Hadamard and then the spectacular discoveries of Fredholm, Hilbert, Riesz, Fréchet and Helly, Banach laid the foundations of this new theory. It was later enriched by Hahn, von Neumann and many others. In addition to its inner beauty, it proved to be very useful in, among other areas, the calculus of variations, the theory of partial differential equations and in quantum mechanics.

Instead of following the historical development,[1] we will try to extend some well-known results of Euclidean geometry to infinite-dimensional spaces:

- if K is a non-empty convex, closed set in $\mathbb{R}^N$, then K has a closest point to each $x \in \mathbb{R}^N$;
- for every proper subspace[2] M of $\mathbb{R}^N$ there exists a point x such that $\text{dist}(x, M) = |x| - 1$;
- two non-empty disjoint convex sets of $\mathbb{R}^N$ may always be separated by an affine hyperplane;
- every bounded convex polytope is the convex hull of its vertices;
- every bounded sequence in $\mathbb{R}^N$ has a convergent subsequence.

This road will lead in a natural way to many deep theorems but also to surprising counterexamples.

The more general the space, the more counter-intuitive the phenomena that appear. We start our investigations with *Hilbert spaces*, the closest to $\mathbb{R}^N$. We follow with the wider class of *Banach spaces*. Then we shortly investigate the still more general *locally convex spaces*: they play an important role in the theory of distributions, the basic framework for the study of linear partial differential

---

[1] The last two chapters of this book are devoted mostly to the Lebesgue integral and its applications.

[2] In this book by a *subspace* without adjective we always mean a *linear* subspace. In case of metric or topological subspaces we will always write *metric subspace* or *topological subspace*.

equations. We end our tour by exhibiting some strange properties of general *topological vector spaces*.

From the immense literature we mention for further studies the classical monographs of Banach [24] and Riesz–Sz.-Nagy [394]: after many decades, they still keep their freshness and elegance. Many additional theoretical results can be found in [2, 32, 35, 40, 97, 117, 119, 254, 266, 285, 309, 321, 349, 367, 397, 403, 406, 411, 488], exciting historical aspects are given in [45, 106, 117, 144, 203, 316, 327, 367, 394, 431, 490], and many exercises are contained in [15, 117, 187, 249, 349, 367, 403, 406, 458].

Chapter 1
Hilbert Spaces

The infinite! No other question has ever moved so profoundly the spirit of man.
–D. Hilbert

Stimulated by Fredholm's discovery of an unexpectedly simple and general theory of integral equations in 1900, Hilbert developed a general theory of infinite-dimensional inner product spaces between 1904 and 1906. This allowed him to solve several important problems of mathematical physics. His student Schmidt replaced his algebraic formulation by a more intuitive geometric language, making the theory accessible to a wider public.

We may define the notion of orthogonality, and many results of plane geometry, such as Pythagoras' theorem, remain valid. Hilbert spaces appear today in almost all branches of mathematics and theoretical physics: since the fundamental works of von Neumann,[1] they have formed the mathematical framework of quantum mechanics.

We give here an introduction to this theory.

1.1 Definitions and Examples

Let X be a real vector space. We recall some basic definitions and properties. By a *norm*[2] in X we mean a function $\|\cdot\| : X \to \mathbb{R}$ satisfying for all $x, y, z \in X$ and $\lambda \in \mathbb{R}$ the following properties:

- $\|x\| \geq 0,$

- $\|x\| = 0 \iff x = 0,$

---

[1] von Neumann [334, 337].
[2] Riesz [383]. Notation of Schmidt [416].

© Springer-Verlag London 2016
V. Komornik, *Lectures on Functional Analysis and the Lebesgue Integral*,
Universitext, DOI 10.1007/978-1-4471-6811-9_1

Fig. 1.1 Triangle inequality

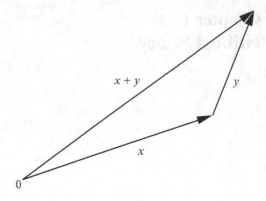

- $\|\lambda x\| = |\lambda| \cdot \|x\|$,
- $\|x + y\| \leq \|x\| + \|y\|$.

The last property is called the *triangle inequality*; see Fig. 1.1.

By a *normed space* we mean a vector space endowed with a norm. The norm is continuous with respect to the corresponding topology.

By a *scalar product* in X we mean a function $(\cdot, \cdot) : X \times X \to \mathbb{R}$ satisfying for all $x, y, z \in X$ and $\alpha, \beta \in \mathbb{R}$ the following properties:

- $(\alpha x + \beta y, z) = \alpha(x, z) + \beta(y, z)$,
- $(x, y) = (y, x)$,
- $(x, x) \geq 0$,
- $(x, x) = 0 \Longleftrightarrow x = 0$.

By a *Euclidean* or *prehilbert space* we mean a vector space endowed with a scalar product.

Every Euclidean space has a natural norm: $\|x\| := (x, x)^{1/2}$. This norm satisfies the *Cauchy–Schwarz inequality*:

$$|(x, y)| \leq \|x\| \cdot \|y\|$$

and the *parallelogram identity*:

$$\|x + y\|^2 + \|x - y\|^2 = 2\|x\|^2 + 2\|y\|^2.$$

Finally, the scalar product is continuous with respect to the corresponding topology:

$$\text{if } x_n \to x \text{ and } y_n \to y, \text{ then } (x_n, y_n) \to (x, y).$$

Definition By a *Hilbert space*[3] we mean a *complete* Euclidean space.

Examples

- We recall from topology that $\mathbb{R}^N$ is a Euclidean space with respect to the natural scalar product

$$(x, y) := x_1 y_1 + x_2 y_2 + \cdots + x_N y_N.$$

Since every finite-dimensional normed space is complete, $\mathbb{R}^N$ is a Hilbert space.
- The set ℓ^2 of sequences $x = (x_n)$ of real numbers satisfying the condition $\sum |x_n|^2 < \infty$ is a Hilbert space with respect to the scalar product

$$(x, y) := \sum_{n=1}^{\infty} x_n y_n.$$

First of all, the inequalities

$$\sum_{n=1}^{\infty} |x_n y_n| \leq \frac{1}{2} \sum_{n=1}^{\infty} |x_n|^2 + \frac{1}{2} \sum_{n=1}^{\infty} |y_n|^2 < \infty,$$

and

$$\sum_{n=1}^{\infty} |\alpha x_n + \beta y_n|^2 \leq 2|\alpha|^2 \sum_{n=1}^{\infty} |x_n|^2 + 2|\beta|^2 \sum_{n=1}^{\infty} |y_n|^2 < \infty$$

(for arbitrary $\alpha, \beta \in \mathbb{R}$) imply that ℓ^2 is a vector space, and that (x, y) is a correctly defined scalar product.

Now let $(x_n^1), (x_n^2), \ldots$ be a Cauchy sequence in ℓ^2. For every fixed $\varepsilon > 0$ there exists a k_0 such that

$$\sum_{n=1}^{\infty} |x_n^k - x_n^\ell|^2 < \varepsilon \tag{1.1}$$

for all $k, \ell \geq k_0$. In particular, (x_n^ℓ) is a Cauchy sequence for every fixed n, and therefore converges to some real number x_n.

Letting $\ell \to \infty$ we deduce from (1.1) the inequality

$$\sum_{n=1}^{N} |x_n^k - x_n|^2 \leq \varepsilon$$

---

[3] Hilbert [208], von Neumann [334], Löwig [312], and Rellich [368].

for every $k \geq k_0$ and $N \geq 1$. Letting $N \to \infty$ this yields $(x_n) \in \ell^2$ and $(x_n^k) \to (x_n)$ in ℓ^2.

Many metric and topological properties of finite-dimensional normed spaces remain valid in all Hilbert spaces. But we have to be careful: *there are important exceptions*. Before giving some examples, we recall some compactness results in finite-dimensional spaces.

We recall from topology that a subset K of a normed (or metric) space is *compact* if every sequence $(x_k) \subset K$ has a subsequence, converging to some element of K. For example, every finite set is compact.

Theorem 1.1

(a) *(Kürschák)*[4] *Every sequence of real numbers has a monotone subsequence.*
(b) *(Bolzano–Weierstrass)*[5] *Every bounded sequence of real numbers has a convergent subsequence.*

Proof

(a) An element of the sequence (x_k) is called a *peak* if it is larger than all later elements: $x_k > x_m$ for all $m > k$.

If there are infinitely many peaks, then they form a decreasing subsequence. Otherwise, there exists an index N such that no element x_k with $k \geq N$ is a peak. This allows us to define by induction a non-decreasing subsequence.
(b) There exists a bounded and monotone subsequence by (a). Its convergence follows from the axioms of real numbers. □

Corollary 1.2 *Let X be a finite-dimensional normed space.*

(a) *Every bounded sequence $(x_k) \subset X$ has a convergent subsequence.*
(b) *A subset of X is compact $\iff$ it is bounded and closed.*
(c) *The distance between two non-empty bounded and closed sets of X is always attained.*
(d) *The diameter of a non-empty bounded and closed set of X is always attained.*
(e) *Every (linear) subspace of X is closed.*[6]
(f) *X is complete.*

Sketch of Proof

(a) For $X = \mathbb{R}^N$ endowed with the usual Euclidean norm the results easily follows from the one-dimensional case by observing that convergence in norm is equivalent to component-wise convergence.

---

[4]Kürschák [275]. This elegant result and its combinatorial proof seems to be little known.
[5]Bolzano [54] and Weierstrass [482].
[6]We recall that, in this book, by a subspace without adjective we always mean a *linear* subspace.

The general case hence follows by a theorem of Tychonoff[7]: *on a finite-dimensional vector space all norms are equivalent.*

(b)–(f) easily follow from (a). □

All these properties may fail in infinite dimensions:

Examples We show that properties (a)–(e) fail in $H := \ell^2$.

(a) The vectors

$$e_k = (\overbrace{0,\dots,0}^{k-1}, 1, 0, \dots), \quad k = 1, 2, \dots$$

form a bounded sequence in ℓ^2 because $\|e_k\| = 1$ for all k.

But this sequence has no convergent subsequence. Indeed, we have $\|e_k - e_m\| = \sqrt{2}$ whenever $k \neq m$, so that no subsequence satisfies the Cauchy convergence criterion.

(b) The previous example also shows that the closed unit ball of ℓ^2, although bounded and closed, is not compact.

(c) The subset

$$F := \left\{ \left(\overbrace{0,\dots,0}^{k-1}, \frac{k+1}{k}, 0, \dots \right) : k = 1, 2, \dots \right\}$$

of ℓ^2 is non-empty, bounded and closed, but it has no element of minimal norm, i.e., its distance from 0 is not attained: we have $\mathrm{dist}(0, F) = 1$, but $\|y\| > 1$ for every $y \in F$.

(d) The subset

$$K := \left\{ x \in \ell^2 : \sum_{n=1}^{\infty} \left(1 + \frac{1}{n} \right)^2 |x_n|^2 \leq 1 \right\}$$

of ℓ^2 is non-empty, convex, bounded and closed,[8] but it has no element of *maximal* norm. Moreover, the diameter of K is not attained: we have $\mathrm{diam}\, K = 2$, but $\|x - y\| < 2$ for all $x, y \in K$.

(e) The proper subspace

$$M := \left\{ x \in \ell^2 : \sum_{n=1}^{\infty} x_n = 0 \right\}$$

of ℓ^2 is dense.

----

[7] Tychonoff [454].
[8] Observe that K is the inverse image of the closed unit ball by a continuous linear map.

For the proof we fix an arbitrary ball $B_r(x)$. We choose first a large positive integer m such that

$$\|(0,\ldots,x_{m+1},x_{m+2},\ldots)\| < r/2,$$

and then a large positive integer k such that $|x_1 + \cdots + x_m| < \sqrt{k}r/2$. Then the vector

$$y := \Big(x_1,\ldots,x_m,\overbrace{c,\ldots,c}^{k},0,0,\ldots\Big), \quad c = -\frac{x_1 + \cdots + x_m}{k}$$

belongs to M, and

$$\|x - y\| \leq \|(0,\ldots,x_{m+1},x_{m+2},\ldots)\|$$
$$+ \Big\|\Big(\overbrace{0,\ldots,0}^{m},\overbrace{c,\ldots,c}^{k},0,0,\ldots\Big)\Big\| < r.$$

Corollary 1.2 (f) may also fail in infinite dimensions:

Examples

(a) Consider the subspace X spanned by the vectors e_k of the first example above: the elements (x_n) of X have at most a finite number of non-zero components. The formula $u_k := \sum_{n=1}^{k} n^{-1}e_n$ defines a Cauchy sequence (u_k) in X because

$$\|u_k - u_m\|^2 = \sum_{n=m+1}^{k} \frac{1}{n^2} \leq \sum_{n=m+1}^{\infty} \frac{1}{n^2} \to 0$$

as $k > m \to \infty$.

But (u_k) does not converge to any point $x \in X$. Indeed, each $x = (x_n) \in X$ has a zero element $x_n = 0$. Therefore

$$\|u_k - x\|^2 \geq \frac{1}{n^2}$$

for all $k \geq n$, so that $\|u_k - x\| \not\to 0$.

(b) A more natural example is given if we take a non-degenerate compact interval I, and we endow the vector space $C(I)$ of continuous functions $x : I \to \mathbb{R}$ with the scalar product $(x, y) := \int_I xy\, dt$.

To prove that this space is *not* complete, we assume for simplicity that $I = [0, 2]$, and we consider the functions

$$x_n(t) := \text{med}\,\{0, n(t-1), 1\}, \quad 0 \leq t \leq 2, \quad n = 1, 2, \ldots,$$

Fig. 1.2 Graph of x_n

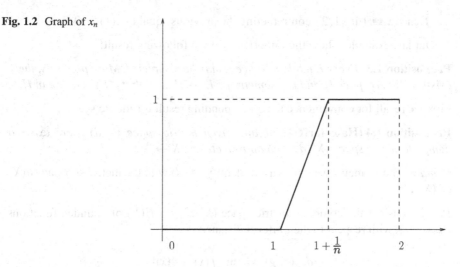

(see Fig. 1.2), where med $\{x, y, z\}$ denotes the middle number among x, y and z.
For $x \leq z$ we have

$$\text{med}\,\{x, y, z\} = \max\,\{x, \min\,\{y, z\}\}\,.$$

If $m > n \to \infty$, then

$$\|x_m - x_n\|^2 = \int_1^{(n+1)/n} |x_m(t) - x_n(t)|^2 \; dt \leq \frac{1}{n} \to 0,$$

so that (x_n) is a Cauchy sequence.

Assume on the contrary that it converges to some $x \in C(I)$. Since x is
continuous, then we deduce from the estimate

$$\int_0^1 |x(t)|^2 \; dt = \int_0^1 |x(t) - x_n(t)|^2 \; dt \leq \|x - x_n\|^2 \to 0$$

that $x \equiv 0$ in $[0, 1]$; in particular, $x(1) = 0$.

On the other hand, for arbitrary integers $n \geq N \geq 1$ we have

$$\int_{(N+1)/N}^2 |x(t) - 1|^2 \; dt = \int_{(N+1)/N}^2 |x(t) - x_n(t)|^2 \; dt \leq \|x - x_n\|^2\,.$$

Letting $n \to \infty$ and then $N \to \infty$, we get

$$\int_{(N+1)/N}^2 |x(t) - 1|^2 \; dt = 0, \quad \text{and then} \quad \int_1^2 |x(t) - 1|^2 \; dt = 0.$$

Hence $x \equiv 1$ in $[1, 2]$, contradicting the previous equality $x(1) = 0$.

Our last examples show the importance of the following result:

Proposition 1.3 *Every Euclidean space E may be completed. More precisely, there exists a Hilbert space H and an isometry $f : E \to H$ such that $f(E)$ is dense in H.*

First we recall for convenience the corresponding result for metric spaces:

Proposition 1.4 (Hausdorff)[9] *For any given metric space (X, d) there exists a complete metric space (X', d') and an isometry $h : X \to X'$.*

Remark The isometry h enables us to identify (X, d) with the metric subspace $h(X)$ of (X', d').

Proof Consider the complete metric space $(X', d') := \mathcal{B}(X)$ of bounded functions $f : X \to \mathbb{R}$ with respect to the uniform distance

$$d_\infty(f, g) := \sup_{x \in X} |f(x) - g(x)| .$$

Fix an arbitrary point $a \in X$. For each $x \in X$ the formula

$$h_x(y) := d(x, y) - d(a, y), \quad y \in X$$

defines a function $h_x \in \mathcal{B}(X)$, because

$$|h_x(y)| \le d(x, a)$$

for all $y \in X$ by the triangle inequality.
    Since

$$|h_x(z) - h_y(z)| = |d(x, z) - d(y, z)| \le d(x, y)$$

for all $z \in X$, we have

$$d'(h_x, h_y) \le d(x, y)$$

for all $x, y \in X$. In fact, this is an equality, because for $z = y$ we have

$$|h_x(y) - h_y(y)| = d(x, y).$$

$\square$

---

[9]Hausdorff [195]. The short proof given here, based on an idea of Fréchet [157, p. 161], is due to Kuratowski [273]. If the metric d is bounded, then the proof may be further shortened by simply taking $h_x(y) := d(x, y)$.

Proof of Proposition 1.3 Every Euclidean space E is a metric space with respect to the distance

$$d(x,y) := \|x - y\|_E = (x - y, x - y)^{1/2},$$

and thus it can be considered as a dense *metric* subspace of a suitable complete metric space (H, d).

For any fixed $x, y \in H$ and $c \in \mathbb{R}$ we choose two sequences (x_n) and (y_n) in E such that $d(x, x_n) \to 0$ and $d(y, y_n) \to 0$, and then we set

$$x + y := \lim(x_n + y_n),$$

$$cx := \lim cx_n,$$

$$(x, y) := \lim(x_n, y_n).$$

One may readily check that

- the limits exist;
- they do not depend on the particular choice of (x_n) and (y_n);
- H is a Euclidean and thus a Hilbert space with respect to this scalar product;
- $d(x,y) = (x - y, x - y)^{1/2}$ for all $x, y \in H$.   □

Definition We denote by $L^2(I)$ the Hilbert space obtained by the completion of $C(I)$.[10]

Remark The Lebesgue integral will provide a more concrete interpretation of $L^2(I)$.[11]

Henceforth, until the end of this chapter the letter H always denotes a Hilbert space.

1.2 Orthogonality

Definition Let $x, y \in H$ and $A, B \subset H$. We say that

- x and y are *orthogonal* if $(x, y) = 0$;
- x and A are *orthogonal* if $(x, y) = 0$ for all $y \in A$;
- A and B are *orthogonal* if $(x, y) = 0$ for all $x \in A$ and $y \in B$.

We express these relations by the symbols $x \perp y$, $x \perp A$ and $A \perp B$.

Now we solve the first problem of the introduction.

---

[10]As in the case of metric spaces, the proof shows that the completion is essentially (up to isomorphism) unique.

[11]See Proposition 9.5 (b), p. 312.

Theorem 1.5 (Orthogonal Projection)[12] *Let* $K \subset H$ *be a non-empty convex, closed set, and* $x \in H$.

(a) *There exists in* K *a unique closest point* y *to* x. *It is characterized by the following properties:*

$$y \in K, \quad and \quad (x - y, v - y) \leq 0 \quad for\ every \quad v \in K. \tag{1.2}$$

(b) *The formula* $P_K x := y$ *defines a Lipschitz continuous function* $P_K : H \to K$ *with some Lipschitz constant* $L \leq 1$.
(c) *If* K *is a subspace, then* (1.2) *is equivalent to the orthogonality property*

$$x - y \perp K, \tag{1.3}$$

and P_K *is a bounded linear map of norm* ≤ 1.

Definition The point $y = P_K(x)$ is called the *orthogonal projection* of x onto K (see Fig. 1.3).

Proof

Existence. Set $d = \mathrm{dist}(x, K)$, and consider a minimizing sequence $(y_n) \subset K$ satisfying $\|x - y_n\| \to d$. This is a Cauchy sequence. Indeed, by the

Fig. 1.3 Orthogonal projection

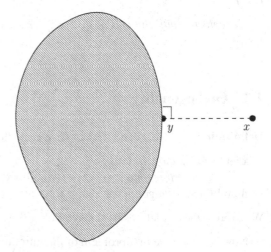

---

[12]Levi [300], Schmidt [416], Nikodým [343] (statement), [344] (proof), and Riesz [389].

parallelogram identity we have

$$\|(x - y_n) - (x - y_m)\|^2 + \|(x - y_n) + (x - y_m)\|^2$$
$$= 2\|x - y_n\|^2 + 2\|x - y_m\|^2.$$

Using the definition of d this implies

$$\|y_m - y_n\|^2 = 2\|x - y_n\|^2 + 2\|x - y_m\|^2 - 4\|x - 2^{-1}(y_m + y_n)\|^2$$
$$\leq 2\|x - y_n\|^2 + 2\|x - y_m\|^2 - 4d^2,$$

because $2^{-1}(y_m + y_n)$ belongs to the convex set K. It remains to observe that the right-hand side tends to zero as $m, n \to \infty$.

The limit y of the sequence belongs to K because K is closed, and we have $\|x - y\| = d$ by the continuity of the norm.

Characterization and uniqueness. Let $y \in K$ be at a minimal distance d from x. For any fixed $v \in K$ the vectors $(1 - t)y + tv = y + t(v - y)$ belong to the convex set K for all $0 < t < 1$, so that

$$0 \geq t^{-1}(\|x - y\|^2 - \|x - y - t(v - y)\|^2) = 2(x - y, v - y) - t\|v - y\|^2.$$

Letting $t \to 0$ this yields (1.2).

Conversely, if (1.2) holds and $v \in K$ is different from y, then

$$\|x - v\|^2 = \|x - y\|^2 + \|y - v\|^2 - 2(x - y, v - y)$$
$$\geq \|x - y\|^2 + \|y - v\|^2$$
$$> \|x - y\|^2.$$

Lipschitz property. If $x, x' \in H$, then writing $y = P_K(x)$ and $y' = P_K(x')$ we have

$$(x - y, y' - y) \leq 0 \quad \text{and} \quad (x' - y', y - y') \leq 0.$$

Summing them we get

$$(x - x' + y' - y, y' - y) \leq 0;$$

hence

$$\|y' - y\|^2 \leq (x' - x, y' - y) \leq \|x' - x\| \cdot \|y' - y\|$$

and therefore

$$\|y' - y\| \le \|x' - x\|.$$

The case when K is a subspace. Let $w \in K$. Applying (1.2) with $v = y \pm w$ we obtain

$$(x - y, \pm w) \le 0,$$

and hence $(x - y, w) = 0$.
Conversely, (1.3) implies $(x - y, v - y) = 0$ because $v - y \in K$.
The linearity of P_K follows from its uniqueness. Indeed, if $y = P_K(x)$, $y' = P_K(x')$ and $\lambda \in \mathbb{R}$, then the relations $x - y \perp K$ and $x' - y' \perp K$ imply

$$(x + x') - (y + y') \perp K \quad \text{and} \quad \lambda x - \lambda y \perp K.$$

$\square$

Example The example of the set F in the preceding section shows that the convexity assumption is necessary also for the *existence* of the orthogonal projection.

In order to state some corollaries we introduce two new notions:

Definitions

- The *orthogonal complement* of a set $D \subset H$ is defined by the formula[13]

$$D^\perp := \{x \in H : x \perp D\}.$$

- The *closed subspace spanned by* a set $D \subset H$ is by definition the intersection of all closed subspaces containing D.[14]

Observe that $D^\perp$ is a closed subspace of H, and that

$$A \subset B \Longrightarrow B^\perp \subset A^\perp, \quad (A \cup B)^\perp = A^\perp \cap B^\perp.$$

Notice also that the closed subspace spanned by D is the closure of the set of all *finite* linear combinations formed by the points of D.

---

[13]For instance, the orthogonal complement of a k-dimensional subspace in $\mathbb{R}^n$ is an $(n - k)$-dimensional subspace.
[14]This is clearly the smallest closed subspace containing D.

Part (b) of the following result solves the second problem of the introduction:

Corollary 1.6

(a) *(Riesz)*[15] *Let $M \subset H$ be a non-empty closed subspace. Every $x \in H$ has a unique decomposition $x = y + z$ with $y \in M$ and $z \in M^\perp$. Consequently, $M = M^{\perp\perp}$.*
(b) *Let $M \subset H$ be a non-empty proper closed subspace. There exists an $x \in H$ such that*

$$\mathrm{dist}(x, M) = \|x\| = 1.$$

(c) *The closed subspace spanned by $D \subset H$ is equal to $D^{\perp\perp}$. Consequently,*

- *if $D^\perp = \{0\}$, then D spans H;*
- *if $M^\perp = \{0\}$ for some* subspace $M \subset H$, *then M is dense in H.*

See Figs. 1.4 and 1.5.

Proof

(a) *Existence.* We have $y := P_M x \in M$ by definition, and $z := x - y \in M^\perp$ by (1.3).
   Uniqueness. If $x = y + z$ and $x = y' + z'$ are two decompositions with $y, y' \in M$ and $z, z' \perp M$, then

$$w := y - y' = z' - z \in M \cap M^\perp.$$

Hence $(w, w) = 0$, thus $w = 0$, and therefore $x = x'$ and $y = y'$.

Fig. 1.4 Orthogonal decomposition

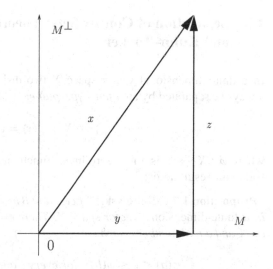

[15]Riesz [389].

Fig. 1.5 $\text{dist}(x, M) = \|x\|$

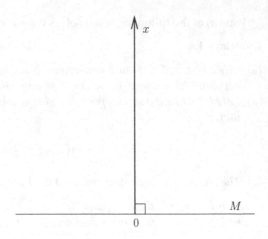

If $x \in M$, then x is orthogonal to every $z \in M^\perp$, i.e., $x \in M^{\perp\perp}$. Conversely, if $x \in M^{\perp\perp}$ and $x = y + z$ is its decomposition with $y \in M$ and $z \in M^\perp$, then $x - y = z$ belongs to $M^\perp$ but also to $M^{\perp\perp}$ because $M \subset M^{\perp\perp}$. Hence $x - y = z = 0$, and therefore $x = y \in M$.

(b) Choosing $y \in H \setminus M$ arbitrarily, $x := (y - P_M y)/\|y - P_M y\|$ has the required property.

(c) The closed subspace M spanned by D satisfies $D^\perp = M^\perp$ and thus $D^{\perp\perp} = M^{\perp\perp}$. Using (a) we conclude that $D^{\perp\perp} = M$. $\qquad\square$

1.3 Separation of Convex Sets: Theorems of Riesz–Fréchet and Kuhn–Tucker

In a finite-dimensional vector space X two disjoint non-empty convex sets may always be separated by an *affine hyperplane*, i.e., by a set of the form

$$\{x \in X \ : \ \varphi(x) = c\},$$

where $\varphi : X \to \mathbb{R}$ is a non-zero linear functional, and $c \in \mathbb{R}$. More precisely, the following result holds:

*Proposition 1.7 (Minkowski)[16] *Let A and B be two disjoint non-empty convex sets in a* finite-dimensional *vector space X. There exist a non-zero linear functional φ on X and a real number c such that*

$$\varphi(a) \leq c \leq \varphi(b) \quad \text{for every} \quad a \in A \quad \text{and} \quad b \in B. \tag{1.4}$$

---

[16]Minkowski [324, 325].

First we establish a weaker property that holds in all Hilbert spaces. We recall that we denote by X' the dual space of a *normed space* X, i.e., the space of *continuous* linear functionals on X.[17]

Theorem 1.8 (Tukey)[18] *Let A and B be two disjoint non-empty convex,* closed *sets in H. If at least one of them is* compact, *then there exist* $\varphi \in H'$ *and* $c_1, c_2 \in \mathbb{R}$ *such that*

$$\varphi(a) \leq c_1 < c_2 \leq \varphi(b) \quad \text{for all} \quad a \in A \quad \text{and} \quad b \in B. \tag{1.5}$$

(See Fig. 1.6.) In particular, for two distinct points $a, b \in H$ *there exists a* $\varphi \in H'$ *such that* $\varphi(a) \neq \varphi(b)$.

Proof The set

$$C := B - A = \{b - a \; : \; a \in A, \, b \in B\}$$

is non-empty convex, closed, and $0 \notin C$. The only nontrivial property is its closedness: we have to show that if a sequence of the form $(b_n - a_n)$ converges

Fig. 1.6 Separation of convex sets

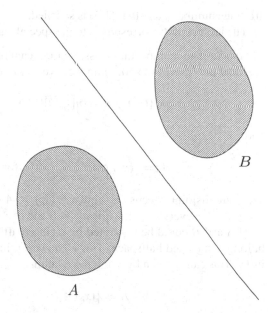

---

[17]The terminology of *bounded linear maps* and *bounded linear functionals* is frequently used instead of *continuous linear maps* and *continuous linear functionals*.
[18]Tukey [460].

to some point x in H, then $x \in C$. Assuming for example that A is compact, there exists a convergent subsequence $a_{n_k} \to a \in A$. Then we have

$$b_{n_k} = (b_{n_k} - a_{n_k}) + a_{n_k} \to x + a.$$

Since B is closed, $x + a \in B$, and therefore $x = (x + a) - a \in B - A = C$.

Let us denote by y the orthogonal projection of 0 to C; then $y \neq 0$ (because $0 \notin C$), and

$$(0 - y, b - a - y) \leq 0 \quad \text{for all} \quad a \in A \quad \text{and} \quad b \in B,$$

i.e.,

$$\|y\|^2 + (a, y) \leq (b, y) \quad \text{for all} \quad a \in A \quad \text{and} \quad b \in B.$$

The formula $\varphi(x) := (x, y)$ defines a bounded linear functional $\varphi \in H'$ by the Cauchy–Schwarz inequality. Since A and B are non-empty, we infer from the just obtained inequality that

$$c_1 := \sup_{a \in A} (a, y), \quad \text{and} \quad c_2 := \inf_{b \in B} (b, y)$$

are finite numbers, and that (1.5) is satisfied.

The last property corresponds to the special case $A := \{a\}$ and $B := \{b\}$.   □

Example The compactness assumption cannot be omitted.[19] To see this we consider in $H := \ell^2$ the non-empty convex, closed sets

$$A := \left\{ (x_n) \in \ell^2 \ : \ n|x_n^{-2/3}| \leq x_1 \quad \text{for every} \quad n \geq 2 \right\}$$

and

$$B := \left\{ (x_n) \in \ell^2 \ : \ x_n = 0 \quad \text{for every} \quad n \geq 2 \right\}.$$

They are disjoint because a sequence $(x_n) \in A \cap B$ should satisfy the inequality $x_1 \geq n^{1/3}$ for every $n \geq 2$, while $x_n \to 0$ and $n^{1/3} \to \infty$.

If A and B could be separated by a closed affine hyperplane, then $A - B$ would belong to a closed halfspace. This is, however, impossible, because $A - B$ is *dense* in ℓ^2. This can be seen by using the relation

$$A - B = \left\{ (x_n) \in \ell^2 \ : \ x_n^{-2/3} = O(1/n) \right\}.$$

---

[19]Tukey [460].

For any fixed $(z_n) \in \ell^2$ and $\varepsilon > 0$ choose a large m such that

$$\sum_{n>m} |z_n|^2 < \varepsilon^2/4 \quad \text{and} \quad \sum_{n>m} n^{-4/3} < \varepsilon^2/4.$$

Then the formula

$$x_n := \begin{cases} z_n & \text{if} \quad n \leq m, \\ n^{-2/3} & \text{if} \quad n > m \end{cases}$$

defines a sequence $(x_n) \in A - B$ for which

$$\left(\sum_{n=1}^{\infty} |x_n - z_n|^2 \right)^{1/2} \leq \left(\sum_{n>m} n^{-4/3} \right)^{1/2} + \left(\sum_{n>m} |z_n|^2 \right)^{1/2} < \varepsilon.$$

The bounded linear functional φ obtained in the proof of Theorem 1.8 is *represented* by a vector $y \in H$. Next we establish the very important fact that *every* bounded linear functional on H has this form.

If $y \in H$, then the formula

$$\varphi_y(x) := (x, y)$$

defines a bounded linear functional $\varphi_y \in H'$ for which $\|\varphi_y\| \leq \|y\|$, because

$$|\varphi_y(x)| \leq \|y\| \cdot \|x\|$$

for every $x \in H$ by the Cauchy–Schwarz inequality. Setting $j(y) := \varphi_y$ we obtain therefore a map j of H into H'. This map is linear by the bilinearity of the scalar product.

Theorem 1.9 (Riesz–Fréchet)[20] *The map j is an isometric isomorphism of H onto H'.*

It follows from the theorem that H' is also a Hilbert space; using the theorem, H' is often identified with H.

Proof We already know that $\|\varphi_y\| \leq \|y\|$ for every y. The equality $|\varphi_y(y)| = \|y\|^2$ implies the converse inequality $\|\varphi_y\| \geq \|y\|$. Hence j is an isometry; it remains to prove the surjectivity.

---

[20]Riesz [373], Fréchet [155, 156] for L^2, Riesz [389] for the general case.

The *kernel*

$$M = N(\varphi) := \{x \in H \; : \; \varphi(x) = 0\}$$

of any $\varphi \in H'$ is a closed subspace. If $M = H$, then $\varphi = \varphi_y$ with $y = 0$.

If $M \neq H$, then applying Corollary 1.6 (p. 15) we may fix a unit vector e, orthogonal to M. We have $\varphi(e)x - \varphi(x)e \in M$ for every $x \in H$ because

$$\varphi\left(\varphi(e)x - \varphi(x)e\right) = \varphi(e)\varphi(x) - \varphi(x)\varphi(e) = 0.$$

By the choice of e this implies

$$0 = (\varphi(e)x - \varphi(x)e, e) = \varphi(e)(x, e) - \varphi(x)(e, e) = (x, \varphi(e)e) - \varphi(x),$$

i.e., $\varphi = \varphi_y$ with $y = \varphi(e)e$. $\square$

Let us return to Minkowski's theorem.

Proof of Proposition 1.7 Let us endow X with a Euclidean norm. As a finite-dimensional space, X is separable, hence the metric subspaces A and B are separable, too. We may therefore fix a dense sequence (a_n) in A and a dense sequence (b_n) in B. Let us denote by A_n and B_n the convex hulls of $a_1, \ldots, a_n$ and $b_1, \ldots, b_n$, for $n = 1, 2, \ldots$.

The sets A_n, B_n are compact because they are the images of the compact[21] *simplex*

$$\{(t_1, \ldots, t_n) \in \mathbb{R}^n \; : \; t_1 \geq 0, \ldots, \, t_n \geq 0, \; t_1 + \cdots + t_n = 1\}$$

by the continuous (linear) maps $f, g : \mathbb{R}^n \to X$, defined by

$$f(t_1, \ldots, t_n) := t_1 a_1 + \cdots + t_n a_n \quad \text{and} \quad g(t_1, \ldots, t_n) := t_1 b_1 + \cdots + t_n b_n.$$

Since $A_n \subset A$ and $B_n \subset B$ are disjoint, by Theorem 1.8 there exists a non-zero functional $\varphi_n \in X'$ such that

$$\varphi_n(a) \leq \varphi_n(b) \quad \text{for all} \quad a \in A_n \quad \text{and} \quad b \in B_n. \tag{1.6}$$

Multiplying by a suitable constant we may assume that $\|\varphi_n\| = 1$.

---

[21] We recall that the finite-dimensional bounded closed sets are compact.

Since X' is finite-dimensional, there exists a convergent subsequence $\varphi_{n_k} \to \varphi$. Then we have $\|\varphi\| = 1$, so that φ is a non-zero functional. We claim that

$$\varphi(a) \le \varphi(b) \quad \text{for all} \quad a \in A \quad \text{and} \quad b \in B;$$

this will yield the proposition with

$$c := \inf \{\varphi(b) \ : \ b \in B\}.$$

Thanks to the density of the sequences (a_n), (b_n) it is sufficient to show that

$$\varphi(a_k) \le \varphi(b_m)$$

for all $k, m = 1, 2, \ldots$. For any fixed k, m, we have

$$\varphi_n(a_k) \le \varphi_n(b_m)$$

for all $n \ge \max \{k, m\}$ by (1.6). We conclude by letting $n \to \infty$.   □

Example Proposition 1.7 does not hold in infinite dimensions.[22] To show this we consider the vector space X of the polynomials and we denote by A the set of polynomials having a (strictly) positive leading coefficient. Then A and $B := \{0\}$ are disjoint non-empty convex sets in X. We claim that if (1.4) is satisfied for some linear functional φ, then $\varphi \equiv 0$.

Indeed, for any fixed polynomial x choose a positive integer $k > \deg x$, and consider the polynomial $e_k(t) := t^k$. Then $\lambda x + e_k \in A$, and thus $\lambda \varphi(x) + \varphi(e_k) \le c$ for all $\lambda \in \mathbb{R}$. Hence $\varphi(x) = 0$.

As an application of Minkowski's theorem we consider a finite number of convex functions $f_0, \ldots, f_n : K \to \mathbb{R}$ defined on a convex subset of a vector space X, and we investigate the minima of the restriction of f_0 to the convex subset

$$\Gamma := \{x \in K \ : \ f_i(x) \le 0, \quad i = 1, \ldots, n\}.$$

We are going to prove the following version of the Lagrange multiplier theorem[23]:

---

[22]Dieudonné [105].

[23]See the books on *differential calculus*.

---

*Theorem 1.10 (Kuhn–Tucker)[24]

(a) *If $f_0|_\Gamma$ has a minimum in a,[25] then there exist $\lambda_0, \ldots, \lambda_n \in \mathbb{R}$, not all zero, such that*

$$\text{the function } \lambda_0 f_0 + \cdots + \lambda_n f_n : K \to \mathbb{R} \text{ has a minimum in } a; \tag{1.7}$$

$$\lambda_0, \ldots, \lambda_n \geq 0; \tag{1.8}$$

$$\lambda_i f_i(a) = 0 \quad \text{for all} \quad i \neq 0. \tag{1.9}$$

(b) *Conversely, let $a \in \Gamma$ and $\lambda_0, \ldots, \lambda_n$ satisfy (1.8)–(1.7). If $\lambda_0 \neq 0$, then $f_0|_\Gamma$ has a minimum in a.*

(c) *If there exist $a, b \in K$ such that*

$$f_i(b) < 0 \quad \text{for all} \quad i \neq 0, \tag{1.10}$$

then (1.7)–(1.9) imply that either $\lambda_0 > 0$ or $\lambda_0 = \cdots = \lambda_n = 0$.

Since a differentiable convex function has a minimum in $a \iff$ its derivative vanishes in a, hence we deduce the following

*Corollary 1.11 *Let K be a convex open subset of a normed space; and let $f_0, \ldots, f_n : K \to \mathbb{R}$ be convex, differentiable functions. Assume that there exist $a, b \in K$ satisfying (1.10).*

Then $f_0|_\Gamma$ has a minimum at some point $a \iff$ there exist real numbers $\lambda_1, \ldots, \lambda_n \geq 0$ satisfying

$$f_0'(a) + \lambda_1 f_1'(a) + \cdots + \lambda_n f_n'(a) = 0$$

and

$$\lambda_i f_i(a) = 0 \quad \text{for all} \quad i.$$

Proof of the Theorem We denote by $x \cdot y$ the usual scalar product of $\mathbb{R}^{n+1}$ and we introduce the canonical unit vectors

$$e_0 = (1, 0, \ldots, 0), \ e_1 = (0, 1, 0, \ldots, 0), \ldots, \ e_n = (0, \ldots, 0, 1).$$

---

[24] Karush 1939, Kuhn–Tucker 1951.

[25] We recall from *differential calculus* that every local minimum of a convex function is also a global minimum.

(a) The formula

$$C := \{c \in \mathbb{R}^{n+1} \; : \; \exists x \in K : f_0(x) < f_0(a) + c_0$$

$$\text{and} \quad f_i(x) \le c_i, \; i = 1, \ldots, n\}$$

defines a non-empty convex set in $\mathbb{R}^{n+1}$ with $0 \notin C$. Applying Proposition 1.7 with $A = \{0\}$ and $B = C$, there exists a non-zero vector $\lambda = (\lambda_0, \ldots, \lambda_n) \in \mathbb{R}^{n+1}$ such that $\lambda \cdot x \ge 0$ for all $x \in C$. By the continuity of the scalar product this yields

$$\lambda \cdot c \ge 0 \quad \text{for all} \quad c \in \overline{C}. \tag{1.11}$$

Observe that

$$\{c \in \mathbb{R}^{n+1} \; : \; \exists x \in K : f_0(x) \le f_0(a) + c_0$$

$$\text{and} \quad f_i(x) \le c_i, \; \forall i \ge 1\} \subset \overline{C}. \tag{1.12}$$

Indeed, if c belongs to the first set, then $(c_0 + \delta, c_1, \ldots, c_n) \in C$ for every $\delta > 0$, and we conclude by letting $\delta \to 0$.

For each fixed i, choosing $x = a$ in (1.12) we get $e_i \in \overline{C}$, whence $\lambda_i \ge 0$ by (1.11).

For $i \ge 1$ this choice also shows that $e_i \in \overline{C}$, whence $\lambda_i f_i(a) \ge 0$ by (1.11). Since $\lambda_i \ge 0$ and $f_i(a) \le 0$ (because $a \in \Gamma$), we conclude that in fact $\lambda_i f_i(a) = 0$.

Finally we observe that

$$c := (f_0(x) - f_0(a), f_1(x), \ldots, f_n(x)) \in \overline{C}$$

for every $x \in K$ by (1.12). Applying (1.11) again, we get

$$\lambda \cdot f(x) - \lambda_0 f_0(a) = \lambda \cdot c \ge 0.$$

Since we already know that $\lambda \cdot f(a) = \lambda_0 f_0(a)$, we conclude that

$$\lambda \cdot f(x) \ge \lambda_0 f_0(a) = \lambda \cdot f(a)$$

for all $x \in K$.

(b) For any fixed $x \in \Gamma$, applying consecutively (1.8)–(1.7) and the property $f_i(x) \le 0$ $(i \ge 1)$, we obtain that

$$\lambda_0 f_0(a) = \lambda \cdot f(a) \le \lambda \cdot f(x) \le \lambda_0 f_0(x).$$

Since $\lambda_0 > 0$, this implies $f_0(a) \le f_0(x)$.

(c) If $\lambda_0 = 0$, then (1.9) and (1.7) imply

$$\sum_{i=1}^{n} \lambda_i f_i(b) = \lambda \cdot f(b) \geq \lambda \cdot f(a) = \lambda_0 f_0(a) = 0.$$

Since $\lambda_i \geq 0$ and $f_i(b) < 0$ for all $i \geq 1$ by (1.8) and (1.10), hence we conclude that $\lambda_1 = \cdots = \lambda_n = 0$. $\square$

1.4   Orthonormal Bases

Hilbert spaces provide an ideal framework for the study of Fourier series.

Definition By an *orthonormal sequence* we mean a sequence of pairwise orthogonal unit vectors.[26]

Examples

- The vectors

$$e_k = (\overbrace{0,\ldots,0}^{k-1},1,0,\ldots), \quad k = 1,2,\ldots$$

  form an orthonormal sequence in ℓ^2.
- (*Trigonometric system*) For any interval I of length 2π the functions

$$e_0 = \frac{1}{\sqrt{2\pi}}, \quad \text{and} \quad e_{2k-1} = \frac{\sin kt}{\sqrt{\pi}}, \quad e_{2k} = \frac{\cos kt}{\sqrt{\pi}}, \quad k = 1,2,\ldots$$

  form an orthonormal sequence in $L^2(I)$.
- The functions $\sqrt{2/\pi}\,\sin kt$ $(k = 1,2,\ldots)$ form an orthonormal sequence in $L^2(0,\pi)$.[27]
- The functions $1/\sqrt{\pi}$ and $\sqrt{2/\pi}\,\cos kt$ $(k = 1,2,\ldots)$ form an orthonormal sequence in $L^2(0,\pi)$.

Lemma 1.12 *If the vectors $x_1,\ldots,x_n$ are pairwise orthogonal, then*

$$\|x_1 + \cdots + x_n\|^2 = \|x_1\|^2 + \cdots + \|x_n\|^2.$$

---

[26]Gram [173] and Schmidt [416].
[27]We write $L^2(0,\pi)$ instead of $L^2([0,\pi])$ for brevity.

Proof Since $(x_j, x_k) = 0$ if $j \neq k$, we have

$$\|x_1 + \cdots + x_n\|^2 = \sum_{j=1}^{n}\sum_{k=1}^{n}(x_j, x_k) = \sum_{j=1}^{n}(x_j, x_j) = \|x_1\|^2 + \cdots + \|x_n\|^2.$$

$\square$

Proposition 1.13 *Let (e_j) be an orthonormal sequence in H.*

(a) *The orthogonal projection P_{M_n} onto $M_n := \text{Vect}\{e_1, \ldots, e_n\}$[28] is given by the explicit formula*

$$P_{M_n}x = \sum_{j=1}^{n}(x, e_j)e_j, \quad x \in H.$$

Consequently,[29]

$$\text{dist}(x, M_n) = \left\| x - \sum_{j=1}^{n}(x, e_j)e_j \right\|. \tag{1.13}$$

(b) *(Bessel's equality)*[30] *The equality*

$$\left\| x - \sum_{j=1}^{m}(x, e_j)e_j \right\|^2 = \|x\|^2 - \sum_{j=1}^{m}|(x, e_j)|^2 \tag{1.14}$$

holds for all $x \in H$ and $m = 1, 2, \ldots$. (See Fig. 1.7.)
(c) *(Bessel's inequality)*[31] *We have*

$$\sum_{j=1}^{\infty}|(x, e_j)|^2 \leq \|x\|^2 \tag{1.15}$$

for all $x \in H$. In particular, the series on the left-hand side is convergent.
(d) *If (c_j) is a sequence of real numbers, then*

$$\sum_{j=1}^{\infty}c_j e_j \quad \text{is convergent in} \quad H \iff \sum_{j=1}^{\infty}|c_j|^2 < \infty.$$

---

[28] The *linear hull* M_n is finite-dimensional, hence closed.
[29] Toepler [455].
[30] Bessel [41, 42]. Figure 1.7 shows that this is a generalization of Pythagoras' theorem.
[31] Bessel [41, 42].

Fig. 1.7 Bessel's equality for $m = 1$

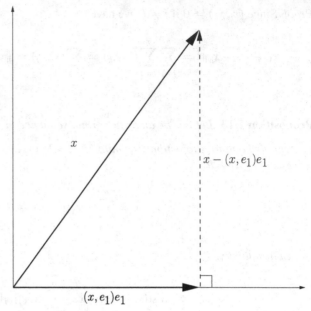

Remarks

- The case $m = 1$ of Bessel's inequality follows from the Cauchy–Schwarz inequality.
- The quantities (x, e_j) are called the *Fourier coefficients* of x.[32]

Proof

(a) It suffices to observe that the vector on the right-hand side belongs to M_n, and that the differences of the two sides is orthogonal to M_n, because it is orthogonal to each of the vectors $e_1, \ldots, e_n$ that span M_n:

$$\left(x - \sum_{j=1}^{n}(x, e_j)e_j, e_k\right) = (x, e_k) - \sum_{j=1}^{n}(x, e_j)(e_j, e_k)$$

$$= (x, e_k) - (x, e_k) = 0, \quad k = 1, \ldots, n.$$

---

[32] Clairaut [88, pp. 546–547], Euler [131], and Fourier [148].

(b) Since

$$x - P_{M_n}x = x - \sum_{j=1}^{n}(x, e_j)e_j$$

is orthogonal to M_n by the properties of the orthogonal projection, the $n + 1$ vectors on the right-hand side of the equality

$$x = \left(x - \sum_{j=1}^{n}(x, e_j)e_j\right) + \sum_{j=1}^{n}(x, e_j)e_j$$

are pairwise orthogonal. Applying the lemma, (1.14) follows.
(c) By Bessel's equality $\|x\|^2$ is an upper bound of all partial sums of this series of nonnegative terms.
(d) Since

$$\left\| \sum_{j=m+1}^{n} c_j e_j \right\|^2 = \sum_{j=m+1}^{n} |c_j|^2$$

for all $n > m$, the Cauchy criteria are the same for the two series. $\square$

Let us investigate the case of equality in Bessel's inequality:

Proposition 1.14 *Let (e_j) be an orthonormal sequence in H. The following four properties are equivalent:*

(a) *(Fourier series)[33] we have $\sum_{j=1}^{\infty}(x, e_j)e_j = x$ for all $x \in H$;*
(b) *the subspace[34] $M := \text{Vect}\{e_1, e_2, \ldots\}$ is dense in H;*
(c) *(Parseval's equality)[35] we have $\sum_{j=1}^{\infty}|(x, e_j)|^2 = \|x\|^2$ for all $x \in H$;*
(d) *if $y \in H$ and $(y, e_j) = 0$ for all j, then $y = 0$.*

Proof (a) $\Longleftrightarrow$ (b). Setting $M_m := \text{Vect}\{e_1, \ldots, e_m\}$, (a) and (b) are equivalent to the conditions

$$\left\| x - \sum_{j=1}^{m}(x, e_j)e_j \right\| \to 0 \quad \text{and} \quad \text{dist}(x, M_m) \to 0$$

for all $x \in H$. We conclude by applying the equality (1.13).

---

[33] Fourier [148].
[34] The *linear hull* M is by definition the set of all *finite* linear combinations of the vectors e_j.
[35] Parseval [352].

(a) $\Longleftrightarrow$ (c) follows from the Bessel equality because the two sides of (1.14) tend to zero at the same time.

(a) $\Longrightarrow$ (d). We have $y = \sum_{j=1}^{\infty}(y, e_j)e_j = \sum_{j=1}^{\infty} 0 = 0$.

(d) $\Longrightarrow$ (a). Set $y := x - \sum_{k=1}^{\infty}(x, e_k)e_k \in H$: the series converges by parts (c) and (d) of the proposition. Since

$$(y, e_j) = (x, e_j) - \sum_{k=1}^{\infty}(x, e_k)(e_k, e_j) = (x, e_j) - (x, e_j) = 0$$

for all j, using (d) we conclude that $y = 0$.[36] $\square$

Definition An orthonormal sequence (e_j) is *complete* if the equivalent conditions (a)–(d) are satisfied. In this case we also say that (e_j) is an *orthonormal basis*.

Examples

- The orthonormal sequence $e_1, e_2, \ldots$ of ℓ^2, given above, is complete because $(x, e_j) = x_j$ for all j for every $x = (x_j) \in \ell^2$, so that Parseval's equality follows from the *definition* of the norm.
- The three other orthonormal sequences given above are complete as well.[37] Applying Parseval's equality for the trigonometric system on the interval $I = [-\pi, \pi]$ and for the function $x(t) \equiv t$ we obtain by an easy computation a famous result of Euler[38]:

$$\sum_{k=1}^{\infty} \frac{1}{k^2} = \frac{\pi^2}{6}.$$

If (e_j) is an orthonormal basis in H, then the finite linear combinations of the vectors e_j with *rational* coefficients form a countable, dense set in H, so that H is *separable*. Conversely, we have the following

Proposition 1.15 *Every separable Hilbert space has an orthonormal basis.*

Proof Let (y_n) be a dense sequence in a Hilbert space H. Let n_k be the first index for which $y_1, \ldots, y_{n_k}$ span a k-dimensional subspace. Then the sequence $y_{n_1}, y_{n_2}, \ldots$ is *linearly independent*; furthermore,

$$y_1, \ldots, y_{n_k} \quad \text{and} \quad y_{n_1}, \ldots, y_{n_k}$$

span the same subspace M_k for each k.

---

[36]The completeness of H was used only in this step, so that (a), (b) and (c) are equivalent in non-complete Euclidean spaces as well. See Exercise 1.12.

[37]See Corollary 9.6, p. 314.

[38]Euler [128] (heuristic proof), [129] (§ 167).

Writing $x_k := y_{n_k}$ for brevity, the formulas[39]

$$e_1 = \frac{x_1}{\|x_1\|} \quad \text{and} \quad e_k := \frac{x_k - P_{M_{k-1}}x_k}{\|x_k - P_{M_{k-1}}x_k\|}, \qquad k = 2, 3, \ldots$$

define a sequence of unit vectors satisfying $e_1, \ldots, e_{k-1} \in M_{k-1}$, $e_k \perp M_{k-1}$ and

$$\text{Vect}\{e_1, \ldots, e_{k-1}\} = \text{Vect}\{x_1, \ldots, x_{k-1}\}$$

for all $k \geq 2$. Hence (e_k) is an orthonormal sequence, and

$$\overline{\text{Vect}\{e_1, e_2, \ldots\}} = \overline{\text{Vect}\{x_1, x_2, \ldots\}} = \overline{\text{Vect}\{y_1, y_2, \ldots\}} = H.$$

$\square$

Remark The convergence and the sum of an orthogonal series do not depend on the order of its terms. Therefore the results of this section may be extended to arbitrary non-separable Hilbert spaces, by considering *orthonormal families* instead of orthonormal sequences.[40]

1.5 Weak Convergence: Theorem of Choice

The examples at the end of Scct. 1.1 show that the Bolzano–Weierstrass theorem fails in infinite-dimensional Hilbert spaces: bounded, closed sets are not always compact. A simple counterexample is provided by the closed balls of infinite-dimensional Hilbert spaces[41]:

Example Every orthonormal sequence (e_n) is bounded, but it does not have any convergent subsequence because $\|e_n - e_m\| > 1$ for all $n \neq m$.

However, Hilbert succeeded in generalizing the Bolzano–Weierstrass theorem for all Hilbert spaces by a suitable weakening of the notion of convergence. The idea comes from the following elementary observation:

Proposition 1.16 *Let $e_1, \ldots, e_k$ be an orthonormal basis in a* finite-dimensional *Hilbert space H. Then the following properties are equivalent:*

(a) $x_n \to x$;
(b) $(x_n, y) \to (x, y)$ *for each fixed* $y \in H$;
(c) $(x_n, e_j) \to (x, e_j)$ *for* $j = 1, \ldots, k$.

---

[39]Gram–Schmidt orthogonalization [173], [415].
[40]See, e.g., Halmos [185].
[41]It suffices to consider unit balls by a similarity argument.

Proof The equivalence (a) $\Longleftrightarrow$ (c) follows from the identity

$$\|x_n - x\|^2 = \left\| \sum_{j=1}^{k}(x_n - x, e_j)e_j \right\|^2 = \sum_{j=1}^{k} |(x_n - x, e_j)|^2 .$$

Property (c) implies the formally stronger property (b) because we have $y = \sum_{j=1}^{k} c_j e_j$ with suitable coefficients c_j, and then

$$(x_n, y) - (x, y) = \sum_{j=1}^{k} c_j\big((x_n, e_j) - (x, e_j)\big) \to 0.$$

$\square$

Remark For the usual orthonormal basis of $H = \mathbb{R}^k$ the equivalence (a) $\Longleftrightarrow$ (c) means that the convergence of a vector sequence is equivalent to its coordinate-wise or component-wise convergence.

Definition The sequence (x_n) converges *weakly*[42] to x in H if $(x_n, y) \to (x, y)$ for each fixed $y \in H$.[43] We express this by writing $x_n \rightharpoonup x$.

Example In *infinite* dimensions every orthonormal sequence (e_n) converges weakly to zero. Indeed, the numerical series $\sum |(y, e_n)|^2$ converges for each $y \in H$ by Bessel's inequality (Proposition 1.13, p. 25), and therefore its general term tends to zero: $(y, e_n) \to 0 = (y, 0)$.

We recall that (e_n) is not norm-convergent.

Let us establish the basic properties of weak convergence:

Proposition 1.17

(a) *A sequence has at most one weak limit.*
(b) *If $x_n \rightharpoonup x$, then $x_{n_k} \rightharpoonup x$ for every (x_{n_k}) subsequence, too.*
(c) *If $x_n \rightharpoonup x$ and $y_n \rightharpoonup y$, then $x_n + y_n \rightharpoonup x + y$.*
(d) *If $x_n \rightharpoonup x$ in H and $\lambda_n \to \lambda$ in $\mathbb{R}$, then $\lambda_n x_n \rightharpoonup \lambda x$ in H.*
(e) *Let $K \subset H$ be a convex closed set and $(x_n) \subset K$. If $x_n \rightharpoonup x$, then $x \in K$.*
(f) *If $\|x_n\| \leq L$ for all n and $x_n \rightharpoonup x$, then $\|x\| \leq L$.[44]*
(g) *The following equivalence holds:*

$$x_n \to x \quad \Longleftrightarrow \quad x_n \rightharpoonup x \quad and \quad \|x_n\| \to \|x\| .$$

---

[42]Hilbert [209].
[43]We often write the last relation in the equivalent form $(x_n - x, y) \to 0$.
[44]Equivalently $\|x\| \leq \lim \inf \|x_n\|$.

Proof

(a) If $x_n \rightharpoonup x$ and $x_n \rightharpoonup y$, then $(x_n, x - y) \to (x, x - y)$ and $(x_n, x - y) \to (y, x - y)$. By the uniqueness of the limit of numerical sequences we conclude $(x, x - y) = (y, x - y)$, i.e., $(x - y, x - y) = 0$, and thus $x - y = 0$.

   (b), (c), (d) follow by definition from the corresponding properties of numerical sequences. For example, (d) may be shown in the following way: we have

$$(\lambda_n x_n, y) = \lambda_n (x_n, y) \to \lambda(x, y) = (\lambda x, y)$$

   for each $y \in H$, i.e., $\lambda_n x_n \rightharpoonup \lambda x$.

(e) Denoting by y the orthogonal projection of x onto K, we have

$$(x_n - y, x - y) \le 0$$

   for all n by Theorem 1.5 (p. 12). Since $x_n \rightharpoonup x$, taking the limit we find $(x - y, x - y) \le 0$. Hence $\|x - y\|^2 \le 0$ and therefore $x = y \in K$.

(f) We apply (e) with $K := \{z \in H \ : \ \|z\| \le L\}$.

(g) If $x_n \to x$, i.e., if $\|x_n - x\| \to 0$, then

$$|(x_n, y) - (x, y)| \le \|x_n - x\| \cdot \|y\| \to 0$$

   for each $y \in H$ by the Cauchy–Schwarz inequality, and

$$|\|x_n\| - \|x\|| \le \|x_n - x\| \to 0$$

   by the triangle inequality.

   Conversely, if $x_n \rightharpoonup x$ and $\|x_n\| \to \|x\|$, then the right-hand side of the identity

$$\|x_n - x\|^2 = \|x_n\|^2 + \|x\|^2 - 2(x_n, x)$$

   tends to zero, so that $x_n \to x$.   □

Remarks

- The convexity condition cannot be omitted in (e): every orthonormal sequence belongs to the closed unit sphere, but its weak limit, the null vector, does not.
- Norm convergence is also called *strong convergence* because it implies weak convergence by (g).

Every weakly convergent sequence is bounded. For the proof of this deeper property we recall Baire's lemma from topology[45]:

Proposition 1.18 *If a complete metric space is covered by countably many closed sets, then at least one of them has a non-empty interior.*

Proposition 1.19

(a) *Every weakly convergent sequence is bounded.*
(b) *If $x_n \to x$ and $y_n \rightharpoonup y$, then $(x_n, y_n) \to (x, y)$.*

Example Part (b) expresses a strengthened continuity property of the scalar product. If (e_n) is an orthonormal sequence, then the example $x_n = y_n := e_n$ shows that it cannot be strengthened further: the relations $x_n \rightharpoonup x$ and $y_n \rightharpoonup y$ do not imply $(x_n, y_n) \to (x, y)$ in general.

Proof

(a) If $x_n \rightharpoonup x$ in H, then the *numerical sequence* $n \mapsto (x_n, y)$ is convergent for each $y \in H$, and hence it is bounded. Consequently, the *closed* sets

$$F_k := \{y \in H : |(x_n, y)| \le k \quad \text{for all} \quad n\}, \quad k = 1, 2, \ldots$$

cover H. By Baire's lemma, one of them, say F_k, contains a ball $B_{2r}(y)$.
If $x_n \ne 0$, then

$$y + r\,\|x_n\|^{-1} x_n \in B_{2r}(y) \subset F_k,$$

and hence

$$|(x_n, y + r\,\|x_n\|^{-1} x_n)| \le k.$$

Since $y \in F_k$, this yields

$$r\,\|x_n\| = |(x_n, r\,\|x_n\|^{-1} x_n)| \le k + |(x_n, y)| \le 2k,$$

i.e., the boundedness of (x_n).

---

[45]Osgood [350], Baire [17], Kuratowski [272], Banach [23]. The usefulness of Baire's lemma in functional analysis was recognized by Saks: see Banach and Steinhaus [28]. See also the self-contained proofs of the more general Theorem 2.23 and Proposition 2.24 below (pp. 81–82), without using Baire's lemma.

(b) Since (y_n) is bounded, we have

$$|(x_n, y_n) - (x, y)| \leq |(x_n - x, y_n)| + |(x, y_n - y)|$$
$$\leq \|x_n - x\| \cdot \|y_n\| + |(x, y_n) - (x, y)|$$
$$\to 0$$

as $n \to \infty$. □

The following lemma simplifies the verification of weak convergence:

Lemma 1.20 *Let (x_n) be a bounded sequence in H and $x \in H$. The set*

$$Y := \{y \in H \ : \ (x_n, y) \to (x, y)\}$$

is a closed subspace of H.

Proof Y is a subspace by the linearity of the scalar product. For the closedness we show that if $(y_k) \subset Y$ and $y_k \to y \in H$, then $y \in Y$. Fixing $\varepsilon > 0$ arbitrarily, we have to find an integer N such that $|(x_n - x, y)| < \varepsilon$ for all $n \geq N$.

Choose a large number L such that $\|x\| < L$, and $\|x_n\| < L$ for all n, and then choose a large index k satisfying $\|y_k - y\| < \varepsilon/3L$. Since $y_k \in Y$, there exists an N such that $|(x_n - x, y_k)| < \varepsilon/3$ for all $n \geq N$.

Then the required inequality holds for all $n \geq N$ because

$$|(x_n - x, y)| \leq |(x_n - x, y - y_k)| + |(x_n - x, y_k)|$$
$$< \|x_n - x\| \cdot \|y - y_k\| + \frac{\varepsilon}{3}$$
$$\leq 2L\frac{\varepsilon}{3L} + \frac{\varepsilon}{3} = \varepsilon.$$

□

Example The sequence $x^1 = (x_k^1)$, $x^2 = (x_k^2), \ldots$ converges weakly to $x = (x_k)$ in $\ell^2 \iff$ it is bounded, and $x_k^n \to x_k$ for each k (*component-wise convergence*).

Indeed, writing $x_k^n \to x_k$ in the equivalent form $(x^n, e_k) \to (x, e_k)$, the necessity of this condition follows from the proposition. The sufficiency follows from Lemma 1.20 because (e_k) spans ℓ^2.

Now we are ready to generalize the Bolzano–Weierstrass theorem:

Theorem 1.21 (Theorem of Choice)[46] *In a Hilbert space every bounded sequence has a weakly convergent subsequence.*

---

[46]Hilbert [209], Schmidt [416], and von Neumann [336].

Proof Let (x_n) be a bounded sequence in H, and fix a constant L such that $\|x_n\| < L$ for all n. Let us denote by M the closed linear hull of (x_n). Observe that M is separable.

If M is finite-dimensional, then (x_n) has even a strongly convergent subsequence by the classical Bolzano–Weierstrass theorem. Henceforth assume that M is infinite-dimensional, and fix an orthonormal basis (e_k) of M by Proposition 1.15 (p. 28).

The numerical sequence $n \mapsto (x_n, e_1)$ is bounded. By the Bolzano–Weierstrass theorem there exist a subsequence $(x_n^1) \subset (x_n)$ and $c_1 \in \mathbb{R}$ such that $(x_n^1, e_1) \to c_1$.

Next, since the numerical sequence $n \mapsto (x_n^1, e_2)$ is also bounded, there exist a subsequence $(x_n^2) \subset (x_n^1)$ and $c_2 \in \mathbb{R}$ such that $(x_n^2, e_2) \to c_2$.

Continuing by recursion we construct an infinite sequence of subsequences

$$(x_n) \supset (x_n^1) \supset (x_n^2) \supset \cdots$$

and real numbers c_k such that

$$(x_n^k, e_k) \to c_k$$

for each fixed $k = 1, 2, \ldots$. Applying *Cantor's diagonal method*,[47] the formula $z_n := x_n^n$ defines a subsequence $(z_n) \subset (x_n)$ converging weakly to $\sum_{k=1}^{\infty} c_k e_k$.

For the proof first we notice that for each fixed k, the truncated subsequence $z_k, z_{k+1}, \ldots$ of (z_n) is also a subsequence of $(x_n^k)_{n=1}^{\infty}$, and hence $(z_n, e_k) \to c_k$.

Next we claim that the orthogonal series $\sum_{k=1}^{\infty} c_k e_k$ converges strongly to some point $z \in M$ of norm $\leq L$. For the convergence it suffices to check by Proposition 1.13 that $\sum_{k=1}^{m} |c_k|^2 \leq L^2$ for each fixed m. We have

$$\sum_{k=1}^{m} |(z_n, e_k)|^2 \leq \|z_n\|^2 < L^2$$

for all n by Bessel's inequality, and the required assertion follows by letting $n \to \infty$. Finally, the inequality $\|z\| \leq L$ follows from the continuity of the norm.

We already know that $(z_n, e_k) \to c_k = (z, e_k)$ for all k. Applying Lemma 1.20 we conclude that $(z_n, y) \to (z, y)$ for all $y \in M$, too.

We prove finally that $(z_n, y) \to (z, y)$ for all $y \in H$. Denoting by u the orthogonal projection of y onto M, we already know that $(z_n, u) \to (z, u)$. Furthermore, we have $y - u \perp M$, so that $(z_n - z, y - u) = 0$ for all n. We conclude that

$$(z_n, y) - (z, y) = (z_n - z, u) + (z_n - z, y - u) = (z_n - z, u) \to 0.$$

$\square$

---

[47]Cantor [75].

1.6 Continuous and Compact Operators

For brevity a linear map $A : H \to H$ is also called an *operator*. Its continuity may also be characterized by weak convergence:

Proposition 1.22 *For an operator $A : H \to H$ the following properties are equivalent:*

(a) *there exists a constant M such that $\|Ax\| \leq M \|x\|$ for all $x \in H$;*
(b) *A sends bounded sets into bounded sets;*
(c) *A sends totally bounded sets into totally bounded sets;*
(d) *$x_n \to x \Longrightarrow Ax_n \to Ax$;*
(e) *$x_n \rightharpoonup x \Longrightarrow Ax_n \rightharpoonup Ax$;*
(f) *$x_n \to x \Longrightarrow Ax_n \rightharpoonup Ax$.*

Remark It suffices to check (d), (e) and (f) for $x = 0$ by linearity. The same remark applies to Proposition 1.24 below.

For the proof we introduce *adjoint operators*:

Proposition 1.23 *For each operator $A \in L(H, H)$ there exists a unique operator $A^* \in L(H, H)$ such that*

$$(Ax, y) = (x, A^*y) \quad \text{for all} \quad x, y \in H. \tag{1.16}$$

Definition A^* is called the *adjoint* of A.[48]

Remark It follows from the proposition that $A^{**} = A$ for every A.

Proof For any fixed $y \in H$ the formula $\psi_y(x) := (Ax, y)$ defines a bounded linear functional $\psi_y \in H'$. Applying the Riesz–Fréchet theorem there exists a unique vector $y^* \in H$ satisfying

$$(Ax, y) = (x, y^*) \quad \text{for all} \quad x, y \in H.$$

Hence y^* is the unique possible candidate for A^*y. On the other hand, defining $A^*y := y^*$ the condition (1.16) is satisfied indeed.

For any $y_1, y_2 \in H$ and $\lambda \in \mathbb{R}$ it follows from the definitions of y_1^*, y_2^* and from the bilinearity of the scalar product that

$$(Ax, y_1 + y_2) = (x, A^*y_1 + A^*y_2) \quad \text{and} \quad (Ax, \lambda y) = (x, \lambda A^*y)$$

for all $x, y \in H$. In view of the uniqueness of the vectors $A^*(y_1 + y_2)$ and $A^*(\lambda y)$ the linearity of A^* follows.

---

[48]Lagrange [279, p. 471] and Riesz [379, 382] (in L^2 and ℓ^2).

Applying (1.16) with $x = A^*y$ we get for every $y \in H$ the estimate

$$\|A^*y\|^2 = (AA^*y, y) \leq \|AA^*y\| \cdot \|y\| \leq \|A\| \cdot \|A^*y\| \cdot \|y\|\,;$$

this shows that A^* continuous, and $\|A^*\| \leq \|A\|$.   □

Proof of Proposition 1.22 The implications (a) $\Longleftrightarrow$ (b), (a) $\Longleftrightarrow$ (c), (a) $\Longrightarrow$ (d) and (e) $\Longrightarrow$ (f) follows from the definitions.

(d) $\Longrightarrow$ (e). We have

$$(Ax_n - Ax, y) = (x_n - x, A^*y) \to 0$$

for any fixed $y \in H$ because $x_n \rightharpoonup x$.

(f) $\Longrightarrow$ (a). If (a) is not satisfied, then there exists a sequence (x_n) such that $\|x_n\| = 1/n$ and $\|Ax_n\| > n$ for every n. Then $x_n \to 0$, while (Ax_n) is unbounded and hence does not converge weakly.   □

Let us strengthen the continuity:

Proposition 1.24 *For an operator $A : H \to H$ the following properties are equivalent:*

(a) *(x_n) is bounded $\Longrightarrow$ (Ax_n) has a (strongly) convergent subsequence;*
(b) *A sends bounded sets into totally bounded sets;*
(c) *$x_n \rightharpoonup x \Longrightarrow Ax_n \to Ax$.*

For the proof we need the following result of Cantor:

Lemma 1.25 (Cantor)[49] *In a topological space a sequence x_n converges to $x \Longleftrightarrow$ every subsequence (x_n') of (x_n) has a subsequence (x_n'') converging to x.*

Proof If $x_n \to x$, then $x_n' \to x$, so that we can choose $x_n'' := x_n'$. On the other hand, if $x_n \not\to x$, then there exist a neighborhood V of x and a subsequence (x_n') of (x_n) such that $x_n' \notin V$ for all n. Then (x_n') has no subsequence converging to x.   □

Proof of Proposition 1.24 (a) $\Longrightarrow$ (b) If (b) does not hold, then there exists a bounded set B such that $A(B)$ is not totally bounded. It means that there exists an $r > 0$ such that $A(B)$ cannot be covered by finitely many balls of radius r. Using this property we may recursively construct a sequence $(x_n) \subset B$ such that $\|Ax_n - Ax_k\| \geq r$ for all $n \neq k$. Then (x_n) is a bounded sequence, but (Ax_n) has no convergent subsequence because the Cauchy criterion is not satisfied. Hence (a) does not hold either.

(b) $\Longrightarrow$ (c) In view of the lemma it is sufficient to show that every subsequence (Ax_n') of (Ax_n) has a subsequence (Ax_n'') converging to Ax.

[49]Cantor [69, p. 89]

Since the sequence (x'_n) is weakly convergent and hence bounded, by property (b) the image sequence (Ax'_n) belongs to a totally bounded set. Since the closure of a totally bounded set is compact,[50] there exists a suitable subsequence $Ax''_n \to y$. It remains to show that $y = Ax$.

Since $x_n \rightharpoonup x$ implies $x''_n \rightharpoonup x$, and since A is continuous by (b) and by Proposition 1.22, we have $Ax''_n \rightharpoonup Ax$. On the other hand, $Ax''_n \to y$ implies $Ax''_n \rightharpoonup y$, so that $y = Ax$ by the uniqueness of the weak limit.

(c) $\Longrightarrow$ (a) Every bounded sequence (x_n) has a weakly convergent subsequence $x'_n \rightharpoonup x$ by Theorem 1.21. Then we have $Ax'_n \to Ax$ by (c). $\qquad\qquad\square$

Definition An operator $A : H \to H$ is *compact* or *completely continuous*,[51] if it satisfies one of the equivalent properties of Proposition 1.24.

Examples

- If H is finite-dimensional, then every operator $A : H \to H$ is continuous, and hence compact.
- The identity map $I : H \to H$ is *not* compact if H is infinite-dimensional. Indeed, we have $e_n \rightharpoonup 0$ for every orthonormal sequence, but $Ie_n = e_n \nrightarrow 0$ in H.

We establish some basic properties of compact operators:

Proposition 1.26

(a) *Every compact operator is continuous.*
(b) *Every continuous operator of* finite rank[52] *is compact.*
(c) *If $A, B \in L(H, H)$ and A is compact, then AB and BA are compact.*
(d) *The compact operators form a* closed *subspace in $L(H, H)$.*

Proof (a), (b) and (c) follow from Propositions 1.22 and 1.24 and from the equivalence of weak and strong convergence in finite-dimensional spaces.

(d) Only the closedness is not obvious. Let A_1, A_2, ... be compact operators satisfying $A_n \to A$ in $L(H, H)$. We have to show that A is compact. If (x_k) is a bounded sequence in H, then repeating the proof of Theorem 1.21 we may construct a subsequence (z_k) such that the image sequences $(A_n z_k)$ are convergent for each fixed n. It is sufficient to show that (Az_k) is a Cauchy sequence.

Fix a constant L such that $\|x_n\| < L$ for all n. For each fixed $\varepsilon > 0$ choose n such that

$$\|A - A_n\| < \frac{\varepsilon}{3L},$$

---

[50]We recall that we are working in a Hilbert space, which is complete by definition.
[51]Hilbert [209] and Riesz [383].
[52]An operator has *finite rank* if its range $R(A)$ is finite-dimensional.

and then choose N such that

$$\|A_n z_k - A_n z_\ell\| < \frac{\varepsilon}{3} \quad \text{for all} \quad k, \ell \geq N.$$

Then

$$\|A z_k - A z_\ell\| \leq \|(A - A_n) z_k\| + \|A_n z_k - A_n z_\ell\| + \|(A_n - A) z_\ell\| < \varepsilon$$

for all $k, \ell \geq N$.   □

An important example of a compact operator is the following:

Proposition 1.27 (Hilbert–Schmidt Operators)[53] *Let (e_n) be an orthonormal basis in H. If $(a_{mn}) \subset \mathbb{R}$ satisfies*

$$\sum_{m,n=1}^{\infty} |a_{mn}|^2 < \infty,$$

then the formula

$$A\Big(\sum_{n=1}^{\infty} x_n e_n\Big) := \sum_{m=1}^{\infty} \Big(\sum_{n=1}^{\infty} a_{mn} x_n\Big) e_m$$

defines a compact operator on H.

Example Intuitively, we may view (a_{mn}) as an infinite square matrix. For example, the diagonal matrix

$$\begin{pmatrix} \lambda_1 & 0 & \dots \\ 0 & \lambda_2 & \dots \\ \vdots & \vdots & \ddots \end{pmatrix}$$

represents a Hilbert–Schmidt operator if $\sum |\lambda_n|^2 < \infty$.

In fact, the weaker condition $\lambda_n \to 0$ is already sufficient, although we do not have a Hilbert–Schmidt operator in that case.

Proof If

$$x = \sum_{n=1}^{\infty} x_n e_n \in H,$$

---

[53]Hilbert [209] and Schmidt [415].

then

$$\|Ax\|^2 = \sum_{m=1}^{\infty}\left|\sum_{n=1}^{\infty} a_{mn}x_n\right|^2 \le \left(\sum_{m,n=1}^{\infty}|a_{mn}|^2\right)\left(\sum_{n=1}^{\infty}|x_n|^2\right)$$

by the Cauchy–Schwarz inequality. Hence A is a bounded operator, and

$$\|A\| \le \left(\sum_{m,n=1}^{\infty}|a_{mn}|^2\right)^{1/2}.$$

Similarly, the formula

$$A_N\left(\sum_{n=1}^{\infty}x_n e_n\right) := \sum_{m=1}^{N}\left(\sum_{n=1}^{\infty}a_{mn}x_n\right)e_m$$

defines a bounded operator of finite rank ($\le N$), hence A_N is a compact operator in H. Since for $N \to \infty$ we have

$$\|A - A_N\| \le \left(\sum_{m>N}|a_{mn}|^2\right)^{1/2} \to 0$$

by an analogous computation, applying the proposition we conclude that A is compact.   □

1.7 Hilbert's Spectral Theorem

We know from linear algebra that every symmetric matrix is diagonalizable. We extend this to infinite-dimensional Hilbert spaces.

Definition An operator $A \in L(H, H)$ is *symmetric*[54] or *self-adjoint* if $A^* = A$, i.e., if

$$(Ax, y) = (x, Ay) \quad \text{for all} \quad x, y \in H.$$

Example A Hilbert–Schmidt operator is self-adjoint if $a_{mn} = a_{nm}$ for all m, n.

---

[54]Hilbert [208] and Schmidt [415].

The main result of this section is the following:

Theorem 1.28 (Hilbert)[55] *Let A be a compact, self-adjoint operator in a separable Hilbert space $H \neq \{0\}$. There exist an orthonormal basis (e_k) in H and a sequence $(\lambda_k) \subset \mathbb{R}$ such that*

$$Ae_k = \lambda_k e_k \quad \text{for all} \quad k.$$

Furthermore, in the infinite-dimensional case we also have

$$\lambda_k \to 0.$$

Remarks

* It follows from the property $\lambda_k \to 0$ that the non-zero eigenvalues of A have a finite multiplicity, i.e., the corresponding eigensubspaces are finite-dimensional.
* Using orthonormal *families* instead of orthonormal sequences the theorem may be extended to the non-separable case as well.[56]

The following proof is due to F. Riesz.[57] For each real λ we denote by $N(A - \lambda I)$ the *kernel* of $A - \lambda I$, i.e., the *eigensubspace* of A associated with the *eigenvalue* λ:

$$N(A - \lambda I) := \{x \in H \ : \ (A - \lambda I)x = 0\} = \{x \in H \ : \ Ax = \lambda x\}.$$

If A is continuous, then its eigensubspaces are closed. The non-zero elements of the eigensubspaces are called *eigenvectors*.

Lemma 1.29 *Let $A \in L(H, H)$ be a self-adjoint operator.*

(a) *The eigensubspaces of A are pairwise orthogonal.*
(b) *If $e_1, e_2, \ldots, e_k$ are eigenvectors of A, then*

$$H_k := \{x \in H \ : \ x \perp e_1, \ldots, x \perp e_k\}$$

is a closed invariant subspace of A, i.e.,

$$x \in H_k \quad \Longrightarrow \quad Ax \in H_k.$$

Consequently, the restriction of A to H_k is a self-adjoint operator in $L(H_k, H_k)$.

---

[55]Hilbert [208, 209], Schmidt [415], and Rellich [368].
[56]See, e.g., Halmos [185].
[57]Riesz [379].

(c) *The norm of A may be determined from the associated quadratic form:*

$$\|A\| = \sup \{|(Ax, x)| \; : \; \|x\| \leq 1\}. \tag{1.17}$$

Proof

(a) If $Ae = \lambda e$, $Af = \mu f$ and $\lambda \neq \mu$, then

$$\lambda(e, f) = (Ae, f) = (e, Af) = (e, \mu f) = \mu(e, f),$$

whence $(e, f) = 0$, i.e., $e \perp f$.

(b) If $Ae_j = \lambda_j e_j$ for $j = 1, \ldots, k$ and $x \in H_k$, then

$$(Ax, e_j) = (x, Ae_j) = (x, \lambda_j e_j) = \lambda_j(x, e_j) = 0, \quad j = 1, \ldots, k,$$

so that $Ax \in H_k$.

(c) Let us denote temporarily by N_A the right-hand side of (1.17), then

$$|(Ax, x)| \leq N_A \|x\|^2 \quad \text{for all} \quad x \in H$$

by homogeneity arguments.

The obvious estimate

$$\|x\| \leq 1 \Longrightarrow |(Ax, x)| \leq \|Ax\| \cdot \|x\| \leq \|A\| \cdot \|x\|^2 \leq \|A\|$$

shows that $N_A \leq \|A\|$. For the converse inequality first we observe that, thanks to the identity

$$(A^2 x, x) = (Ax, Ax),$$

the following estimate holds for all $\lambda > 0$:

$$4 \|Ax\|^2 = (A(\lambda x + \lambda^{-1} Ax), \lambda x + \lambda^{-1} Ax)$$
$$- (A(\lambda x - \lambda^{-1} Ax), \lambda x - \lambda^{-1} Ax)$$
$$\leq N_A \|\lambda x + \lambda^{-1} Ax\|^2 + N_A \|\lambda x - \lambda^{-1} Ax\|^2$$
$$= 2N_A (\lambda^2 \|x\|^2 + \lambda^{-2} \|Ax\|^2).$$

If $Ax \neq 0$, then $x \neq 0$, and choosing $\lambda^2 = \frac{\|Ax\|}{\|x\|}$ we get

$$4 \|Ax\|^2 \leq 4N_A \|Ax\| \cdot \|x\|;$$

hence

$$\|Ax\| \leq N_A \|x\|.$$

The last inequality also holds if $Ax = 0$, so that $\|A\| \leq N_A$.   □

Lemma 1.30 *If* $A \in L(H, H)$ *is a* compact, self-adjoint *operator and* $H \neq \{0\}$, *then* A *has an eigenvalue* λ *satisfying* $|\lambda| = \|A\|$.

Proof If $A = 0$, then $\lambda = 0$ is an eigenvalue of A. Assume henceforth that $A \neq 0$. By the lemma there exists a sequence $(x_n) \subset H$ satisfying $\|x_n\| \leq 1$ and $|(Ax_n, x_n)| \to \|A\|$. Taking a subsequence and multiplying A by a suitable constant if necessary, we may also assume that $(Ax_n, x_n) \to \|A\| = 1$, and that (here we use the compactness of A) $Ax_n \to x$ for some $x \in H$. Then we have

$$0 \leq \|Ax_n - x_n\|^2 = \|Ax_n\|^2 - 2(Ax_n, x_n) + \|x_n\|^2 \leq 2 - 2(Ax_n, x_n) \to 0,$$

whence $\lim x_n = \lim Ax_n = x$, and thus $Ax = \lim Ax_n = x$. We complete the proof by observing that

$$\|x\|^2 = (x, x) = (\lim Ax_n, \lim x_n) = \lim(Ax_n, x_n) = 1,$$

i.e., $\|x\| = 1$.   □

Proof of Theorem 1.28 First we assume that A is also one-to-one. We define recursively an orthonormal sequence $e_1, e_2, \ldots$ and $(\lambda_k) \subset \mathbb{R}$ satisfying $Ae_k = \lambda_k e_k$ for all k, and the inequalities $|\lambda_1| \geq |\lambda_2| \geq \cdots$.

By the above lemmas there exist a unit vector e_1 and $\lambda_1 \in \mathbb{R}$ with

$$Ae_1 = \lambda_1 e_1 \quad \text{and} \quad |\lambda_1| = \|A\| > 0.$$

If $e_1, \ldots, e_k$ and $\lambda_1, \ldots, \lambda_k$ are already defined for some $k \geq 1$, then we consider the restriction of A to H_k. If $H_k \neq \{0\}$, then applying the lemmas again, there exist a unit vector $e_{k+1} \in H_k$ and $\lambda_{k+1} \in \mathbb{R}$ such that

$$Ae_{k+1} = \lambda_{k+1} e_{k+1} \quad \text{and} \quad |\lambda_{k+1}| = \|A|_{H_k}\| > 0.$$

We have $|\lambda_k| \geq |\lambda_{k+1}|$ because $H_k \subset H_{k-1}$ $(H_0 := H)$.

If $\dim H = n < \infty$, then we get an orthonormal basis of H after n steps, and it satisfies the requirements of the theorem.

In case $\dim H = \infty$ it remains to prove that $\lambda_k \to 0$, and that the orthonormal sequence (e_k) is complete.

Assume on the contrary that $\lambda_k \not\to 0$. Then $\inf |\lambda_k| > 0$, and therefore $(x_k) := (\lambda_k^{-1} e_k)$ is a bounded sequence. This contradicts the compactness of A because the image sequence $(Ax_k) = (e_k)$ is orthonormal, and hence it cannot have a (strongly) convergent subsequence. This proves the relation $\lambda_k \to 0$.

For the completeness of (e_k) we show that if $x \in H$ is orthogonal to every e_k, then $x = 0$. For this we observe that $x \in H_k$ for all k, i.e.,

$$\|Ax\| = \|A|_{H_k} x\| \le |\lambda_{k+1}| \cdot \|x\|$$

for all k. Since $\lambda_k \to 0$, this yields $Ax = 0$, and hence $x = 0$ because A is one-to-one.

If A is not one-to-one, then we may apply the above proof to the restriction of A to $N(A)^\perp$.[58] Since $N(A)$ is a closed subspace of H by the continuity of A, and therefore H is the direct sum of the orthogonal closed subspaces $N(A)$ and $N(A)^\perp$, we complete the proof by completing the orthonormal basis (e_k) of $N(A)^\perp$ by an arbitrarily chosen orthonormal basis (f_m) of the kernel $N(A)$; each f_m is an eigenvector associated with the eigenvalue 0.[59]   □

*Remark Using the spectral theorem we may define *continuous* functions of compact, self-adjoint operators as follows. We define the *spectrum*[60] of A by the formula

$$\sigma(A) := \{\lambda_k\} \cup \{0\};$$

observe that it is compact. If $f \in C(\sigma(A))$, then the formula

$$f(A)\left(\sum x_k e_k\right) := \sum f(\lambda_k) x_k e_k$$

defines a bounded operator $f(A) \in L(H, H)$.

One can show that the map $f : C(\sigma(A)) \to L(H, H)$ is a linear isometry, and that $(fg)(A) = f(A)g(A)$ for all $f, g \in C(\sigma(A))$. In particular, the definition reduces to the usual one for polynomials $p(z) = a_n z^n + \cdots + a_1 z + a_0$ with real coefficients:

$$p(A) := a_n A^n + \cdots + a_1 A + a_0 I.$$

This remark shows the intimate relationship between the spectral theorem and the theory of *Banach algebras* that we cannot investigate here.[61]

Let us consider the *linear non-homogeneous* equation

$$x - Ax = y \tag{1.18}$$

---

[58] This is also an A-invariant subspace by Lemma 1.29 (b).

[59] We obtain an orthonormal basis of H satisfying the conditions of the theorem by taking $f_1, \ldots, f_m, e_1, e_2, \ldots$ if $\dim N(A) = m < \infty$ and $e_1, f_1, e_2, f_2, \ldots$ if $\dim N(A) = \infty$.

[60] Hilbert [209].

[61] See, e.g., Berberian [34], Dunford–Schwartz [117], Halmos [185], Neumark [341], Rudin [406], and Sz.-Nagy [447].

and the associated *linear homogeneous* equation

$$z - Az = 0, \tag{1.19}$$

with a given operator A in H. The following result is of great importance in the theory of partial differential equations[62]:

Proposition 1.31 (Fredholm Alternative)[63] *Let A be a compact, self-adjoint operator on a Hilbert space H.*

(a) *The solutions of (1.19) form a finite-dimensional subspace M.*
(b) *The Eq. (1.18) is solvable $\Longleftrightarrow y \perp M$.*
(c) *If $y \perp M$, then the solutions of (1.18) form a translate M_y of M.*

Remark There are thus two mutually exclusive possibilities: either (1.19) has a nontrivial solution, or (1.18) has a unique solution for every $y \in H$.

Proof Assume for simplicity that H is infinite-dimensional and separable.[64]

(a) Since $\lambda_n \to 0$ by Theorem 1.28, the eigensubspaces of A are finite-dimensional for every non-zero λ. In particular, $N(A - I)$ is finite-dimensional.
(b) Using the sequences (e_n) and (λ_n) of Theorem 1.28 and using the Fourier series

$$x = \sum_{n=1}^{\infty} x_n e_n \quad \text{and} \quad y = \sum_{n=1}^{\infty} y_n e_n,$$

(1.18) takes the following form:

$$(1 - \lambda_n)x_n = y_n, \quad n = 1, 2, \ldots. \tag{1.20}$$

If it has a solution, then $y_n = 0$ for all n with $\lambda_n = 1$. In other words, we have $y \perp M$ because M is the subspace spanned by $\{e_n : \lambda_n = 1\}$.
    Conversely, if $y \perp M$ the formula

$$x_n := \begin{cases} (1 - \lambda_n)^{-1} y_n & \text{if} \quad \lambda_n \neq 1, \\ \text{arbitrary} & \text{if} \quad \lambda_n = 1 \end{cases}$$

gives a solution of (1.20). Since $(y_n) \in \ell^2$, and since the numerical sequence $(1 - \lambda_n)^{-1}$ is bounded (because converges to 1), the relation $(x_n) \in \ell^2$ holds, too. Consequently, $x := \sum x_n e_n$ is a solution of (1.18).
(c) We have $M_y = x + M$ for any fixed solution x of (1.18). $\square$

---

[62] See, e.g., Riesz and Sz.-Nagy [394], §81.
[63] Fredholm [150, 151].
[64] The proof may be easily adapted to the general case. The finite-dimensional case is well known from linear algebra.

1.8   * The Complex Case

Most results of this chapter may be easily adapted to the complex case. Let us briefly indicate the necessary modifications. We recall that every complex vector space may also be considered as a real vector space, by allowing only multiplication by real numbers. For example, $\mathbb{C}^N$ is isomorphic to $\mathbb{R}^{2N}$ as a real vector space.

Let X and Y be complex vector spaces. We say that the map $A : X \to Y$ is *linear* if

$$A(x + y) = A(x) + A(y) \quad \text{and} \quad A(\lambda x) = \lambda A(x)$$

for all $x, y \in X$ and $\lambda \in \mathbb{C}$, and *antilinear* if

$$A(x + y) = A(x) + A(y) \quad \text{and} \quad A(\lambda x) = \overline{\lambda} A(x)$$

for all $x, y \in X$ and $\lambda \in \mathbb{C}$.

Section 1.1. By a *norm* defined on a complex vector space X we mean a real-valued function $\|\cdot\|$ satisfying for all $x, y, z \in X$ and $\lambda \in \mathbb{C}$ the same properties and in the real case[65]:

- $\|x\| \geq 0,$

- $\|x\| = 0 \quad \Longleftrightarrow \quad x = 0,$

- $\|\lambda x\| = |\lambda| \cdot \|x\|,$

- $\|x + y\| \leq \|x\| + \|y\|.$

The last property is still called the *triangle inequality*. A *normed space* is a vector space endowed with a norm. A norm induces a metric in the usual way, and the norm function is continuous with respect to the corresponding topology.

A complex-valued function $(\cdot, \cdot) : X \times X \to \mathbb{C}$ defined on a complex vector space X is called a *scalar product* if it satisfies for all $x, y, z \in X$ and $\alpha, \beta \in \mathbb{C}$ the following properties:

- $(\alpha x + \beta y, z) = \alpha(x, z) + \beta(y, z),$

- $(x, y) = \overline{(y, x)},$

- $(x, x) \geq 0,$

- $(x, x) = 0 \quad \Longleftrightarrow \quad x = 0.$

---

[65] Wiener [487].

A *Euclidean space* is a vector space endowed with a scalar product. A scalar product induces a norm in the usual way, which satisfies the Cauchy–Schwarz inequality and the parallelogram identity. The scalar product is continuous with respect to the norm topology.

A complete Euclidean space is called a *Hilbert space*.[66] For example, $\mathbb{C}^N$ is a Hilbert space with respect to the scalar product

$$(x, y) := x_1\overline{y_1} + x_2\overline{y_2} + \cdots + x_N\overline{y_N},$$

and the complex numerical sequences $x = (x_n)$ satisfying the condition $\sum |x_n|^2 < \infty$ form a Hilbert space with respect to the scalar product

$$(x, y) := \sum x_n\overline{y_n}.$$

On the other hand, the continuous, complex-valued functions defined on a non-degenerate compact interval form a *non-complete* Euclidean space with respect to the scalar product

$$(f, g) := \int_I f\overline{g}\, dx.$$

Section 1.2. Condition (1.2) of Theorem 1.5 (p. 12) has to be changed to

$$y \in K, \text{ and } \Re(x - y, v - y) \le 0 \text{ for all } v \in K$$

(the letter $\Re$ stands for the real part), and we have to write $\Re(\cdot, \cdot)$ instead of $(\cdot, \cdot)$ everywhere in the proof.

Section 1.3. We have to write $\Re\varphi(a)$ and $\Re\varphi(b)$ instead of $\varphi(a)$ and $\varphi(b)$ in formulas (1.4) and (1.5) of Proposition 1.7 and Theorem 1.8.

In the Riesz–Fréchet theorem (p. 19) the map j is *antilinear* in the complex case.

Section 1.4. Everything remains valid with one modification: we have to change (x, e_k) to $\overline{(x, e_k)}$ in the proof of Bessel's equality.

The trigonometric system takes a more elegant form: the exponential functions $(2\pi)^{-1/2}e^{ikt}$, where k runs over *all* integers, form an orthonormal basis in $L^2(I)$ for every interval I of length 2π.

Section 1.5. Everything remains valid with one modification: in the proof of Proposition 1.17 (e) (p. 31) we have to write $\Re(x_n - y, x - y) \le 0$ instead of $(x_n - y, x - y) \le 0$.

Section 1.6. No modification is needed; Proposition 1.27 of Hilbert–Schmidt (p. 38) remains valid for complex numbers a_{mn}, too.

---

[66]Hilbert [208], von Neumann [334], Löwig [312], and Rellich [368].

Section 1.7. Everything remains valid with one remark: if we also consider complex numbers a_{mn}, then the self-adjointness of the Hilbert–Schmidt operator is ensured by the condition $a_{mn} = \overline{a_{nm}}$ instead of $a_{mn} = a_{nm}$.

In the complex case the spectral theorem may be generalized beyond self-adjoint operators. Let us state the results:[67]

Definition An operator $A \in L(H, H)$ is *normal*[68] if $AA^* = A^*A$.

Examples

- Every self-adjoint operator is normal.
- Every unitary operator is normal. (An operator $A \in L(H, H)$ is *unitary* if it is invertible and $A^{-1} = A^*$, i.e., if $AA^* = A^*A = I$.)
- The operator in $\mathbb{C}^2$ given by the matrix $\begin{pmatrix} 0 & 1 \\ 0 & 0 \end{pmatrix}$ is *not* normal.

Theorem 1.32 (Spectral Theorem of Normal Operators)[69] *Let A be a compact, normal operator in a separable, complex Hilbert space H. There exist an orthonormal basis (e_k) in H and a sequence $(\lambda_k) \subset \mathbb{C}$ such that*

$$Ae_k = \lambda_k e_k \quad \text{for all} \quad k.$$

Furthermore, if H is infinite-dimensional, then

$$\lambda_k \to 0.$$

1.9 Exercises

Exercise 1.1 Prove that the sequences $x = (x_n) \subset \mathbb{R}$ satisfying $\sum |x_n| < \infty$ form a normed space with respect to the norm $\|x\| := \sum |x_n| < \infty$, and that this norm is not Euclidean.

Exercise 1.2 Let (x_n) and (y_n) be two sequences in the closed unit ball of a Euclidean space. Prove that if $(x_n, y_n) \to 1$, then $\|y_n - x_n\| \to 0$.

---

[67]A proof similar to that of Sect. 1.7 is given in Bernau and Smithies [36]. Another proof is given in Halmos [185].

[68]Frobenius [162, p. 391] in finite dimensions, Toeplitz [456].

[69]Frobenius [162, p. 391] in finite dimensions, Toeplitz [456] in the general case. Von Neumann [336] generalized the theorem for *unbounded* normal operators.

Exercise 1.3 Is ℓ^2 a Hilbert space with respect to the new scalar product

$$(x, y) = \sum_{k=1}^{\infty} \frac{x_k y_k}{k^2}?$$

Exercise 1.4 Let (x_j), (y_j) be two *biorthogonal* sequences in a Euclidean space E, satisfying $(x_i, y_j) = \delta_{ij}$.[70] Prove that both sequences are linearly independent.

Exercise 1.5 Consider the subspace $E := \mathrm{Vect}\,\{e_1, e_2, \ldots\}$ of ℓ^2 with the induced scalar product and norm. Prove that the formula

$$M := \left\{ x = (x_n) \in E \ : \ \sum \frac{x_n}{n} = 0 \right\}$$

defines a proper closed subspace of E satisfying $M^{\perp} = \{0\}$. Does this contradict Corollary 1.6 (a)?

Exercise 1.6 Consider the Euclidean space E of continuous functions $f : [-1, 1] \to \mathbb{R}$ with the scalar product $(f, g) := \int_{-1}^{1} fg \, dt$. Let M denote the subspace of functions $f \in E$ vanishing in $[0, 1]$.

(i) Prove that M is a closed subspace of E.
(ii) Determine the closed subspace $M^{\perp}$.
(iii) Do we have $E = M \oplus M^{\perp}$? Why?

Exercise 1.7 Consider the Euclidean space of continuous functions $f : [-1, 1] \to \mathbb{R}$ with the scalar product $(f, g) := \int_{-1}^{1} fg \, dt$. Determine the first three functions obtained by the Gram–Schmidt orthogonalization of the sequence of polynomials $f_n(t) = t^n$, $n = 0, 1, 2, \ldots$.[71]

Henceforth the letter H denotes a Hilbert space.

Exercise 1.8 Let $M, N \in H$ and assume that every $x \in H$ has a unique decomposition $x = u + v$ with $u \in M$ and $v \in N$. Are M and N linear subspaces of H?

Exercise 1.9 (Lax–Milgram Lemma) Let $a(\cdot, \cdot)$ be a continuous bilinear form on H, satisfying for some positive constant α the inequality

$$|a(x, x)| \geq \alpha \, \|x\|^2$$

---

[70]We use the Kronecker symbol: $\delta_{ij} = 1$ and $i = j$, and $\delta_{ij} = 0$ otherwise.
[71]Legendre polynomials.

for all $x \in H$. Prove that the *variational equality*

$$a(x, y) = \varphi(y) \quad \text{for all} \quad y \in H$$

has a unique solution $x \in H$ for each $\varphi \in H'$.[72]

Exercise 1.10 Assume that H is separable and let M be a dense subspace of H. Prove that H has an orthonormal basis formed by vectors belonging to M.

Exercise 1.11 Consider in ℓ^2 the set

$$M = \left\{ x = (x_k) \in \ell^2 \ : \ \sum_{k=1}^{\infty} x_k = 0 \right\}.$$

(i) Show that M is a dense subspace of ℓ^2.
(ii) Find a linearly independent sequence in M whose orthogonalization leads to an orthonormal basis of ℓ^2.

Exercise 1.12 Let $e_1, e_2, \ldots$ be an orthonormal sequence in H and consider the (linear) subspace E spanned by

$$f_1 := \sum_{n=1}^{\infty} \frac{e_n}{n} \quad \text{and} \quad e_2, e_3, \ldots.$$

Show that the truncated orthonormal sequence $e_2, e_3, \ldots$ satisfies property (d) of Proposition 1.14 (p. 27) in the subspace E instead of H, but not the other three. Explain.

Exercise 1.13 We recall that every Euclidean norm satisfies the parallelogram identity. The purpose of this exercise is to prove the converse.[73] We consider a norm in a vector space X satisfying the parallelogram identity, and we set

$$(x, y) = 4^{-1} \left(\|x + y\|^2 - \|x - y\|^2 \right)$$

for all $x, y \in X$. Prove the following assertions for all $x, y, z \in X$:

(i) $(x, z) + (y, z) = 2 \left(\frac{x+y}{2}, z \right)$;
(ii) $(x, z) = 2 \left(\frac{x}{2}, z \right)$;
(iii) $(x, z) + (y, z) = (x + y, z)$;
(iv) $(\alpha x, y) = \alpha(x, y)$ for all $\alpha \in \mathbb{Q}$;
(v) the maps $\alpha \mapsto \|\alpha x \pm y\|$ are continuous;

---

[72] If $a(\cdot, \cdot)$ is symmetric, then this follows from the Riesz–Fréchet theorem.
[73] Jordan and von Neumann [233]. We follow Yosida [488, p. 39].

(vi) $(\alpha x, y) = \alpha(x, y)$ for all $\alpha \in \mathbb{R}$;
(vii) (x, y) is a scalar product associated with our norm.

Exercise 1.14 Prove the following propositions:

(i) Every decreasing sequence of non-empty bounded closed convex sets in a Hilbert space has a non-empty intersection.
(ii) The hypothesis "bounded" cannot be omitted.
(iii) The hypothesis "convex" may be omitted in finite dimensions, but not in general.

Exercise 1.15 Let $P \in L(H, H)$ be a *projection*, i.e., satisfying the equality $P^2 = P$. Show that the following conditions are equivalent:

(i) P is an orthogonal projector;
(ii) P is self-adjoint: $P^* = P$;
(iii) P is normal: $PP^* = P^*P$;
(iv) $(Px, x) = \|Px\|^2$ for all $x \in H$.

Exercise 1.16 Prove that the *Hilbert cube*

$$\left\{ x = (x_n) \in \ell^2 \; : \; |x_n| \leq 1/n \quad \text{for all} \quad n \right\}$$

is compact.

Exercise 1.17 Let P be the orthogonal projection of a Hilbert space onto a closed subspace M. Show that

$$P \quad \text{is compact} \quad \Longleftrightarrow \quad \dim M < \infty.$$

Exercise 1.18 Consider in the Hilbert space ℓ^2 the following operators, where we use the notation $x = (x_1, x_2, \ldots) \in \ell^2$:

$$Ax = (0, x_1, x_2, \ldots);$$

$$Bx = \left(x_1, \frac{x_2}{2}, \frac{x_3}{3}, \ldots \right);$$

$$Cx = \left(0, x_1, \frac{x_2}{2}, \frac{x_3}{3}, \ldots \right).$$

Are they compact?

Exercise 1.19 Let (e_n) be an orthonormal basis in H, and (λ_n) a sequence of real numbers, converging to 0. Prove that the formula

$$Ax = \sum_{n=1}^{\infty} \lambda_n (x, e_n) e_n$$

defines a compact operator A in H.

Exercise 1.20 Let (e_n) be an orthonormal basis in H and $A \in L(H, H)$. Assume that

$$\sum_{n=1}^{\infty} \|Ae_n\|^2 < \infty.$$

Show that A is compact.

Exercise 1.21 Let $T \in L(H, H)$.

(i) Prove that TT^* and T^*T are self-adjoint.
(ii) Prove the following equalities:

$$\|TT^*\| = \|T^*T\| = \|T\|^2 = \|T^*\|^2.$$

(iii) Let $A \in L(H, H)$ be a self-adjoint operator. Does there exist a $T \in L(H, H)$ such that $A = T^*T$?

Exercise 1.22 We define the *spectral radius* of an operator $A \subset L(H, H)$ by the formula

$$\rho(A) := \inf_{n=1,2,\dots} \|A^n\|^{1/n}.$$

Prove the following:

(i) $|\lambda| \leq \rho(A)$ for all eigenvalues of A;
(ii) if $\dim H < \infty$ and $A^* = A$, then there exists an eigenvalue satisfying $|\lambda| = \rho(A)$;
(iii) the following equalities hold[74]:

$$\|A\| = \sqrt{\rho(A^*A)} = \sqrt{\rho(AA^*)}.$$

Exercise 1.23 Let $A \in L(H, H)$. Prove that H is the orthogonal direct sum of $\overline{R(A)}$ and $N(A^*)$.

---

[74]Glazman and Ljubic [170, p. 199].

Exercise 1.24 Let $T \in L(H,H)$ satisfy $\|T\| \le 1$. Prove that

$$N(I - T) = N(I - T^*).$$

Exercise 1.25 (Mean Ergodic Theorem)[75] Let $T \in L(H,H)$ satisfy $\|T\| \le 1$. Prove the relation

$$S_n(x) := \frac{1}{n}(x + Tx + \cdots + T^{n-1}x) \to Px, \quad n \to \infty$$

for all $x \in H$, where P denotes the orthogonal projector onto the invariant subspace $N(I - T)$ of T, by establishing the following facts:

  (i) $N(I - T)$ is a closed subspace of H;
 (ii) $N(I - T) = R(I - T)^{\perp}$;
(iii) $S_n(x) \to x$ for all $x \in N(I - T)$;
 (iv) $S_n(x) \to 0$ for all $x \in R(I - T)$;
  (v) $S_n(x) \to 0$ for all $x \in \overline{R(I - T)}$;
 (vi) conclude.

Exercise 1.26

  (i) Let $u_n \rightharpoonup 0$ in H. Construct a subsequence (u_{n_k}) satisfying

$$\left|(u_{n_k}, u_{n_j})\right| < \frac{1}{k} \quad \text{for all} \quad k > j.$$

 (ii) Show that

$$\left\| \frac{1}{p} \sum_{j=1}^{p} u_{n_j} \right\| \to 0$$

    as $p \to \infty$.
(iii) Prove that every bounded sequence $(v_n) \subset H$ has a subsequence (v_{n_k}) for which

$$\left(\frac{1}{p} \sum_{j=1}^{p} v_{n_j} \right)_{p=1}^{\infty}$$

    is strongly convergent.

---

[75]Riesz 1938. Use the preceding two exercises.

Exercise 1.27 Let (e_n) be an orthonormal sequence in a Hilbert space and (c_n) a bounded sequence of real numbers. Set

$$u_n := \frac{1}{n} \sum_{i=1}^{n} c_i e_i, \quad n = 1, 2, \ldots.$$

(i) Show that $u_n \to 0$.
(ii) Show that $\sqrt{n} u_n \rightharpoonup 0$.
(iii) Give an example such that $\sqrt{n} u_n \not\to 0$.

Exercise 1.28 Let $x_n \rightharpoonup x$ in H.

(i) Show that $n^{-1}(x_1 + \cdots + x_n) \rightharpoonup x$.
(ii) Show that if (x_n) belongs to a compact subset of H, then $x_n \to x$.

Exercise 1.29 Fix a bounded sequence $\alpha_1, \alpha_2, \ldots$ of real numbers, and set

$$Tx := (\alpha_1 x_2, \alpha_2 x_3, \ldots), \quad x = (x_1, x_2, \ldots) \in H := \ell^2.$$

(i) Show that $T \in L(H, H)$ and compute $\|T\|$.
(ii) Show that T is compact $\iff \alpha_n \to 0$.

Henceforth assume that $\alpha_n = 1$ if n is odd, and $\alpha_n = 2$ if n is even.

(iii) Show that each $\lambda \in (-\sqrt{2}, \sqrt{2})$ is an eigenvalue of T, and determine the associated eigensubspaces.
(iv) Compute $\|T^n\|$ for $n = 1, 2, \ldots$, and determine $\lim \|T^n\|^{1/n}$.
(v) Determine the adjoint operator T^*.

Exercise 1.30 Let $A \in L(H, H)$ be an isometric, non-surjective operator.

(i) Prove that there exists a unit vector e_0, orthogonal to $R(A)$.
(ii) Show that the formula $e_n := A e_{n-1}$, $n = 1, 2, \ldots$ defines an orthonormal sequence in H.
(iii) Show that $A^* e_0 = 0$ and $A^* e_1 = e_0$.
(iv) Compute $A^* e_n$ for all $n > 1$.
(v) Show that each $\lambda \in (-1, 1)$ is an eigenvalue of A^*.

Exercise 1.31 Consider the left and right shifts in ℓ^2 defined by

$$L(x_1, x_2, \ldots) := (x_2, x_3, \ldots) \quad \text{and} \quad R(x_1, x_2, \ldots) := (0, x_1, x_2, \ldots).$$

Prove the following:

(i) $\|L\| = \|R\| = 1$;
(ii) $L^* = R$;
(iii) the eigenvalues of L form the open interval $(-1, 1)$;
(iv) R has no eigenvalues;
(v)   The spectrum of both L and R is the closed interval $[-1, 1]$.[76]

---

[76]See the definition of the spectrum on p. 108 below.

Chapter 2
Banach Spaces

A mathematician, like a painter or a poet, is a maker of patterns. If his patterns are more
permanent than theirs, it is because they are made with ideas.
–G. Hardy

Hilbert spaces are not suitable for many important situations. For example, the uniform convergence of continuous functions is not associated with any scalar product. For this and many other situations infinite-dimensional normed spaces provide an appropriate framework.

Unlike the finite-dimensional case, *infinite-dimensional normed spaces are not always complete*, and non-complete normed spaces have many pathological properties. On the other hand, Banach and his colleagues discovered in the 1920s that by adding the completeness, many general deep results hold, despite the great variety of these spaces. In particular, although we cannot define orthogonality any more, many results of the preceding chapter remain valid.

In this chapter we give an introduction to this fascinating theory.

In the first four sections, mainly devoted to convexity, arbitrary normed spaces are considered. In the remaining sections the completeness of the spaces plays an essential role.

For the first reading, we advise the reader to skip the results concerning the somewhat particular spaces ℓ^1, ℓ^∞, c_0, and to concentrate on the spaces ℓ^p with $1 < p < \infty$.

We have to be careful: unlike the finite-dimensional case, *the closed balls of infinite-dimensional normed spaces*, although bounded and closed, are *never* compact. Some first basic results are the following:

Proposition 2.1 (Riesz)[1] *Let X be an infinite-dimensional normed space.*

---

[1]Riesz [383].

© Springer-Verlag London 2016
V. Komornik, *Lectures on Functional Analysis and the Lebesgue Integral*,
Universitext, DOI 10.1007/978-1-4471-6811-9_2

(a) *If $M \subset X$ is a proper closed subspace, then there exists a sequence $(x_n) \subset X$ satisfying*

$$\|x_n\| = 1 \quad \text{for all} \quad n, \quad \text{and} \quad \text{dist}(x_n, M) \to 1. \tag{2.1}$$

(b) *If $M \subset X$ is a finite-dimensional subspace, then there exists an $x \in X$ satisfying*

$$\|x\| = 1 \quad \text{and} \quad \text{dist}(x, M) = 1.$$

(c) *There exists a sequence (x_n) of unit vectors satisfying $\|x_m - x_n\| \geq 1$ for all $m \neq n$.*
(d) *The closed balls and the spheres of X are not compact.*[2]

Proof

(a) Choose an arbitrary point $z \in X \setminus M$, and then a *minimizing sequence* $(y_n) \subset M$ satisfying

$$\|z - y_n\| \to \text{dist}(z, M).$$

Since $y_n \in M$ and the subspace property of M imply that

$$\text{dist}(z, M) = \text{dist}(z - y_n, M) = \|z - y_n\| \, \text{dist}\left(\frac{z - y_n}{\|z - y_n\|}, M \right),$$

the unit vectors $x_n := (z - y_n)/\|z - y_n\|$ satisfy the relation $\text{dist}(x_n, M) \to 1$.
(b) Since M is finite-dimensional, the above sequence (y_n) has a convergent subsequence $y_{n_k} \to y$. Then $x := (z - y)/\|z - y\|$ has the required properties.
(c) By a repeated application of property (b) we may construct a sequence of unit vectors x_n such that $\text{dist}(x_n, \text{Vect}\{x_1, \ldots, x_{n-1}\}) = 1$ for all $n \geq 2$. This implies $\|x_n - x_m\| \geq 1$ for all $n > m$.
(d) By similarity it suffices to consider the closed unit ball B and the unit sphere $S \subset B$. The sequence constructed in (c) belongs to them but none of its subsequences has the Cauchy property.

$\square$

Remarks

• The finite-dimensional assumption cannot be omitted in (b): see the counterexample following Proposition 2.31, p. 95.

---

[2]By *spheres* we mean the boundaries of the balls.

- Kottman[3] proved that we may even require the strict inequalities $\|x_m - x_n\| > 1$ in (c). We recall that if (x_n) is an orthonormal sequence in a Euclidean space, then $\|x_m - x_n\| = \sqrt{2}$ for all $m \neq n$.

2.1 Separation of Convex Sets

Theorem 1.8 (p. 17) on the separation of convex sets remains valid in all normed spaces, and this has many important applications. However, a different proof is needed: even the existence of non-zero continuous linear functionals is a nontrivial result.

First we investigate the hyperplanes of vector spaces.

Definitions Let X be a vector space.

- By a *linear functional* on X we mean a linear map $\varphi : X \to \mathbb{R}$. They form a set X^* having a natural vector space structure.[4]
- By a *hyperplane* of X we mean a maximal proper subspace. In other words, a proper subspace H of X is a hyperplane if $\text{Vect}\,\{H, a\} = X$ for every $a \in X \setminus H$, where $\text{Vect}\,\{H, a\}$ denotes the subspace generated by H and a, i.e., the smallest subspace containing H and a.[5]
- By an *affine hyperplane* of X we mean a translate of a hyperplane.

Lemma 2.2 *The hyperplanes of X are the kernels of the non-zero linear functionals of X.*

Proof If $\varphi \in X^*$ and $\varphi \neq 0$, then $H := \varphi^{-1}(0)$ is a proper subspace of X. Furthermore, if $a \in X \setminus H$, then $\text{Vect}\,\{H, a\} = X$, because[6]

$$\varphi\left(x - \frac{\varphi(x)}{\varphi(a)}a\right) = \varphi(x) - \frac{\varphi(x)}{\varphi(a)}\varphi(a) = 0,$$

and hence

$$x - \frac{\varphi(x)}{\varphi(a)}a \in H$$

for every $x \in X$.

---

[3] Kottman [265]. See Diestel [104, p. 7].
[4] It is a (linear) subspace of the vector space of all functions $f : X \to \mathbb{R}$.
[5] We will weaken this definition in the remark following the next lemma.
[6] Note that $\varphi(a) \neq 0$, because $a \notin H$.

Conversely, if H is a hyperplane of X and $a \in X \setminus H$, then every $x \in X$ has a unique decomposition $x = ta + h$ with $t \in \mathbb{R}$ and $h \in H$.[7] Then the formula $\varphi(x) := t$ defines a non-zero linear functional X whose kernel is H.[8]   □

Remark Let H be a proper subspace. If *there exists* a vector $a \in X$ such that Vect$\{H, a\} = X$, then H is the kernel of a non-zero linear functional by the second part of the above proof, and hence H is a hyperplane by the first part of the proof.

The following notion is useful in the study of linear functionals.

Definition A subset U of a vector space X is *balanced* if

$$x \in U, \quad \lambda \in \mathbb{R} \quad \text{and} \quad |\lambda| \le 1 \quad \Longrightarrow \quad \lambda x \in U.$$

Examples

- Every subspace is balanced.
- The intersection of a family of balanced sets is balanced.
- The image of a balanced set by a linear map is balanced.
- The balanced sets of $\mathbb{R}$ are the intervals that are symmetric to 0.
- The open and closed balls centered at 0 of normed spaces are balanced.

Lemma 2.3 *Let U be a balanced set in a vector space X, and $\varphi \in X^*$ a linear functional satisfying $\varphi(a) = 1$. Then*

$$(a + U) \cap \varphi^{-1}(0) = \varnothing \Longleftrightarrow |\varphi| < 1 \quad in \quad U.$$

Proof First we observe the following equivalences:

$$(a + U) \cap \varphi^{-1}(0) = \varnothing \Longleftrightarrow 0 \notin \varphi(a + U) \Longleftrightarrow -\varphi(a) \notin \varphi(U).$$

Since $\varphi(U) \subset \mathbb{R}$ is an interval symmetric to 0 and not containing $-\varphi(a) = -1$, we conclude that $\varphi(U) \subset (-1, 1)$.   □

Next we study the hyperplanes of *normed spaces*.

Lemma 2.4

(a) *A hyperplane in a normed space is either closed or dense.*
(b) *A hyperplane of the form $H = \varphi^{-1}(0)$, $\varphi \in X^*$, is closed $\Longleftrightarrow \varphi$ is continuous.*

Proof

(a) If H is closed, then $\overline{H} = H \neq X$, so that H is not dense.
   If H is not closed, then $\overline{H}$ is a subspace of X satisfying $H \subset \overline{H}$ and $H \neq \overline{H}$. By the maximality of H we conclude that $\overline{H} = X$, i.e., H is dense.

---

[7] The uniqueness follows from the condition $a \notin H$.
[8] The linearity follows from the uniqueness of the decomposition.

(b) If φ is continuous, then $\varphi^{-1}(0)$ is closed. Conversely, if $\varphi^{-1}(0)$ is closed, then we choose a point a with $\varphi(a) = 1$, and then a small number $r > 0$ such that $\varphi \neq 0$ in the ball $U := B_r(a)$. Applying the lemma we conclude that $|\varphi| < 1$ in U, and hence $\|\varphi\| \leq 1/r$.

$\square$

Remarks

- The following proof[9] of part (b) does not use Lemma 2.3. We show that if $H = \varphi^{-1}(0)$ is closed, then φ is continuous. The case $\varphi = 0$ is obvious. If $\varphi \neq 0$, then there exists a point $e \in X$ such that $\varphi(e) = 1$, and then $d := \text{dist}(e, H) > 0$. If $x \in X \setminus H$, then $e - \frac{x}{\varphi(x)} \in H$, and therefore

$$d \leq \left\| e - \left(e - \frac{x}{\varphi(x)} \right) \right\| = \frac{\|x\|}{|\varphi(x)|},$$

whence $|\varphi(x)| \leq d^{-1} \|x\|$. This inequality holds of course for $x \in H$ as well.
- If X is finite-dimensional, then $X^* = X'$ because every linear functional on X is continuous. On the other hand, if X is infinite-dimensional, then X' is a proper subspace of X^*.[10]

We are ready to generalize Theorem 1.8 (p. 17); see Figs. 2.1, 2.2 and 2.3.

Fig. 2.1 Theorem of Mazur

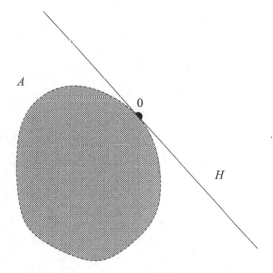

---

[9]Private communication of Z. Sebestyén.
[10]We can define non-continuous linear functionals by using a Hamel basis of X.

Fig. 2.2 Eidelheit's theorem

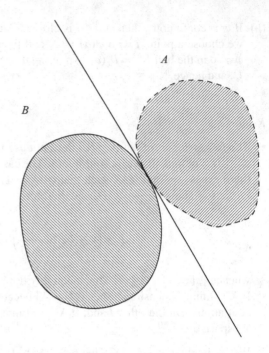

Fig. 2.3 Tukey's theorem

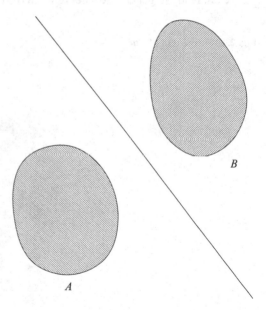

Theorem 2.5 *Let A and B be two disjoint non-empty convex sets in a normed space X.*

(a) *(Mazur)*[11] *If A is open and B is a subspace, then there exists a closed hyperplane H such that*

$$B \subset H \quad \text{and} \quad A \cap H = \varnothing.$$

(b) *(Eidelheit)*[12] *If A is open, then there exist $\varphi \in X'$ and $c \in \mathbb{R}$ such that*

$$\varphi(a) < c \leq \varphi(b) \quad \text{for all} \quad a \in A \quad \text{and} \quad b \in B.$$

(c) *(Tukey)*[13] *If A is closed and B is compact, then there exist $\varphi \in X'$ and $c_1, c_2 \in \mathbb{R}$ such that*

$$\varphi(a) \leq c_1 < c_2 \leq \varphi(b) \quad \text{for all} \quad a \in A \quad \text{and} \quad b \in B. \tag{2.2}$$

Remark Applying (a) with $0 \in \partial A$ and $B = \{0\}$, by translation we obtain that a convex open set has a *supporting affine hyperplane* at each boundary point; see Fig. 2.1.

The following lemma is the core of the proof:

Lemma 2.6 *Let X be a normed space, H a subspace of X, and A a non-empty convex open set in X, disjoint from H. If H is not a hyperplane, then there exists an $x \in X \setminus H$ such that Vect $\{H, x\} = X$ is still disjoint from A.*

Proof If Vect $\{H, x\}$ meets A, then $a = h + sx$ with suitable vectors $a \in A, h \in H$ and a real number s. Since $A \cap H = \varnothing$ implies $s \neq 0$, this yields

$$x = -s^{-1}h + s^{-1}a \in H + \bigcup_{t \in \mathbb{R}} tA.$$

Therefore it is sufficient to show that $H + \bigcup_{t \in \mathbb{R}} tA \neq X$.

Assume on the contrary that

$$H \cup H_+ \cup H_- = X, \tag{2.3}$$

---

[11]Brunn [66] and Minkowski [324, §16, pp. 33–35] in finite dimensions, Ascoli [13, pp. 53–56 and 205] in separable spaces, Mazur [317].

[12]Eidelheit [122].

[13]Tukey [460].

where we use the notations

$$H_+ := H + \bigcup_{t>0} tA \quad \text{and} \quad H_- := -H_+ = H + \bigcup_{t<0} tA.$$

Observe that H_+, H_- are (non-empty) open sets, and that H, H_+, H_- are pairwise disjoint. Indeed, if there were for example a point $x \in H_+ \cap H_-$, then we would have

$$x = h + ta = h' - t'a'$$

with suitable vectors $h, h' \in H$, $a, a' \in A$ and real numbers $t, t' > 0$. This would imply the equality

$$a'' := \frac{ta + t'a'}{t + t'} = \frac{h' - h}{t + t'} \in H.$$

Since $a'' \in A$ by the convexity of A, this contradicts the relation $A \cap H = \varnothing$.

The proof of the relations $H_+ \cap H = \varnothing$ and $H \cap H_- = \varnothing$ is similar: we may repeat the above proof with $t' = 0$ and $t = 0$, respectively.

Now choose a point $a \in H_+$. Since $H \neq X$ and H is not a hyperplane, we have $\text{Vect}\{H, a\} \neq X$. Let $b \in X \setminus \text{Vect}\{H, a\}$, then $b \in H_+ \cup H_-$ by (2.3). Changing b to $-b$ if needed, we may assume that $b \in H_-$.

Observe that $b \notin \text{Vect}\{H, a\}$ implies $[a, b] \cap H = \varnothing$, and hence $[a, b]$ is the union of the disjoint sets $[a, b] \cap H_+$ and $[a, b] \cap H_-$. The latter sets are open in the subspace topology of $[a, b]$. Since $a \in H_+$ and $b \in H_-$, they are non-empty, and this contradicts the connectedness of the interval $[a, b]$.   □

We also need the following equivalent form of the *axiom of choice* in set theory[14]:

Lemma 2.7 (Zorn)[15] *Let $\mathcal{A}$ be a non-empty family of sets satisfying the following condition: every* monotone *subfamily $\mathcal{B}$ has a* majorant *in $\mathcal{A}$.*

In other words, if for any two sets $B_1, B_2 \in \mathcal{B}$ we have either $B_1 \subset B_2$ or $B_2 \subset B_1$, then there exists a set $A \in \mathcal{A}$ containing all $B \in \mathcal{B}$.

Then the family $\mathcal{A}$ has a maximal element, i.e., there exists an $A \in \mathcal{A}$ that is not contained in any other set of $\mathcal{A}$.[16]

We will also use the following simple result:

Lemma 2.8 *Every non-zero linear functional φ on a normed space X is an* open mapping.

---

[14]See, e.g., Kelley [247].

[15]Zorn [492].

[16]There may be several maximal elements in general.

Proof Let x be an arbitrary point of an open set A, and consider the ball $B_r(x) \subset A$. Fix a point $e \in X$ such that $\|e\| = 1$ and $\varphi(e) > 0$. If $-r < t < r$, then $x + te \in B_r(x)$, and hence $\varphi(x + te) \in \varphi(B_r(x)) \subset \varphi(A)$, i.e.,

$$(\varphi(x) - r\varphi(e), \varphi(x) + r\varphi(e)) \subset \varphi(A).$$

This shows that $\varphi(x)$ is an inner point of $\varphi(A)$. □

Proof of Theorem 2.5

(a) We consider the family of subspaces H of X satisfying $B \subset H$ and $A \cap H = \varnothing$. The assumptions of Zorn's lemma are satisfied, hence it has a *maximal* element H. By Lemma 2.6 H is a hyperplane. Since H does not meet the non-empty open set A, H is not dense, but then it is closed by Lemma 2.4 (p. 58).

(b) Applying (a) with $B := \{0\}$ and with $A - B$ instead of A, we obtain $\varphi \in X'$ such that $\varphi(A)$ and $\varphi(B)$ are disjoint, non-empty convex sets in $\mathbb{R}$, i.e., disjoint, non-empty intervals; in particular φ is a non-zero functional. Changing φ to $-\varphi$ if needed, we may assume that

$$\sup_A \varphi \leq \inf_B \varphi.$$

Since A and B are non-empty sets, $c := \inf_B \varphi$ is a (finite) real number, and

$$\varphi(a) \leq c \leq \varphi(b) \quad \text{for all} \quad a \in A \quad \text{and} \quad b \in B.$$

Finally, by Lemma 2.8 the openness of A implies that $\varphi(A)$ is an *open* interval, and hence $\varphi < c$ in A.

(c) We claim that $\text{dist}(A, B) > 0$. For otherwise there exist two sequences $(a_n) \subset A$ and $(b_n) \subset B$ satisfying $\|a_n - b_n\| \to 0$. Since B is compact, there is a convergent subsequence $b_{n_k} \to b \in B$. Then we also have $a_{n_k} \to b$ by the relation $\|a_n - b_n\| \to 0$, and then $b \in A$ by the closedness of A. However, this contradicts the disjointness of A and B.

Fix a real number $0 < r < 2^{-1} \text{dist}(A, B)$ and introduce the following open neighborhoods of A and B:

$$A' := A + B_r(0), \quad B' := B + B_r(0).$$

Applying (b) to the sets A', B', there exist $\varphi \in X'$ and a real number c such that

$$\varphi(a') < c \leq \varphi(b') \quad \text{for all} \quad a' \in A' \quad \text{and} \quad b' \in B'.$$

This yields

$$\varphi(a) + r\,\|\varphi\| \le c \le \varphi(b) - r\,\|\varphi\| \quad \text{for all} \quad a \in A \quad \text{and} \quad b \in B.$$

The theorem follows with $c_1 := c - r\,\|\varphi\|$ and $c_2 := c + r\,\|\varphi\|$. $\square$

Using the above theorem we may generalize Corollary 1.6 (p. 15). Given $D \subset X$ and $\Delta \subset X'$ we define the *orthogonal complements*

$$D^\perp := \{\varphi \in X' \; : \; \varphi(x) = 0 \quad \text{for all} \quad x \in D\}$$

and

$$\Delta^\perp := \{x \in X \; : \; \varphi(x) = 0 \quad \text{for all} \quad x \in \Delta\}.$$

They are closed subspaces. If X is a Hilbert space and we identify X with its dual X', then both definitions reduce to the former one (p. 14).

Corollary 2.9 *(Banach)*[17] *Let X be a normed space and $D \subset X$.*

(a) *We have* $\overline{\text{Vect}(D)} = (D^\perp)^\perp$.
(b) *If $D^\perp = \{0\}$, then* $\overline{\text{Vect}(D)} = X$.
(c) *If $N^\perp = \{0\}$ for some subspace $N \subset X$, then N is dense in X.*

Proof (a) Set $M := \overline{\text{Vect}(D)}$ for brevity. By definition $D \subset (D^\perp)^\perp$ and $(D^\perp)^\perp$ is a closed subspace, so that $M \subset (D^\perp)^\perp$. It remains to prove that if $x \notin M$, then $x \notin (D^\perp)^\perp$.

If $x \notin M$, then we apply part (c) of the theorem with $A = M$ and $B = \{x\}$: there exists a $\varphi \in X'$ satisfying

$$\sup_M \varphi < \varphi(x).$$

As a linear image of a subspace, $\varphi(M)$ is a subspace of $\mathbb{R}$: either $\varphi(M) = \mathbb{R}$ or $\varphi(M) = \{0\}$. Therefore the previous relation implies $\varphi = 0$ on M, and then $\varphi(x) > 0$. Consequently, $\varphi \in D^\perp$ and $x \notin (D^\perp)^\perp$.

(b) and (c) readily follow from (a). $\square$

Remark We will show by some examples at the end of Sect. 2.3 (p. 76) that the role of X and X' cannot be exchanged in the above corollary.[18]

The following result shows that there are many continuous linear functionals on a normed space.

---

[17]Banach [22].
[18]We will give the topological description of $(\Delta^\perp)^\perp$ in Proposition 3.17, p. 137.

Corollary 2.10 *Let X be a normed space.*

(a) *For any two distinct points $a, b \in X$ there exists a $\varphi \in X'$ such that $\varphi(a) \neq \varphi(b)$.*
(b) *If $x_1, \ldots, x_n \in X$ are linearly independent vectors, then there exist linear functionals $\varphi_1, \ldots, \varphi_n \in X'$ such that*

$$\varphi_i(x_j) = \delta_{ij} \quad \text{for all} \quad i, j = 1, \ldots, n.$$

Consequently, $\dim X' \geq \dim X$.

Proof

(a) Apply Theorem 2.5 (c) with $A = \{a\}$ and $B = \{b\}$.
(b) The subspace $A := \text{Vect}\{x_1, \ldots, x_{n-1}\}$ is finite-dimensional, hence closed. Applying Theorem 2.5 (c) with A and $B = \{x_n\}$, there exist $\varphi \in X'$ and real numbers $c_1 < c_2$ such that $\varphi \leq c_1$ on A, and $\varphi(x_n) \geq c_2$.

Since $\varphi(A)$ is a linear subspace of $\mathbb{R}$, hence $\varphi = 0$ on A and then $\varphi(x_n) > 0$. Therefore $\varphi_n := \varphi/\varphi(x_n)$ has the required property. The construction of $\varphi_1, \ldots, \varphi_{n-1}$ is analogous. □

2.2 Theorems of Helly–Hahn–Banach and Taylor–Foguel

The following theorem if one of the most important results of Functional Analysis.

Theorem 2.11 (Helly–Hahn–Banach)[19] *If $\varphi : M \to \mathbb{R}$ is a continuous linear functional on a subspace $M \subset X$, then φ may be extended, by preserving its norm, to a continuous linear functional $\Phi : X \to \mathbb{R}$.*

Because of its fundamental importance, we give two different proofs here. The first is the original one, essentially due to Helly.

The second one deduces the result from Mazur's theorem.[20]

First Proof For $\varphi = 0$ we may take $\Phi := 0$. Otherwise, multiplying φ by a suitable constant we may assume that $\|\varphi\| = 1$.

---

[19] Helly [204] investigated the case $X = C([0, 1])$, but his proof remains valid in all separable normed spaces; in fact, his work paved the way to the introduction of normed spaces some years later. Based on Helly's crucial finite-dimensional construction, Hahn [182] and Banach [22] treated the non-separable case as well, by changing complete induction to transfinite induction. See also Hochstadt [215] on the life of Helly.

[20] Historically it was the converse: Mazur deduced his result from the extension theorem. See, e.g., Brezis [65] or Rudin [406].

First Step. First we show that for any fixed $a \in X \backslash M$, φ may be extended to a continuous linear functional $\psi : \text{Vect}\{M, a\} \to \mathbb{R}$, with preservation of the norm.

For any fixed real number c, the formula

$$\psi(x + ta) := \varphi(x) + tc, \quad x \in M, \quad t \in \mathbb{R}$$

defines a linear extension $\psi : \text{Vect}\{M, a\} \to \mathbb{R}$ of φ. Being an extension of φ, we have obviously $\|\psi\| \geq 1$. We have to show that the inverse inequality $\|\psi\| \leq 1$ also holds for a suitable choice of c.

Since $\psi(-y) = -\psi(y)$, it suffices to find c satisfying

$$\psi(x \pm ta) \leq \|x \pm ta\|$$

for all $x \in M$ and $t \geq 0$. This is obvious for $t = 0$ because we have an extension. Otherwise, dividing by $t > 0$ we obtain the equivalent condition

$$\psi(x' \pm a) \leq \|x' \pm a\| \quad \text{for all} \quad x' \in M;$$

this may be rewritten in the form

$$\varphi(x') - \|x' - a\| \leq c \leq \|x' + a\| - \varphi(x') \quad \text{for all} \quad x' \in M.$$

In order to ensure the existence of c, it is therefore sufficient to establish the inequalities

$$\varphi(x') - \|x' - a\| \leq \|x'' + a\| - \varphi(x'')$$

for all $x', x'' \in M$. This follows by a direct computation:

$$\varphi(x') + \varphi(x'') = \varphi(x' + x'')$$
$$\leq \|x' + x''\|$$
$$= \|(x' - a) + (x'' + a)\|$$
$$\leq \|x' - a\| + \|x'' + a\|.$$

Second Step. If X is finite-dimensional or, more generally, if M has finite co-dimension in X, then the theorem follows by applying the first step a finite number of times.

In the general case we consider the family of all norm-preserving linear extensions ψ of φ to subspaces of X. If we identify the linear functionals with their graphs, then we get a family of sets satisfying the assumptions of Zorn's lemma (p. 62). There exists therefore a maximal norm-preserving linear extension Φ of φ. By the first step of the proof it is defined on the whole space X.   □

Second Proof of Theorem 2.11 In case $\varphi \equiv 0$ we take simply $\Phi \equiv 0$. In the remaining cases we may assume, multiplying φ by a suitable positive number, that $\|\varphi\| = 1$. Denoting by U the open unit ball of X, centered at 0, then we have $|\varphi| \leq 1$ on U. Lemma 2.8 implies that in fact $|\varphi| < 1$ on U.

Fix $a \in M$ with $\varphi(a) = 1$. Since $|\varphi| < 1$ on U, $a + U$ does not meet $\varphi^{-1}(0)$ by Lemma 2.3 (p. 58). Applying Theorem 2.5 (a) (p. 61) there exists a hyperplane H such that $\varphi^{-1}(0) \subset H$ and H does not meet $a + U$. By Lemma 2.2 (p. 57) there exists a (unique) linear functional $\Phi \in X^*$ satisfying $\Phi^{-1}(0) = H$ and $\Phi(a) = 1$. Another application of Lemma 2.3 shows that $|\Phi| < 1$ on U. Hence Φ is continuous, and $\|\Phi\| \leq 1 = \|\varphi\|$.

It remains to prove that Φ is an extension of φ; this will also imply the reverse inequality $\|\Phi\| \geq \|\varphi\|$. If $x \in M$, then

$$\varphi\left(x - \varphi(x)a\right) = \varphi(x) - \varphi(x)\varphi(a) = \varphi(x) - \varphi(x) = 0,$$

so that $x - \varphi(x)a \in \varphi^{-1}(0) \subset H = \Phi^{-1}(0)$. Hence $\Phi\left(x - \varphi(x)a\right) = 0$, i.e., $\Phi(x) = \varphi(x)\Phi(a) = \varphi(x)$. □

Remark There are many generalizations of the theorem for vector valued linear maps.[21]

In general the extension Φ is not unique, except the trivial case where M is dense. The extension is also unique if X is a Hilbert space. In order to formulate a more precise result we need the following notion:

*Definition A normed space X is *strictly convex* if for any two distinct points $x_1, x_2 \in X$ with $\|x_1\| = \|x_2\| = 1$ we have $\|(x_1 + x_2)/2\| < 1$.

*Remarks

- If X is strictly convex and $x_1, x_2 \in X$ are two distinct points with $\|x_1\| = \|x_2\| = c$, then $\|(x_1 + x_2)/2\| < c$ by homogeneity.
- We recall the elementary fact that if $f : \mathbb{R} \to \mathbb{R}$ is a convex function and $f(0) = f(1) = 1$, then $f(t) \leq 1$ for all $0 < t < 1$ and $f(t) \geq 1$ otherwise. Moreover, either $f \equiv 1$ or $f < 1$ everywhere in $(0, 1)$.

  Applying this with $f(t) := \|tx_1 + (1 - t)x_2\|$ we obtain that a normed space X is strictly convex $\Longleftrightarrow$ its unit sphere does not contain any line segment.

  We obtain also that if for any two distinct points with $\|x_1\| = \|x_2\| = 1$ there exists $t \in \mathbb{R}$ such that $\|tx_1 + (1 - t)x_2\| < 1$, then X is strictly convex.

---

[21]See, e.g., Banach–Mazur [27], Fichtenholz–Kantorovich [145], Murray [328], Goodner [172], Nachbin [330], Kelley [246], and a general review in Narici–Beckenstein [331].

*Proposition 2.12 (Taylor–Foguel)[22] *All continuous linear functionals defined on subspaces of a normed space X have a unique norm-preserving extension to X* $\iff$ *the dual space X' of X is strictly convex.*

Proof Assume that X' is strictly convex, and let $\varphi_1, \varphi_2 \in X'$ be two distinct extensions of a linear functional $\varphi : Y \to \mathbb{R}$ such that $\|\varphi_1\| = \|\varphi_2\| = c$. Then $(\varphi_1 + \varphi_2)/2$ is also a linear extension of φ, and therefore

$$\|\varphi\|_{Y'} \le \|(\varphi_1 + \varphi_2)/2\|_{X'} < c$$

by the strict convexity of X', so that the extensions φ_1, φ_2 are not norm-preserving.

Conversely, assume that all norm-preserving extensions are unique and consider two distinct elements φ_1, φ_2 of X' with $\|\varphi_1\| = \|\varphi_2\| = 1$. In view of the above remark it is sufficient to find a real number t satisfying $\|t\varphi_1 + (1-t)\varphi_2\|$.

The common restriction of φ_1 and φ_2 to the hyperplane $Y := \{x \in X : \varphi_1(x) = \varphi_2(x)\}$ has a unique norm-preserving extension $\varphi \in X'$. Since the distinct extensions φ_1, φ_2 cannot both be norm-preserving, we have necessarily $\|\varphi\| < 1$. It remains to show that $\varphi = t\varphi_1 + (1-t)\varphi_2$ for some $t \in \mathbb{R}$.

Fix an arbitrary point $x_0 \in X \setminus Y$. Since $\varphi_1(x_0) \ne \varphi_2(x_0)$, there exists a $t \in \mathbb{R}$ such that

$$\varphi(x_0) = \varphi_2(x_0) + t\,(\varphi_1(x_0) - \varphi_2(x_0)) = t\varphi_1(x_0) + (1-t)\varphi_2(x_0).$$

Then φ and $t\varphi_1 + (1-t)\varphi_2$ coincide on Vect $\{Y, x_0\} = X$, so that $\varphi = t\varphi_1 + (1-t)\varphi_2$ as required. $\square$

Corollary 2.13 (Banach)[23] *Let M be a closed subspace of a normed space X.*

(a) *For every* $x \in X \setminus M$ *there exists a* $\varphi \in X'$ *such that*

$$\|\varphi\| = 1, \quad \varphi = 0 \quad on \quad M, \quad and \quad \varphi(x) = \text{dist}(x, M).$$

(b) *For every* $x \in X$ *there exists a* $\varphi \in X'$ *such that*

$$\|\varphi\| \le 1 \quad and \quad \varphi(x) = \|x\|.$$

(c) *We have*

$$\|x\| = \max_{\|\varphi\| \le 1} |\varphi(x)|$$

for every $x \in X$.

---

[22]Taylor [450] and Foguel [147]. See also Phelps [355], Holmes [216, p. 175], Beesack, Hughes and Ortel [33] and Ciarlet [87, p. 265].
[23]Banach [22]. In part (b) we also have $\|\varphi\| = 1$, except in the degenerate case $X = \{0\}$.

Proof

(a) The formula

$$\psi(tx - y) := t \, \text{dist}(x, M), \quad t \in \mathbb{R}, \, y \in M$$

defines a linear functional on the subspace Vect$\{M, x\}$. (The linearity follows from the uniqueness of the decomposition $tx - y$.) We have obviously $\psi(x) = \text{dist}(x, M)$, and $\psi = 0$ on M. Furthermore, $\|\psi\| \leq 1$, because for $t \neq 0$ and $y \in M$ we have

$$|\psi(tx - y)| = |t \, \text{dist}(x, M)| \leq |t| \cdot \left\| x - \frac{y}{t} \right\| = \|tx - y\|.$$

(This is also true for $t = 0$ because then the left-hand side is zero.)

For the proof of the converse inequality we choose a sequence $(y_n) \subset M$ satisfying $\|x - y_n\| \to \text{dist}(x, M)$. Then $\psi(x - y_n) = \text{dist}(x, M)$ for every n, and therefore

$$\|\psi\| \geq \lim \frac{\psi(x - y_n)}{\|x - y_n\|} = 1.$$

We conclude by extending ψ to X by applying the theorem.

(b) If $x = 0$, then take $\varphi = 0$. If $x \neq 0$, then apply (a) with $M = \{0\}$.

(c) Since $\varphi \in X'$ and $\|\varphi\| \leq 1$ imply $|\varphi(x)| \leq \|x\|$ by the definition of the norm, the result follows from (b).

$\square$

Remark We may compare the formula in (c) with the *definition*

$$\|\varphi\| = \sup_{\|x\| \leq 1} |\varphi(x)|, \quad \varphi \in X'.$$

In the latter we cannot write max in general. We will return to this question later.[24]

2.3 The ℓ^p Spaces and Their Duals

By Lemmas 2.2 and 2.4 (p. 58) knowledge of the closed hyperplanes is equivalent to knowledge of the dual space. The case of Hilbert spaces is easy because X' may be identified with X by the Riesz–Fréchet theorem (p. 19). In this section we show by some examples that X and X' may have different structures for general normed spaces.

---

[24]See Proposition 2.31, p. 91.

Definitions

- The bounded real sequences $x = (x_n)$ form a normed space ℓ^∞ with respect to the norm

$$\|x\|_\infty := \sup |x_n|,$$

  because $\ell^\infty = \mathcal{B}(K)$ with $K := \{1, 2, \ldots\}$.
- The real null sequences form a subspace c_0 of ℓ^∞, and hence a normed space.
- Given a real number $1 \leq p < \infty$, let us denote by ℓ^p the set of real sequences $x = (x_n)$ satisfying $\sum |x_n|^p < \infty$, and set

$$\|x\|_p := \left(\sum |x_n|^p\right)^{1/p}.$$

The following result shows that all ℓ^p spaces are normed spaces.

Proposition 2.14 *Let $p, q \in [1, \infty]$ be conjugate exponents, i.e., satisfying $p^{-1} + q^{-1} = 1$.*

(a) *(Young's inequality)[25] If p and q are finite, then*

$$xy \leq \frac{x^p}{p} + \frac{y^q}{q}$$

  for all nonnegative numbers x and y.

(b) *(Hölder's inequality)[26] If $x \in \ell^p$ and $y \in \ell^q$, then $xy \in \ell^1$ and*

$$\|xy\|_1 \leq \|x\|_p \cdot \|y\|_q.$$

(c) *(Minkowski's inequality)[27] If $x, y \in \ell^p$, then $x + y \in \ell^p$ and*

$$\|x + y\|_p \leq \|x\|_p + \|y\|_p.$$

(d) *ℓ^p is a normed space.*

---

[25]Young [489].
[26]Rogers [399], Hölder [217], and Riesz [382].
[27]Minkowski [323, pp. 115–117] and Riesz [382].

Fig. 2.4 Young's inequality

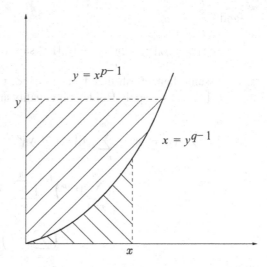

Proof

(a) We may assume by symmetry that $p \geq 2$. Consider the graph of the function $y = x^{p-1}$ or equivalently $x = y^{q-1}$ (see Fig. 2.4). The union of the two shaded regions contains the rectangle of sides x and y. Hence their areas satisfy the inequality

$$xy \leq \int_0^x s^{p-1}\, ds + \int_0^y t^{q-1}\, dt = \frac{x^p}{p} + \frac{y^q}{q}.$$

(b) For $p = 1$ and $q = \infty$ the result follows from the straightforward estimate

$$\|xy\|_1 = \sum_{n=1}^{\infty} |x_n y_n| \leq \left(\sum_{n=1}^{\infty} |x_n| \right) \sup |y_n| = \|x\|_1 \|y\|_\infty .$$

The case of $p = \infty$ and $q = 1$ is similar.

Assume henceforth that $1 < p < \infty$, then $1 < q < \infty$. We may assume by homogeneity that $\|x\|_p = \|y\|_q = 1$, and we have to prove that $\|xy\|_1 \leq 1$. This follows by applying Young's inequality:

$$\left| \sum_{n=1}^{\infty} x_n y_n \right| \leq \sum_{n=1}^{\infty} |x_n| \cdot |y_n| \leq \sum_{n=1}^{\infty} \frac{x_n^p}{p} + \frac{y_n^q}{q} = \frac{1}{p} + \frac{1}{q} = 1.$$

(c) The cases $p = 1$ and $p = \infty$ follow at once from the estimates

$$\sum_{n=1}^{\infty} |x_n + y_n| \leq \sum_{n=1}^{\infty} (|x_n| + |y_n|) = \|x\|_1 + \|y\|_1$$

and

$$\sup |x_n + y_n| \le \sup(|x_n| + |y_n|) \le \sup |x_n| + \sup |y_n| = \|x\|_\infty + \|y\|_\infty .$$

Assume henceforth that $1 < p < \infty$, then $1 < q < \infty$. For each fixed $m = 1, 2, \dots$ we apply Hölder's inequality and the relation $(p-1)q = p$ to get

$$\sum_{n=1}^{m} |x_i + y_i|^p \le \sum_{n=1}^{m} |x_i| \cdot |x_i + y_i|^{p-1} + \sum_{n=1}^{m} |y_i| \cdot |x_i + y_i|^{p-1}$$

$$\le \left(\sum_{n=1}^{m} |x_i|^p \right)^{1/p} \left(\sum_{n=1}^{m} |x_i + y_i|^{(p-1)q} \right)^{1/q}$$

$$+ \left(\sum_{n=1}^{m} |y_i|^p \right)^{1/p} \left(\sum_{n=1}^{m} |x_i + y_i|^{(p-1)q} \right)^{1/q}$$

$$\le \left(\|x\|_p + \|y\|_p \right) \left(\sum_{n=1}^{m} |x_i + y_i|^{(p-1)q} \right)^{1/q}$$

$$= \left(\|x\|_p + \|y\|_p \right) \left(\sum_{n=1}^{m} |x_i + y_i|^p \right)^{1/q} .$$

Since $1/q = 1 - 1/p$, hence

$$\left(\sum_{n=1}^{m} |x_i + y_i|^p \right)^{1/p} \le \|x\|_p + \|y\|_p .$$

Letting $m \to \infty$ we conclude that the left-hand sum converges, and $\|x + y\|_p \le \|x\|_p + \|y\|_p$.

(d) We already know that ℓ^∞ is a normed space; henceforth we assume that $1 \le p < \infty$. Using (c) we see that ℓ^p is a *vector space*[28] and $\|\cdot\|_p$ is a norm.

□

Consider $X = \ell^p$ for some p. If $y = (y_n) \in \ell^q$, where q is the conjugate exponent of p, then the formula

$$\varphi_y(x) := \sum_{n=1}^{\infty} x_n y_n, \quad x = (x_n) \in \ell^p \tag{2.4}$$

---

[28]More precisely, a subspace of the vector space of all real sequences.

defines a continuous linear functional. Indeed, applying Hölder's inequality we see that the definition is correct, and

$$\left|\varphi_y(x)\right| \leq \|y\|_q \cdot \|x\|_p$$

for every x. Consequently,

$$\varphi_y \in (\ell^p)', \quad \text{and} \quad \|\varphi_y\| \leq \|y\|_q \quad \text{for every} \quad y \in \ell^q.$$

Hence the formula $j(y) := \varphi_y$ defines a continuous *linear* map $j : \ell^q \to (\ell^p)'$ (of norm ≤ 1).

Since c_0 is a subspace of ℓ^∞ the same formula also defines a continuous linear map $j : \ell^1 \to (c_0)'$ (of norm ≤ 1).

A special case of a theorem of F. Riesz[29] shows that much more is true:

Proposition 2.15

(a) *If $1 \leq p < \infty$, then $j : \ell^q \to (\ell^p)'$ is an isometric isomorphism.*
(b) *$j : \ell^1 \to (c_0)'$ is an isometric isomorphism.*

*Remarks

- According to the proposition we often identify $(c_0)'$ with ℓ^1, and $(\ell^p)'$ with ℓ^q for $1 \leq p < \infty$.
- We show at the end of this section that $(\ell^\infty)'$ is not isomorphic to ℓ^1.

We need a lemma:

Lemma 2.16 *If $X = \ell^p$, $1 \leq p < \infty$, or if $X = c_0$, then the vectors*

$$e_k = (\overbrace{0, \ldots, 0}^{k-1}, 1, 0, \ldots), \quad k = 1, 2, \ldots$$

generate X. Hence these spaces are separable.

Proof For any given $x = (x_n) \in \ell^p$, $1 \leq p < \infty$, the relation

$$\left\| x - \sum_{n=1}^{k} x_n e_n \right\|_p^p = \sum_{n=k+1}^{\infty} |x_n|^p \to 0 \quad (k \to \infty)$$

shows that the vectors e_k generate ℓ^p.

---

[29]Riesz [382]: see Theorem 9.14, p. 332.

If $x = (x_n) \in c_0$, then

$$\left\| x - \sum_{n=1}^{k} x_n e_n \right\|_\infty = \max\{|x_n| \: : \: n > k\} \to 0,$$

because $x_n \to 0$ by the definition of c_0.

It follows that the finite linear combinations of the vectors e_k with *rational* coefficients form a countable, dense set in X.   □

Remarks

• The lemma does not hold in ℓ^∞ because c_0 is a *proper closed* subspace of ℓ^∞ so that the vectors e_k cannot generate ℓ^∞.
• The space ℓ^∞ is not even separable. For the proof we consider the uncountable set of (open) unit balls, centered at the points $x = (x_n)$ such that $x_n = \pm 1$ for every n. Since they are pairwise disjoint, no countable set D may meet all of them, and therefore D cannot be dense.

Proof of Proposition 2.15

(a) For any fixed $\varphi \in (\ell^p)'$ we have to find a unique sequence $y \in \ell^q$ satisfying $\varphi_y = \varphi$ and $\|y\|_q \le \|\varphi\|$. (The converse inequality $\|\varphi_y\| \le \|y\|_q$ is already known.)

If there exists a $y \in \ell^q$ such that $\varphi_y = \varphi$, then we have necessarily

$$\varphi(e_n) = \varphi_y(e_n) = y_n$$

for every n, whence

$$y_n = \varphi(e_n), \quad n = 1, 2, \ldots.$$

Hence there exists at most one such y.

It remains to show that the above formula indeed defines a suitable sequence. If $p = 1$, then

$$|y_n| = |\varphi(e_n)| \le \|\varphi\|$$

for every n, so that $y \in \ell^\infty$ and $\|y\|_\infty \le \|\varphi\|$.

If $p > 1$ and thus $q < \infty$, then we consider for each fixed $k = 1, 2, \ldots$ the sequence $x = (x_n)$ defined by the formula

$$x_n := \begin{cases} |y_n|^{q-1} \text{ sign } y_n & \text{if } n \le k, \\ 0 & \text{if } n > k. \end{cases}$$

Then $x \in \ell^p$ (because the sequence has only finitely many terms), and

$$\varphi(x) = \sum_{n=1}^{k} |y_n|^q = \|x\|_p^p$$

by a simple computation. Using these equalities we deduce from the estimate

$$|\varphi(x)| \leq \|\varphi\| \cdot \|x\|_p$$

that

$$\sum_{n=1}^{k} |y_n|^q \leq \|\varphi\| \cdot \left(\sum_{n=1}^{k} |y_n|^q \right)^{1/p},$$

and therefore

$$\sum_{n=1}^{k} |y_n|^q \leq \|\varphi\|^q.$$

Letting $k \to \infty$ we conclude that $y \in \ell^q$ and $\|y\|_q \leq \|\varphi\|$.

It remains to prove the equality $\varphi = \varphi_y$. Since the continuous linear functionals φ and φ_y coincide at the points e_n by definition, they also coincide on the closed subspace generated by these points, i.e., on the whole space ℓ^p by the lemma.[30]

(b) For any given $\varphi \in (c_0)'$ we may repeat the proof of (a) with $p = \infty$ and $q = 1$.

$\square$

Let us mention the following result:

*Proposition 2.17 *If X' is separable for some normed space X, then X is also separable.*

Proof We fix a dense sequence (φ_n) in X', and then we choose for each n a vector $x_n \in X$ satisfying

$$\|x_n\| \leq 1 \quad \text{and} \quad |\varphi_n(x_n)| \geq 2^{-1} \|\varphi_n\|.$$

It suffices to prove that (x_n) generates X because then their finite linear combinations with rational coefficients form a countable, dense set in X.

In view of Corollary 2.9 (p. 64) it is sufficient to show that if some functional $\varphi \in X'$ satisfies $\varphi(x_n) = 0$ for every n, then $\varphi = 0$. For this we choose a suitable

---

[30]The assumption $p \neq \infty$ is used only at this last step.

subsequence $\varphi_{n_k} \to \varphi$. Then we have

$$\|\varphi_{n_k}\| \leq 2\,|\varphi_{n_k}(x_{n_k})| = 2\,|(\varphi_{n_k} - \varphi)(x_{n_k})| \leq 2\,\|\varphi_{n_k} - \varphi\|\,.$$

Letting $k \to \infty$ we conclude that $\|\varphi\| \leq 0$, i.e., $\varphi = 0$. $\square$

Remark Since ℓ^∞ is not separable, $(\ell^\infty)'$ is not separable either by the preceding proposition. Since ℓ^1 is separable, it is not isomorphic to $(\ell^\infty)'$.

Now we can give some counterexamples promised on p. 64.

Examples The following examples show that the role of X and X' cannot be exchanged in Corollary 2.9, p. 64.

- Let $X = c_0$ and $X' = \ell^1$. Then

$$\Delta := \left\{ (y_n) \in X' \,:\, \sum y_n = 0 \right\}$$

  is a proper closed subspace of X', while $(\Delta^\perp)^\perp = X'$ because $\Delta^\perp = \{0\}$. For the proof of the latter we observe that if $x = (x_n) \in \Delta^\perp$, then $x \perp e_1 - e_n$ for every n, so that $x_1 = x_2 = \cdots$. But $x \in c_0$ implies that $x_n \to 0$, so that $x = 0$.
- Let $X = \ell^1$ and $X' = \ell^\infty$. Then $\Delta := c_0$ is a proper closed subspace of X', while $(\Delta^\perp)^\perp = X'$ because $\Delta^\perp = \{0\}$. Indeed, if $x = (x_n) \in \Delta^\perp$, then $x \perp e_n$ for every n. In other words, $x_n = 0$ for every n, i.e., $x = 0$.

2.4 Banach Spaces

All finite-dimensional normed spaces are complete. On the other hand, we have already encountered non-complete normed spaces (even Euclidean spaces) in the preceding chapter. The rest of this chapter is devoted to *complete normed spaces*.

Definition A *Banach space* is a complete normed space.[31]

Examples

- Every finite-dimensional normed space is a Banach space.
- Every Hilbert space is a Banach space.
- If K is a non-empty set and X a Banach space, then the vector space $\mathcal{B}(K, X)$ of bounded functions $f : K \to X$ is complete with respect to the norm

$$\|f\|_\infty := \sup_{t \in K} \|f(t)\|_X\,,$$

  and hence it is a Banach space. If $X = \mathbb{R}$, then we write $\mathcal{B}(K)$ for brevity.

---

[31]Riesz [383], Banach [19], Hahn [181], and Wiener [487]. Terminology of Fréchet [161].

- If K is a topological space and X a Banach space, then the bounded continuous functions $f : K \to X$ form a *closed* subspace $C_b(K,X)$ of $\mathcal{B}(K,X)$, and hence a Banach space. If K is compact, then we write simply $C(K,X)$. If $X = \mathbb{R}$, then we write $C_b(K)$ or $C(K)$ instead of $C_b(K,X)$ or $C(K,X)$.
- If I is a non-empty open interval, Y a Banach space and k a natural number, then the C^k functions $f : I \to Y$ for which $f, f', \ldots, f^{(k)}$ are all bounded form a Banach space $C_b^k(I, Y)$ with respect to the norm

$$\|f\|_\infty + \|f'\|_\infty + \cdots + \|f^{(k)}\|_\infty.$$

- The bounded real sequences $x = (x_n)$ form a Banach space ℓ^∞ with respect to the norm

$$\|x\|_\infty := \sup |x_n|,$$

because $\ell^\infty = \mathcal{B}(K)$ with $K := \{1, 2, \ldots\}$.
- The real null sequences form a *closed* subspace c_0 of ℓ^∞, and hence a Banach space.

We give another important example. We recall that if X and Y are normed spaces, then the continuous linear maps $A : X \to Y$ form a normed space $L(X, Y)$ with respect to the norm

$$\|A\| := \sup \{\|Ax\| : \|x\| \le 1\}.$$

Proposition 2.18 *If X is a normed space and Y a Banach space, then $L(X, Y)$ is a Banach space. In particular, the dual of any normed space is a Banach space.*

Proof If (A_n) is a Cauchy sequence in $L(X, Y)$, then $(A_n x)$ is a Cauchy sequence in Y for each fixed $x \in X$, because

$$\|A_n x - A_m x\| \le \|A_n - A_m\| \cdot \|x\| \to 0$$

as $m, n \to \infty$. Since Y is complete, $(A_n x)$ converges to some point $Ax \in Y$.

Since the maps A_n are linear, A is also linear. Since the Cauchy sequence (A_n) is necessarily bounded, there exists a constant M such that

$$\|A_n x\| \le M \|x\|$$

for all n and x. Letting $n \to \infty$ we conclude that $\|A\| \le M$, i.e., $A \in L(X, Y)$.

Finally, for any fixed $\varepsilon > 0$ choose N such that

$$\|A_n - A_m\| \le \varepsilon$$

for all $m, n \geq N$. Then

$$\|A_n x - A_m x\| \leq \varepsilon \|x\|$$

for all $m, n \geq N$ and $x \in X$. Letting $m \to \infty$ we obtain

$$\|A_n x - A x\| \leq \varepsilon \|x\|$$

for all $n \geq N$ and $x \in X$, i.e., $A_n \to A$ in $L(X, Y)$. $\square$

Corollary 2.19 *All ℓ^p spaces are Banach spaces.*

Proof We have seen in the preceding section that all ℓ^p spaces are dual spaces, and hence complete by the preceding proposition.

Alternatively, the completeness of ℓ^p for $1 \leq p < \infty$ may be proved by a simple adaptation of the proof given for ℓ^2 in Sect. 1.1, by changing the exponents 2 to p everywhere. $\square$

Examples

- If U is a non-empty open set in a normed space, Y a Banach space, and k a natural number, then the C^k functions $f : U \to Y$ for which $f, f', \ldots, f^{(k)}$ are all bounded form a Banach space $C_b^k(U, Y)$ with respect to the norm

$$\|f\|_\infty + \|f'\|_\infty + \cdots + \|f^{(k)}\|_\infty,$$

  because the derivative functions map into Banach spaces of the form $L(X, Z)$ by the proposition.[32]
- Let $I = [a, b]$ be a non-degenerate compact interval and $1 \leq p < \infty$. We know that $C(I)$ is a normed space with respect to the norm

$$\|x\|_p := \left(\int_I |x(t)|^p \, dt \right)^{1/p}.$$

This norm is not complete. For $p = 2$ we have already proved this on page 10; the general case follows by changing every exponent 2 to p in that proof.

An easy adaptation of the proof of Proposition 1.3 (p. 10) leads to the following result:

Proposition 2.20 *Every normed space may be completed, i.e., may be considered as a dense subspace of a Banach space.*

Definition We denote by $L^p(I)$, for $1 \leq p < \infty$, the Banach space obtained by completion of $C(I)$ with respect to the norm $\|\cdot\|_p$.

---

[32]See any book on *differential calculus*.

Remark Later we will give a concrete interpretation of these spaces.[33]

We end this section by giving another proof of the last proposition.

Definition By the *bidual* of a normed space X we mean the Banach space $X'' := (X')'$.[34]

Example If $x \in X$, then the formula

$$\Phi_x(\varphi) := \varphi(x), \quad \varphi \in X'$$

defines a continuous linear functional $\Phi_x \in X''$, and $\|\Phi_x\| \leq \|x\|$ because $|\Phi_x(\varphi)| = |\varphi(x)| \leq |\varphi| \cdot \|x\|$ for every $\varphi \in X'$.

Let us look more closely at the correspondence $x \mapsto \Phi_x$:

Corollary 2.21 *(Hahn)*[35] *Let X be a normed space.*

(a) *The formula $J(x) := \Phi_x$ defines a linear isometry $J : X \to X''$.*
(b) *X may be completed: there exist a Banach space Y and a linear isometry $J : X \to Y$ such that $J(X)$ is dense in Y.*

Proof

(a) The linearity of J is straightforward. The isometry follows from Corollary 2.13 (c):

$$\|Jx\| = \sup_{\|\varphi\| \leq 1} |(Jx)(\varphi)| = \sup_{\|\varphi\| \leq 1} |\varphi(x)| = \|x\|.$$

(b) In view of (a) we may choose for Y the closure in X'' of the range $J(X)$ of J: as a closed subspace of the Banach space X'', it is also a Banach space.

$\square$

2.5 Weak Convergence: Helly–Banach–Steinhaus Theorem

Weak convergence proved to be a useful tool in the study of Hilbert spaces. We generalize this notion to normed spaces.

Definition A sequence (x_n) in a normed space X *converges weakly*[36] to $x \in X$ if

$$\varphi(x_n) \to \varphi(x)$$

---

[33]See Proposition 9.5 (b), p. 312.
[34]Hahn [182]. We will investigate these spaces in Sect. 2.6, p. 87.
[35]Hahn [182].
[36]Riesz [380], Banach [24].

for every $\varphi \in X'$. We express this by writing $x_n \rightharpoonup x$.

Remarks

- For Hilbert spaces this reduces to the former notion by the Riesz–Fréchet theorem.
- Norm convergence implies weak convergence by the continuity of the functionals of X'. Therefore norm convergence is also called *strong convergence*.
- In finite-dimensional normed spaces the strong and weak convergences coincide.

Let us collect the elementary properties of weak convergence:

Proposition 2.22

(a) *A sequence has at most one weak limit.*
(b) *If $x_n \rightharpoonup x$, then $x_{n_k} \rightharpoonup x$ for every subsequence (x_{n_k}).*
(c) *If $x_n \rightharpoonup x$ and $y_n \rightharpoonup y$, then $x_n + y_n \rightharpoonup x + y$.*
(d) *If $x_n \rightharpoonup x$ in X and $\lambda_n \to \lambda$ in $\mathbb{R}$, then $\lambda_n x_n \rightharpoonup \lambda x$ in X.*
(e) *Let $K \subset X$ be a convex closed set. If $x_n \rightharpoonup x$, and $x_n \in K$ for every n, then $x \in K$.*
(f) *If $x_n \rightharpoonup x$, and $\|x_n\| \le L$ for every n, then $\|x\| \le L$.*[37]
(g) *If $x_n \to x$, then $x_n \rightharpoonup x$ and $\|x_n\| \to \|x\|$.*

*Remark In contrast to Hilbert spaces the relations $x_n \rightharpoonup x$ and $\|x_n\| \to \|x\|$ do not imply $x_n \to x$ in general.[38] If this holds, then X is said to have the *Radon–Riesz property*.

Proof We may repeat the corresponding proofs given for Hilbert spaces (p. 30), except for (a) and (e); for the proof of (g) we now apply the continuity of $\varphi \in X'$ instead of the Cauchy–Schwarz inequality.

(a) If $x_n \rightharpoonup x$ and $x_n \rightharpoonup y$, then by Corollary 2.13 there exists a $\varphi \in X'$ satisfying $\varphi(x - y) = \|x - y\|$. Since $\varphi(x_n) \to \varphi(x)$ and $\varphi(x_n) \to \varphi(y)$ imply $\varphi(x) = \varphi(y)$, hence

$$\|x - y\| = \varphi(x - y) = \varphi(x) - \varphi(y) = 0,$$

and therefore $x = y$.

(e) Instead of the orthogonal projection we use Tukey's theorem (p. 61). Assume on the contrary that $x \notin K$; then there exist $\varphi \in X'$ and $c_1, c_2 \in \mathbb{R}$ such that

$$\varphi(x) \le c_1 < c_2 \le \varphi(y) \quad \text{for every} \quad y \in K.$$

Then $\varphi(x_n) \ge c_2$ for every n, so that $\varphi(x_n) \not\to \varphi(x)$, i.e., $x_n \not\rightharpoonup x$.   □

---

[37] Equivalently, $\|x\| \le \liminf \|x_n\|$.
[38] We give soon an example. See also Proposition 9.11, p. 328.

Every weakly convergent sequence is bounded. Before proving this deeper result, we establish another essential result of Functional Analysis: the *uniform boundedness theorem*:

Theorem 2.23 (Helly–Banach–Steinhaus)[39] *Consider a family* $\mathcal{A} \subset L(X, Y)$ *of continuous linear maps where X is a* Banach-*space, and Y a normed space. If the sets*

$$\mathcal{A}(x) := \{Ax \in Y \,:\, A \in \mathcal{A}\}, \quad x \in X$$

are all bounded in Y, then $\mathcal{A}$ is bounded in $L(X, Y)$:

$$\sup \, \{\|A\| \,:\, A \in \mathcal{A}\} < \infty.$$

---

*Remark The idea of this theorem had already appeared in Riemann's work.[40]

*Example The theorem fails in non-complete spaces X. Consider for example the subspace X of ℓ^2 formed by the sequences having at most finitely many non-zero elements. The formula

$$\varphi_n(x) := n x_n$$

defines a pointwise bounded but uniformly unbounded sequence of functionals in $L(X, \mathbb{R})$.

Proof It suffices to prove[41] that $\mathcal{A}$ is uniformly bounded in some ball, say

$$\|Ax\| \leq C \quad \text{for every} \quad A \in \mathcal{A} \quad \text{and} \quad x \in B_{2r}(x').$$

This will imply for all $A \in \mathcal{A}$ and $x \in X$, $\|x\| \leq 1$, the relations $x', x' + rx \in B_{2r}(x')$, and therefore the inequalities

$$\|Ax\| = \frac{1}{r} \|A(x' + rx) - Ax'\| \leq \frac{\|A(x' + x)\| + \|Ax'\|}{r} \leq \frac{2C}{r},$$

whence $\|A\| \leq 2C/r$ for every $A \in \mathcal{A}$.

---

[39] Helly [204], and Banach–Steinhaus [28]. See Hochstadt [215] on Helly's contribution. See also Banach [19], Hahn [181], and Hildebrandt [211].

[40] *Condensation of singularities*, Riemann [371], and Hankel [190]. See also Gal [166].

[41] Following a suggestion of Saks, Banach and Steinhaus proved their theorem with the help of Baire's lemma (p. 32). We prefer to adapt, following Riesz–Sz. Nagy [394], an argument of Osgood [350, pp. 163–164], that can also be used to prove Baire's lemma.

Assume on the contrary that $\mathcal{A}$ is not uniformly bounded on any open ball, and fix an arbitrary ball B_0.[42]

By our assumption there exist $A_1 \in \mathcal{A}$ and $x_1 \in B_0$ such that $\|A_1 x_1\| > 1$. By the continuity of A_1 the inequality remains valid in a small ball B_1 centered at x_1. By choosing its radius sufficiently small, we may also assume that diam $B_1 < 1$ and $\overline{B_1} \subset B_0$.

Repeating these arguments, there exist $A_2 \in \mathcal{A}$ and a ball B_2 such that diam $B_2 < 1/2$, $\overline{B_2} \subset B_1$, and $\|A_2 x\| > 2$ for every $x \in B_2$.

Continuing by induction we obtain a sequence $(A_k) \subset \mathcal{A}$ of maps and a sequence (B_k) of balls such that diam $B_k < 1/k$, $\overline{B_k} \subset B_{k-1}$, and $\|A_k x\| > k$ for every $x \in B_k$, $k = 1, 2, \dots$. Applying Cantor's intersection theorem we conclude that $\cap_k \overline{B_k} \neq \varnothing$. If x is a common point of the balls $\overline{B_k}$, then $\|A_k x\| \geq k$ for every k, contradicting the boundedness of $\mathcal{A}(x)$. $\square$

Proposition 2.24 *Let (x_n) be a sequence in a normed space X.*

(a) *If $x_n \rightharpoonup x$, then the sequence (x_n) is bounded.*
(b) *If $x_n \rightharpoonup x$ in X and $\varphi_n \to \varphi$ in X', then $\varphi_n(x_n) \to \varphi(x)$.*
(c) *If $x_n \to x$ in X and $\varphi_n \rightharpoonup \varphi$ in X', then $\varphi_n(x_n) \to \varphi(x)$.*

Proof

(a) We apply Theorem 2.23 for the family $(\Phi_n) \subset X''$ of the functionals

$$\Phi_n \varphi := \varphi(x_n), \quad \varphi \in X', \ n = 1, 2, \dots,$$

and we use the equalities $\|\Phi_n\| = \|x_n\|$ from Corollary 2.21 (a) (p. 79).
(b) The right-hand side of the identity

$$\varphi_n(x_n) - \varphi(x) = (\varphi_n - \varphi)(x_n) + \varphi(x_n - x)$$

tends to zero because $x_n \rightharpoonup x$ implies $\varphi(x_n - x) \to 0$, and because (x_n) is bounded by (a), so that

$$|(\varphi_n - \varphi)(x_n)| \leq \|\varphi_n - \varphi\| \sup \|x_n\| \to 0.$$

(c) Writing $\Phi(\psi) := \psi(x)$ we have $\Phi \in X''$, and the right-hand side of the identity

$$\varphi_n(x_n) - \varphi(x) = \varphi_n(x_n - x) + (\varphi_n - \varphi)x = \varphi_n(x_n - x) + \Phi(\varphi_n - \varphi)$$

---

[42] As usual, all balls are considered to be open.

tends to zero because $\varphi_n \rightharpoonup \varphi$ implies $\Phi(\varphi_n - \varphi) \to 0$, and because (φ_n) is bounded by (a), so that

$$|\varphi_n(x_n - x)| \le \|x_n - x\| \sup \|\varphi_n\| \to 0.$$

$\square$

A simple adaptation of the proof of Lemma 1.20 (p. 33) yields the following results:

Lemma 2.25 *Let (x_k) be a bounded sequence in a normed space X.*

(a) *For each $x \in X$ the set*

$$\{\varphi \in X' : \varphi(x_k) \to \varphi(x)\}$$

is a closed linear subspace of X'.

(b) *The set*

$$\{\varphi \in X' : (\varphi(x_k)) \quad converges\ in \quad \mathbb{R}\}$$

is a closed linear subspace of X'.

*Examples

- Let $X = c_0$ or $X = \ell^p$ for some $1 < p < \infty$. Let $k \mapsto (x_n^k)$ be a *bounded* sequence in X, and let $(x_n) \in X$. Lemmas 2.16 and 2.25 (pp. 73 and 83) yield the following characterizations of weak convergence (*component-wise convergence*):

$$(x_n^k) \rightharpoonup (x_n) \iff x_n^k \to x_n \quad \text{for each} \quad n.$$

- In particular, the sequence of the vectors

$$e_k = (\overbrace{0, \ldots, 0}^{k-1}, 1, 0, \ldots), \quad k = 1, 2, \ldots$$

  converges weakly to zero in the above spaces.
- But this sequence does *not* converge weakly in ℓ^1. Indeed, the formula

$$\varphi(x) := \sum_{n=1}^{\infty} (-1)^n x_n, \quad x = (x_n) \in \ell^1$$

  defines a functional $\varphi \in (\ell^1)'$ for which the numerical sequence of numbers $\varphi(e_n) = (-1)^n$ is divergent.

- Let $x_n = e_1 + e_n$, then $x_n \rightharpoonup e_1$ in c_0 by the first example. Observe that $\|x_n\|_\infty \to \|e_1\|_\infty$, but $\|x_n - e_1\|_\infty \nrightarrow 0$. Hence c_0 does not have the Radon–Riesz property.
- Since c_0 is a subspace of ℓ^∞, the relation $x_n \rightharpoonup e_1$ also holds in ℓ^∞. Hence ℓ^∞ does not have the Radon–Riesz property either.
- On the other hand, it will follow from a later result[43] that ℓ^p has the Radon–Riesz property for all $1 < p < \infty$.
- Our next proposition will imply that ℓ^1 also has the Radon–Riesz property.

The fact that component-wise convergence does not imply weak convergence in ℓ^1 also follows from the next surprising result:

*Proposition 2.26 (Schur)[44] *In ℓ^1 the strong and weak convergences coincide.*

Proof It suffices to prove that if $x^k \rightharpoonup x$ in ℓ^1, then $\|x^k - x\|_1 \to 0$. Changing x^k to $x^k - x$ we may assume that $x = 0$.

Assume on the contrary that $x^k \rightharpoonup 0$ in ℓ^1, but $\|x^k\|_1 \nrightarrow 0$. Denoting the elements of x^k by x_n^k, we have $x_n^k \to 0$ for each fixed n by the definition of weak convergence. Set[45]

$$\varepsilon := \limsup \|x^k\|_1 > 0 \quad \text{and} \quad k_0 = n_0 := 0.$$

Proceeding recursively, if k_{m-1} and n_{m-1} have already been defined for some m, then choose a large index $k = k_m > k_{m-1}$ such that

$$\left\|x^{k_m}\right\|_1 > \frac{\varepsilon}{2} \quad \text{and} \quad \sum_{n=1}^{n_{m-1}} |x_n^{k_m}| < \frac{\varepsilon}{10},$$

and then a large integer $n_m > n_{m-1}$ such that

$$\sum_{n > n_m} |x_n^{k_m}| < \frac{\varepsilon}{10}.$$

The formula

$$y_n := \operatorname{sign} x_n^{k_m} \quad \text{if} \quad n_{m-1} < n \le n_m$$

---

[43]Proposition 9.11, p. 328.

[44]Schur [418].

[45]We apply the *gliding hump method* of Lebesgue [291].

defines a sequence $(y_n) \in \ell^\infty$ of norm ≤ 1, satisfying the following inequalities for each $m = 1, 2, \dots$:

$$\sum_{n=1}^{\infty} x_n^{k_m} y_n \geq \sum_{n_{m-1} < n \leq n_m} |x_n^{k_m}| - \sum_{n \leq n_{m-1}} |x_n^{k_m}| - \sum_{n > n_m} |x_n^{k_m}|$$

$$= \|x^{k_m}\|_1 - 2\sum_{n \leq n_{m-1}} |x_n^{k_m}| - 2\sum_{n > n_m} |x_n^{k_m}|$$

$$> \frac{\varepsilon}{2} - \frac{4\varepsilon}{10}$$

$$= \frac{\varepsilon}{10}.$$

Hence $x^{k_m} \nrightarrow 0$, and thus $x^k \nrightarrow 0$, contradicting our hypothesis. $\qquad\square$

Finally we prove an interesting converse of Hölder's inequality:

*Proposition 2.27 (Hellinger–Toeplitz)[46] *Let (y_n) be a real sequence and $p, q \in [1, \infty]$ two conjugate exponents. If the series $\sum x_n y_n$ converges for every $(x_n) \in \ell^p$, then $y \in \ell^q$.*

Proof [47] The formula

$$\varphi_k(x) := \sum_{n=1}^{k} x_n y_n, \quad x \in \ell^p, \quad k = 1, 2, \dots$$

defines a sequence (φ_k) in $(\ell^p)'$.[48] By assumption the sequence $(\varphi_k(x))$ is convergent, and hence bounded, for every $x \in \ell^p$. Applying the Banach–Steinhaus theorem there exists therefore a constant C such that

$$\left|\sum_{n=1}^{k} x_n y_n\right| \leq C \|x\|_p \quad \text{for every} \quad x \in \ell^p, \quad k = 1, 2, \dots .$$

If $q = \infty$ and thus $p = 1$, then choosing $x = e_k$ we deduce that $|y_k| \leq C$ for all k, and hence $y \in \ell^\infty$.

If $1 \leq q < \infty$, then introducing for each k the sequence

$$x_n := \begin{cases} |y_n|^{q-1} \operatorname{sign} x_n & \text{if} \quad n \leq k, \\ 0 & \text{if} \quad n > k, \end{cases}$$

---

[46]Hellinger–Toeplitz [201] and Landau [282].

[47]See also a short elementary proof of Riesz [382, pp. 47–48] by the *gliding hump method*.

[48]The continuity of the functionals is evident because we have only finite sums here.

similarly to the proof of Proposition 2.15 we obtain that

$$\sum_{n=1}^{k} |y_n|^q \le C^p.$$

Letting $k \to \infty$ we conclude that $y \in \ell^q$ and $\|y\|_q \le C$.   □

Our next objective is to generalize the Bolzano–Weierstrass theorem to Banach spaces. Unfortunately, there are counterexamples even for the weak convergence:

Examples

- In ℓ^1 the bounded sequence (e_n) has no weakly convergent subsequence. Indeed, such a subsequence would also converge strongly by Schur's theorem (p. 84). But this is impossible because no subsequence has the Cauchy property: $\|e_m - e_n\| = 2$ for all $m \ne n$.

  We can avoid the use of Schur's theorem as follows. If (e_{n_k}) is an arbitrary subsequence of (e_n), then the formula

$$\varphi(x) := \sum_{k=1}^{\infty} (-1)^k x_{n_k}, \quad x = (x_n) \in \ell^1$$

  defines a functional $\varphi \in (\ell^1)'$. Since $\varphi(e_{n_k}) = (-1)^k$ does not converge as $k \to \infty$, the subsequence (e_{n_k}) does not converge weakly.
- In c_0 the bounded sequence $(e_1 + \cdots + e_n)$ has no weakly convergent subsequence. Indeed, if we had $e_1 + \cdots + e_{n_k} \rightharpoonup a$ for some subsequence, then we would also have $\varphi(e_1 + \cdots + e_{n_k}) \to \varphi(a)$ for every $\varphi \in c_0'$. Applying this for each fixed $m = 1, 2, \ldots$ to the functional $\varphi(y) := y_m$, we would get the equality $a = (1, 1, \ldots)$. But this is impossible because the last sequence does not belong to c_0.
- The bounded sequence $(e_1 + \cdots + e_n)$ has no weakly convergent subsequence in ℓ^∞ either. Indeed, the previous reasoning shows again that the only possible weak limit is $a = (1, 1, \ldots)$. But this is impossible because a does not belong to c_0, which is the closed subspace generated by the sequence $(e_1 + \cdots + e_n)$: see Proposition 2.22 (e), p. 80.

Nevertheless, we will see later[49] that the above sequences converge in a natural, even weaker sense.

In spite of these counterexamples, we prove in the next section that the weak convergence version of the Bolzano–Weierstrass theorem remains valid in a large class of Banach spaces.

---

[49] See the examples on p. 136.

2.6 Reflexive Spaces: Theorem of Choice

Let X be a normed space. We recall from Corollary 2.21 (p. 79) that the formula

$$\Phi_x(\varphi) := \varphi(x), \quad \varphi \in X'$$

defines a functional $\Phi_x \in X''$ for each $x \in X$, where X'' denotes the bidual of X.

In certain spaces *every* element of X'' has this form:

Definition A normed space X is *reflexive*[50] if for each $\Phi \in X''$ there exists an $x \in X$ such that

$$\Phi(\varphi) = \varphi(x) \quad \text{for all} \quad \varphi \in X'.$$

Before giving many examples, we discuss some consequences of the definition. We recall from Corollary 2.21 that the formula

$$(Jx)(\varphi) := \varphi(x), \quad x \in X, \quad \varphi \in X'$$

defines a linear isometry $J : X \to X''$.

Proposition 2.28 *(Hahn)*[51] *Let X be a normed space.*

(a) *X is reflexive $\iff$ J is an isometric isomorphism between X and X''.*
(b) *If X is reflexive, then it is complete, i.e., a Banach space.*

Proof

(a) We already know that J is a linear isometry. By definition, J is surjective $\iff$ X is reflexive.
(b) X is isomorphic to $X'' = (X')'$, and every dual space is complete.

$\square$

Remark Reflexive Banach spaces are often identified with their bidual by the map J.

Now we turn to the examples.

Proposition 2.29

(a) *Every finite-dimensional normed space is reflexive.*
(b) *Every Hilbert space is reflexive.*
(c) *The spaces ℓ^p spaces are reflexive for all $1 < p < \infty$.*

---

[50]Hahn [182].
[51]Hahn [182].

Proof

(a) We recall from linear algebra that $\dim X = \dim X^*$ for every finite-dimensional vector space X. Hence we have $\dim X \geq \dim X''$ for every finite-dimensional normed space X.[52] Therefore the linear isometry $J : X \to X''$ must be onto (and $\dim X = \dim X''$).

(b) Let H be a Hilbert space and consider the Riesz–Fréchet isomorphism (Theorem 1.9, p. 19) $j : H \to H'$ defined by the formula

$$(jy)(x) = (x, y), \quad x, y \in H. \tag{2.5}$$

For each $\Phi \in H''$, $\Phi \circ j$ is a continuous linear functional on H. Applying the Riesz–Fréchet theorem again, there exists an $x \in H$ such that

$$\Phi(jy) = (y, x) \quad \text{for all} \quad y \in H.$$

Using (2.5) this implies

$$\Phi(jy) = (jy)(x) \quad \text{for all} \quad y \in H.$$

Since $j : H \to H'$ is onto, we conclude that

$$\Phi(\varphi) = \varphi(x) \quad \text{for all} \quad \varphi \in H'.$$

(c) Consider the Riesz isomorphism $j : \ell^q \to (\ell^p)'$ (Proposition 2.15, p. 73) defined by the formula

$$(jy)(x) = \sum y_n x_n, \quad x \in \ell^p, \ y \in \ell^q. \tag{2.6}$$

For each $\Phi \in (\ell^p)''$, $\Phi \circ j$ is a continuous linear functional on ℓ^q. Applying Proposition 2.15 again, there exists an $x \in \ell^p$ such that

$$\Phi(jy) = \sum x_n y_n \quad \text{for all} \quad y \in \ell^q.$$

Using (2.6) this implies

$$\Phi(jy) = (jy)(x) \quad \text{for all} \quad y \in \ell^q.$$

---

[52]In fact we have equality because every linear functional is continuous on finite-dimensional normed spaces.

Since $j : \ell^q \to (\ell^p)'$ is onto, we conclude that

$$\Phi(\varphi) = \varphi(x) \quad \text{for all} \quad \varphi \in (\ell^p)'.$$

$\square$

Now we give some examples of non-reflexive Banach spaces.

*Examples

- c_0 is not reflexive: the formula

$$\Phi(\varphi) := \sum_{n=1}^{\infty} \varphi_n, \quad \varphi = (\varphi_n) \in \ell^1$$

defines a functional $\Phi \in c_0'' = (\ell^1)'$ which is not represented by any $(x_n) \in c_0$.
  Indeed, if such a sequence (x_n) existed, then choosing $\varphi := e_k$ in the corresponding equality

$$\sum_{n=1}^{\infty} \varphi_n = \sum_{n=1}^{\infty} x_n \varphi_n$$

we would get $x_k = 1$ for every k. But the constant sequence $(1, 1, \ldots)$ does not belong to c_0.
  Let us give another proof. Since c_0' is isomorphic to ℓ^1, and $(\ell^1)'$ is isomorphic to ℓ^∞, c_0'' is isomorphic to ℓ^∞. Consequently, c_0'' is not separable. Since c_0 is separable, it cannot be isomorphic to c_0''.

- ℓ^1 is not reflexive. For the proof we consider the subspace c of ℓ^∞ formed by the convergent sequences.
  Applying Theorem 2.11 theorem we extend the continuous linear functional $(y_n) \mapsto \lim y_n$, given on c, to a functional $\Phi \in (\ell^\infty)' = (\ell^1)''$. We claim that Φ is not represented by any sequence $(x_n) \in \ell^1$.
  Indeed, if such a sequence (x_n) existed, then choosing $y := e_k$ in the corresponding equality

$$\Phi(y) = \sum_{n=1}^{\infty} x_n y_n$$

we would get $x_k = 0$ for every k, i.e., $\Phi = 0$. But this is impossible because for $x = (1, 1, \ldots)$ we have $\Phi(x) = \lim 1 = 1$.

- We will give further proofs for the non-reflexivity of c_0, ℓ^1 and ℓ^∞ at the end of this section and in Sect. 3.6 (p. 144).

One of the most important properties of reflexive spaces is the following:

Theorem 2.30 (Theorem of Choice)[53] *In a reflexive Banach space every bounded sequence has a weakly convergent subsequence.*

Remark The converse of this theorem also holds: see Theorem 3.21, p. 140.

Proof Let (x_k) be a bounded sequence in a reflexive Banach space X. We identify X with its bidual X'', so that for every set $\Delta \subset X'$ we have

$$\Delta^{\perp} := \left\{\Phi \in X'' \; : \; \Phi(\varphi) = 0 \quad \text{for all} \quad \varphi \in \Delta\right\}$$
$$= \{x \in X \; : \; \varphi(x) = 0 \quad \text{for all} \quad \varphi \in \Delta\}.$$

Let us arrange the finite linear combinations of the vectors x_k with rational coefficients into a sequence (y_n). Applying Corollary 2.13 (b) (p. 68) we fix for each n a functional $\varphi_n \in X'$ satisfying

$$\|\varphi_n\| \le 1 \quad \text{and} \quad |\varphi_n(y_n)| = \|y_n\|.$$

Applying Cantor's diagonal method similarly to the proof of Theorem 1.21 (p. 33), we obtain a subsequence (z_k) of (x_k) such that the numerical sequence

$$k \mapsto \varphi_n(z_k)$$

converges for each fixed n. Since for $\varphi \perp \{z_k\}$ the numerical sequence $(\varphi(z_k))$ vanishes identically, $(\varphi(z_k))$ converges for every

$$\varphi \in \Delta := \{\varphi_n\} \cup \{z_k\}^{\perp}.$$

Assume temporarily that Δ generates X'. Then $(\varphi(z_k))$ converges for every $\varphi \in X'$ by Lemma 2.25 (p. 83), so that the formula

$$\Phi(\varphi) := \lim \varphi(z_k)$$

defines a map $\Phi : X' \to \mathbb{R}$. This map is clearly linear. Letting $k \to \infty$ in the inequalities

$$|\varphi(z_k)| \le \|z_k\| \cdot \|\varphi\| \le \sup_k \|z_k\| \cdot \|\varphi\|$$

---

[53]Riesz [379, 380] and Pettis [357].

we obtain

$$|\Phi(\varphi)| \leq \sup_k \|x_k\| \cdot \|\varphi\|$$

for every $\varphi \in X'$. Since (x_k) is bounded, we conclude that Φ is continuous and $\|\Phi\| \leq \sup_k \|x_k\|$. Since X is reflexive, $\Phi \in X''$ may be represented by a vector $x \in X$:

$$\Phi(\varphi) = \varphi(x)$$

for all $\varphi \in X'$. In view of the definition of Φ this yields $\varphi(z_k) \to \varphi(x)$ for all $\varphi \in X'$, i.e., $z_k \rightharpoonup x$.

It remains to show that Δ generates X'. By Corollary 2.9 (p. 64) it is sufficient to show that $\Delta^\perp = \{0\}$.

For any given $y \in \Delta^\perp$ we have $\varphi_n(y) = 0$ for all n by the definition of Δ, and y belongs to the closed subspace $\{z_k\}^{\perp\perp}$ generated by $\{z_k\}$. (We apply Corollary 2.9 again.) Choose a subsequence $y_{n_k} \to y$, then

$$\|y_{n_k}\| = |\varphi_{n_k}(y_{n_k})| = |\varphi_{n_k}(y_{n_k} - y)| \leq \|y_{n_k} - y\|.$$

Letting $k \to \infty$ we conclude that $\|y\| \leq 0$, i.e., $y = 0$.   □

Examples We have seen in the previous section that ℓ^1, ℓ^∞ and c_0 have bounded sequences without convergent subsequences. Applying the theorem we conclude again that these spaces are not reflexive.

2.7 Reflexive Spaces: Geometrical Applications

Using Theorem 2.30 (p. 90) we may generalize several results of plane geometry, mentioned in the introduction, to arbitrary reflexive Banach spaces.

Proposition 2.31 *If X is a normed space, then the properties below satisfy the following implications:*

$$(a) \Longrightarrow (b) \Longrightarrow (c) \Longrightarrow (d) \Longrightarrow (e).$$

(a) *X is reflexive.*
(b) *(Tukey)*[54] *Let A and B be disjoint non-empty convex, closed sets in X. If at least one of them is* bounded, *then there exist a functional $\varphi \in X'$ and real numbers*

---

[54]Tukey [460].

c_1, c_2 such that

$$\varphi(a) \le c_1 < c_2 \le \varphi(b) \quad \text{for all} \quad a \in A \quad \text{and} \quad b \in B. \tag{2.7}$$

(c) *If $K \subset X$ is a non-empty convex, closed set and $x \in X$, then there exists a point $y \in K$ at a minimal distance from x:*

$$\|x - y\| \le \|x - z\| \quad \text{for all} \quad z \in K.$$

(d) *If $M \subset X$ is a proper non-empty closed subspace, then there exists an $x \in X$ satisfying*

$$\|x\| = 1 \quad \text{and} \quad \text{dist}(x, M) = 1.$$

(e) *If $\varphi \in X'$ is a non-zero functional, then there exists an $x \in X$ satisfying*

$$\|x\| = 1 \quad \text{and} \quad |\varphi(x)| = \|\varphi\|.$$

*Remarks

- Let us compare property (b) with Theorem 2.5 (c) (p. 61): We recall[55] that *every* infinite-dimensional normed space contains bounded and closed, but *noncompact* sets.
- Klee[56] proved the converse implication (b) $\implies$ (a): he constructed in every non-reflexive normed space two disjoint non-empty convex, bounded and closed sets, that cannot be separated in the sense of (2.7).
- Property (c) is the generalization of the orthogonal projection Theorem 1.5 (p. 12). In strictly convex spaces[57] the point y is unique. Indeed, if y_1, y_2 are two distinct points in K with $c := \|x - y_1\| = \|x - y_2\|$, then $c > 0$ (for otherwise $y_1 = x = y_2$), and $(y_1 + y_2)/2 \in K$ is closer to x:

$$\left\| x - \frac{y_1 + y_2}{2} \right\| = \left\| \frac{(x - y_1) + (x - y_2)}{2} \right\| < c.$$

See also Proposition 9.10, p. 326.
- It is interesting to compare (d) with Proposition 2.1 (b), p. 55.
- In Hilbert spaces property (d) is equivalent to the existence of a unit vector, orthogonal to M.

---

[55] See Proposition 2.1, p. 55.
[56] Klee [250].
[57] See p. 67.

- Property (e) shows that in a reflexive space X we have

$$\|\varphi\| = \max_{\|x\|\leq 1} |\varphi(x)|$$

for every functional $\varphi \in X'$, i.e., we may write max instead of sup.
- James[58] also established the implication (e) $\implies$ (a) so that the above five properties are in fact *equivalent*.

Proof

(a) $\implies$ (b). We may repeat the proof of Theorem 2.5 (c) (p. 61), except the proof of the inequality $\text{dist}(A, B) > 0$. Now we can proceed as follows:

If $\text{dist}(A, B) = 0$, then there exist two sequences $(a_n) \subset A$ and $(b_n) \subset B$ satisfying $\|a_n - b_n\| \to 0$. If for example A is bounded (the other case is analogous), then there exists a *weakly* convergent subsequence $a_{n_k} \rightharpoonup a$. Since $a_n - b_n \rightharpoonup 0$, this implies that $b_{n_k} \rightharpoonup a$. Since A and B are *convex, closed* sets, $a \in A$ and $a \in B$, contradicting the disjointness of A and B.

(b) $\implies$ (c). We may assume by translation that $x = 0$. It is sufficient to show that every non-empty convex, closed set K has an element of minimal norm. The case $0 \in K$ is obvious. Henceforth we assume that $0 \notin K$; then $r := \text{dist}(0, K) > 0$ by the closedness of K.

Assume on the contrary that K has no element of minimal norm. Then we may apply property (b) to the sets

$$A := \{x \in X : \|x\| \leq r\}$$

and $B := K$ to get $\varphi \in X'$ and $c_1, c_2 \in \mathbb{R}$ satisfying (2.7).

Let (y_n) be a sequence in K satisfying $\|y_n\| \to r$. Then

$$c_2 \leq \varphi(y_n) = \frac{\|y_n\|}{r}\varphi\left(\frac{ry_n}{\|y_n\|}\right) \leq \frac{\|y_n\|}{r}c_1 \to c_1,$$

contradicting the inequality $c_1 < c_2$.

(c) $\implies$ (d). For any fixed $z \in X \setminus M$ there exists by (c) a closest point $y \in M$ to z:

$$\|z - y\| \leq \|z - u\| \quad \text{for all} \quad u \in M.$$

Since $z - y \neq 0$ (because $z \notin M$ and $y \in M$), this may be rewritten as

$$1 \leq \left\| \frac{z - y}{\|z - y\|} - \frac{u - y}{\|z - y\|} \right\| \quad \text{for all} \quad u \in M,$$

---

[58] James [226]. See, e.g., Diestel [103] or Holmes [216].

or, using the unit vector $x := (z - y)/\|z - y\|$, as

$$1 \le \left\| x - \frac{u - y}{\|z - y\|} \right\| \quad \text{for all} \quad u \in M.$$

If u runs over the subspace M, then $\frac{u-y}{\|z-y\|}$ also runs over M, so that $\text{dist}(x, M) \ge 1$. Since $0 \in M$, the converse inequality is obvious.

(d) $\implies$ (e). Applying (d) to the kernel $M := \varphi^{-1}(0)$ of φ, there exists an $x \in X$ satisfying

$$\|x\| = 1 = \text{dist}(x, M).$$

It suffices to show that

$$|\varphi(z)| \le |\varphi(x)| \cdot \|z\|$$

for all $z \in X$, because this will imply $\|\varphi\| \le |\varphi(x)|$; since $\|x\| = 1$, the converse inequality is obvious.

The required inequality is obvious if $\varphi(z) = 0$. If $\varphi(z) \ne 0$, then the equality

$$\varphi \left(x - \frac{\varphi(x)}{\varphi(z)} z \right) = \varphi(x) - \frac{\varphi(x)}{\varphi(z)} \varphi(z) = 0$$

implies $x - \frac{\varphi(x)}{\varphi(z)} z \in M$, and hence

$$1 \le \left\| x - \left(x - \frac{\varphi(x)}{\varphi(z)} z \right) \right\| = \frac{|\varphi(x)|}{|\varphi(z)|} \|z\| ,$$

i.e., $|\varphi(z)| \le |\varphi(x)| \cdot \|z\|$. $\square$

Examples We show that properties (b)–(e) may fail in non-reflexive spaces. Let $X = \ell^1$, and fix a positive, strictly increasing sequence (α_n) converging to one, for example $\alpha_n := n/(n + 1)$.

- The formula $\varphi(x) := \sum \alpha_n x_n$ defines a functional of norm 1. Indeed, on the one hand we have

$$|\varphi(x)| \le \sum |x_n| = \|x\|_1$$

for all $x \in \ell^1$, whence $\|\varphi\| \le 1$.
  On the other hand, we have

$$\|\varphi\| \ge |\varphi(e_n)| = |\alpha_n|$$

for all n, and $|\alpha_n| \to 1$.

But the norm $\|\varphi\| = 1$ is not attained because

$$|\varphi(x)| < \|x\|_1 \quad \text{for all} \quad x \neq 0.$$

Indeed, there is at least one non-zero component x_k of x, and then

$$|\varphi(x)| \leq |\alpha_k| \cdot |x_k| + \sum_{n \neq k} |\alpha_n| \cdot |x_n|$$

$$\leq |\alpha_k| \cdot |x_k| + \sum_{n \neq k} |x_n|$$

$$< \sum |x_n| = \|x\|_1 .$$

Hence property (e) is not satisfied.

- The kernel $M := \varphi^{-1}(0)$ of the above functional is a closed hyperplane. We show that $\text{dist}(x, M) < \|x\|$ for all $x \neq 0$, so that property (d) is not satisfied.

  We already know that $|\varphi(x)| < \|x\|_1$ if $x \neq 0$, and hence $|\varphi(x)| < \alpha_k \|x\|_1$ for all sufficiently large k. Then

$$z := x - \frac{\varphi(x)}{\alpha_k} e_k \in M$$

because

$$\varphi(z) = \varphi(x) - \frac{\varphi(x)}{\alpha_k} \varphi(e_k) = 0,$$

and hence

$$\text{dist}(x, M) \leq \|x - z\|_1 = \left| \frac{\varphi(x)}{\alpha_k} \right| < \|x\|_1 .$$

- Consider the above hyperplane M. If $x \in X \setminus M$, then the distance $\text{dist}(x, M)$ is not attained, so that property (c) is not satisfied for $K = M$.

  Indeed, if we had $\text{dist}(x, M) = \|x - z\|_1$ for some $z \in M$, then we would also have $\text{dist}(x - z, M) = \|x - z\|_1$ because $\text{dist}(x - z, M) = \text{dist}(x, M)$. But this would contradict our previous result because $x \notin M$ and therefore $x - z \neq 0$.

- We have just seen that the distance $r := \text{dist}(x, M)$ is not attained for any $x \in X \setminus M$. Therefore the above *proof* of the implication (b) $\Longrightarrow$ (c) shows that $A := B_r(x)$ and $B := M$ cannot be separated in the sense of (2.7).

Remark Similar examples may be given in $X = c_0$ by using the linear functional $\varphi(x) := \sum 2^{-n} x_n$.

2.8   * Open Mappings and Closed Graphs

The results of this section play an important role in the theory of partial differential equations.[59]

Theorem 2.32 *Let X and Y be two Banach spaces.*

(a) *(Open mapping theorem)*[60] *If $A \in L(X, Y)$ is* onto, *then A maps every open set of X onto an open set of Y.*

(b) *(Inverse mapping theorem)*[61] *If $A \in L(X, Y)$ is* bijective, *then its inverse A^{-1} is continuous.*

(c) *(Equivalent norms)*[62] *Let $\|\cdot\|_1$ and $\|\cdot\|_2$ be two* complete *norms on a vector space Z. If there exists a constant c_1 such that $\|z\|_2 \leq c_1 \|z\|_1$ for all $z \in Z$, then there also exists a constant c_2 such that $\|z\|_1 \leq c_2 \|z\|_2$ for all $z \in Z$.*

(d) *(Closed graph theorem)*[63] *If the linear map $A : X \to Y$ has a closed graph*

$$\{(x, Ax) \ : \ x \in X\}$$

in $X \times Y$, then A is continuous.

Remark The converse of the last property always holds: if X, Y are topological spaces and $f : X \to Y$ is continuous function, then its graph $\{(x, f(x)) \ : \ x \in X\}$ is closed in $X \times Y$.

All these theorems are based on the following key lemma. For simplicity we denote by B_r the unit ball of radius r, centered at 0 in both spaces X and Y.

Lemma 2.33 *Let $A \in L(X, Y)$, where X and Y are Banach spaces. If A is* onto, *then there exists an $r > 0$ such that $B_r \subset A(B_1)$.*

Proof First we prove that there exists an $r > 0$ such that

$$B_{2r} \subset \overline{A(B_1)}. \tag{2.8}$$

Since A is onto,

$$Y = \bigcup_{k=1}^{\infty} A(B_k) = \bigcup_{k=1}^{\infty} \overline{A(B_k)}.$$

---

[59]See, e.g., Hörmander [218, 219].
[60]Schauder [413].
[61]Banach [22].
[62]Banach [22].
[63]Banach [24].

By Baire's lemma (p. 32) at least one of the sets $\overline{A(B_k)}$ contains a ball, say $B_s(y) \subset \overline{A(B_k)}$.[64] Then we have

$$B_s(-y) \subset -\overline{A(B_k)} = \overline{A(B_k)}.$$

If $x \in B_s$, then $x \pm y \in B_s(\pm y) \subset \overline{A(B_k)}$; using the convexity of $\overline{A(B_k)}$, this yields

$$x = \frac{(x+y) + (x-y)}{2} \in \overline{A(B_k)}.$$

We thus have $B_s \subset \overline{A(B_k)}$, and (2.8) follows by homogeneity with $r := s/2k$.

Now we fix an arbitrary point $y \in B_r$. We seek $x \in B_1$ satisfying $Ax = y$. For this we observe that (2.8) implies by similarity the more general relations

$$B_{2^{1-n}r} \subset \overline{A(B_{2^{-n}})}, \quad n = 1, 2, \ldots.$$

Using them we may construct recursively a sequence $x_1, x_2, \ldots$ in X such that

$$\|x_n\| < \frac{1}{2^n} \quad \text{and} \quad \|y - A(x_1 + \cdots + x_n)\| < \frac{r}{2^n}$$

for all n. Then the series $\sum x_n$ converges to some point $x \in B_1$. Using the continuity of A we conclude that $Ax = \sum Ax_n = y$. $\qquad\square$

Proof of Theorem 2.32

(a) Given an open set U in X and a point $x \in U$, we have to find $s > 0$ such that $B_s(Ax) \subset A(U)$. Fix $\varepsilon > 0$ satisfying $B_\varepsilon(x) \subset U$; then the choice $s := r\varepsilon$ is suitable. Indeed, applying the lemma we have

$$B_s(Ax) = Ax + B_s = Ax + \varepsilon B_r \subset Ax + \varepsilon A(B_1) = A(B_\varepsilon(x)) \subset A(U).$$

(b) follows from (a) by using the characterization of continuity by open sets.
(c) The identity map is continuous from $(Z, \|\cdot\|_1)$ to $(Z, \|\cdot\|_2)$ by assumption. Applying (b) we conclude that it is an isomorphism.
(d) The formula

$$\|x\|_1 := \|x\| + \|Ax\|$$

defines a complete norm on X by our assumption. Since we have obviously $\|\cdot\| \le \|\cdot\|_1$, by (c) there exists a constant c_2 such that $\|\cdot\|_1 \le c_2 \|\cdot\|$. Hence A is continuous (and $\|A\| \le c_2 - 1$).

$\qquad\square$

---

[64]Of course, then all the others also contain some balls by homogeneity, but we do not need this here.

The above proofs may be simplified for reflexive spaces.[65] We show this for the inverse mapping theorem:

Proof of the Inverse Mapping Theorem if X is Reflexive Since A is onto, the sets

$$F_k := \{Ax \ : \ \|x\| \le k\} = A(\overline{B_k}), \quad k = 1, 2, \ldots$$

cover Y. Assume for a moment that these sets are closed. Then at least one of them contains a ball by Baire's theorem, say $B_r(y) \subset F_k$. Hence

$$\|A^{-1}x\| \le k \quad \text{for all} \quad x \in B_r(y),$$

and therefore

$$\|A^{-1}x\| \le k + \|A^{-1}y\| \quad \text{for all} \quad x \in B_r(0).$$

Consequently, $\|A^{-1}\| \le r^{-1}(k + \|A^{-1}y\|)$.

It remains to prove the closedness of the sets F_k. If $\|x_n\| \le k$ and $Ax_n \to y \in Y$, then there exists a weakly convergent subsequence $x_{n_k} \rightharpoonup x$ by the reflexivity of X, and $\|x\| \le k$ by a basic property of the weak convergence. Then[66] $Ax_n \rightharpoonup Ax$ by the continuity of A, and therefore $y = Ax \in F_k$ by the uniqueness of the weak limit.
□

We give only one application here:

Proposition 2.34 (Hellinger–Toeplitz)[67] *Let $A, B : H \to H$ be two linear maps on a Hilbert space H. If*

$$(Ax, y) = (x, By)$$

for all $x, y \in H$, then A and B are continuous.

Proof For the continuity of A (the case of B is analogous) it suffices to show that $x_n \to x$ and $Ax_n \to z$ imply $z = Ax$. Indeed, then we may conclude by applying the closed graph theorem.

Letting $n \to \infty$ in the equality $(Ax_n, y) = (x_n, By)$ we get

$$(z, y) = (x, By), \quad \text{i.e.,} \quad (Ax - z, y) = 0$$

for all $y \in H$. Choosing $y := Ax - z$ this yields the required equality $Ax = z$.   □

---

[65]Private communication of O. Gebuhrer.

[66]See Proposition 2.36 below

[67]Hellinger–Toeplitz [202, pp. 321–327] and Stone [439].

Remarks

- Instead of the closed graph theorem we may also apply the Banach–Steinhaus theorem here. Indeed, denote by F the closed unit ball of H and introduce for each $y \in F$ the linear functional φ_y by the formula

$$\varphi_y(x) := (x, By);$$

we clearly have $\|\varphi_y\| = \|By\|$.
  The family $\{\varphi_y\}$ is pointwise bounded because for each fixed $x \in H$ we have

$$\left|\varphi_y(x)\right| = |(Ax, y)| \leq \|Ax\| \cdot \|y\| \leq \|Ax\|$$

for all $y \in F$. Then the family is uniformly bounded by the Banach–Steinhaus theorem, and thus

$$\|B\| = \sup_{y \in F} \|By\| = \sup_{y \in F} \|\varphi_y\| < \infty.$$

2.9 * Continuous and Compact Operators

As in the case of Hilbert spaces, the introduction of the adjoint operator helps to clarify the relationship between continuity and weak convergence.

Definition Let X and Y be normed spaces and $A \in L(X, Y)$. By the *adjoint*[68] of A we mean the linear map $A^* : Y' \to X'$ defined by the formula

$$A^*\varphi := \varphi A, \quad \varphi \in Y'.$$

Remarks

- If $X = Y$ is a Hilbert space, then this definition reduces to that of the preceding chapter if we identify X' with X by the Riesz–Fréchet theorem.
- In order to emphasize the analogy with the scalar product we often write $\langle \varphi, x \rangle$ instead of $\varphi(x)$; then the definition of the adjoint takes the following form:

$$\langle A^*\varphi, x \rangle = \langle \varphi, Ax \rangle \quad \text{for all} \quad x \in X.$$

Proposition 2.35 *Let X, Y and Z be normed spaces.*

(a) *If $A \in L(X, Y)$, then $A^* \in L(Y', X')$ and $\|A^*\| = \|A\|$.*
(b) *The map $A \mapsto A^*$ is a linear isometry.*

---

[68]Riesz [379, 380], Banach [22], and Schauder [414].

(c) *If $B \in L(X, Y)$ and $A \in L(Y, Z)$, then $(AB)^* = B^* A^*$.*
(d) *If $A \in L(X, Y)$ is bijective, then $A^* \in L(Y', X')$ is also bijective, and*

$$(A^*)^{-1} = (A^{-1})^*.$$

Proof Only the equality $\|A^*\| = \|A\|$ requires a proof.[69] The inequality $\|A^*\| \leq \|A\|$ follows from obvious estimate

$$\|A^* \varphi\| = \|\varphi A\| \leq \|\varphi\| \cdot \|A\|,$$

valid for all $\varphi \in X'$.

For the proof of the converse inequality we choose for each $x \in X$ a functional $\varphi \in X'$ satisfying $\|\varphi\| \leq 1$ and $\varphi(Ax) = \|Ax\|$.[70] Then we have

$$\|Ax\| = \varphi(Ax) = (A^* \varphi)x \leq \|A^*\| \cdot \|\varphi\| \cdot \|x\| \leq \|A^*\| \cdot \|x\|,$$

and hence $\|A\| \leq \|A^*\|$.   □

Now we generalize the characterization of continuous and completely continuous operators.

Proposition 2.36 *Let X, Y be normed spaces and $A : X \to Y$ a linear map. The following properties are equivalent:*

(a) *there exists a constant M such that $\|Ax\| \leq M \|x\|$ for all $x \in X$;*
(b) *A sends bounded sets into bounded sets;*
(c) *A sends totally bounded sets into totally bounded sets;*
(d) *if $x_n \to x$, then $Ax_n \to Ax$;*
(e) *if $x_n \rightharpoonup x$, then $Ax_n \rightharpoonup Ax$;*
(f) *if $x_n \to x$, then $Ax_n \rightharpoonup Ax$.*

Proof Using Propositions 2.24 (a) (p. 82) and 2.35 we may repeat word for word the proof of Proposition 1.22 (p. 35).   □

*Example The embeddings $i : \ell^p \to \ell^q$ are continuous for all $1 \leq p \leq q \leq \infty$. For this we show that $\|x\|_p \leq 1$ implies $\|x\|_q \leq 1$.

If $\|x\|_p \leq 1$, then $|x_n| \leq 1$ for all n; the case $q = \infty$ hence already follows. If $q < \infty$, then the inequalities $|x_n| \leq 1$ imply that

$$\|x\|_q^q = \sum_{n=1}^{\infty} |x_n|^q \leq \sum_{n=1}^{\infty} |x_n|^p = \|x\|_p^p \leq 1.$$

---

[69] Property (d) follows from (c) applied with $B = A^{-1}$.
[70] We apply Corollary 2.13, p. 68.

We have also shown here that $\|i\| \leq 1$. Since $\|x\|_p = \|x\|_q > 0$ for every vector having exactly one non-zero component, we have in fact $\|i\| = 1$.

Definition Let X and Y be Banach spaces. A linear map $A : X \rightarrow Y$ is called *completely continuous* or *compact*[71] if one of the following two equivalent conditions hold:

(a) for every bounded sequence (x_n) in X, (Ax_n) has a convergent subsequence in Y;
(b) A sends bounded sets into totally bounded sets.

Remark The equivalence of the conditions follows from the completeness of Y: see the proof of Proposition 1.24, p. 36.

Let us list some basic properties:

Proposition 2.37 *Let X, Y, Z be Banach spaces.*

(a) *Every completely continuous linear map is continuous.*
(b) *If* $\dim Y < \infty$, *then every* $A \in L(X, Y)$ *map is completely continuous.*
(c) *Let* $B \in L(X, Y)$ *and* $A \in L(Y, Z)$. *If A or B is completely continuous, then AB is completely continuous.*
(d) *The completely continuous linear maps* $A : X \rightarrow Y$ *form a* closed *subspace of* $L(X, Y)$.

Proof (a) We use the fact that every totally bounded set is bounded.
    (b) We observe that the bounded and totally bounded sets are the same in Y.
    (c) and (d) The corresponding proofs of Proposition 1.26 (p. 37) remain valid.
                                                                            □

Examples

• If X is infinite-dimensional, then the identity map $I : X \rightarrow X$ is *not* completely continuous by Proposition 2.1.
• The embeddings $i : \ell^p \rightarrow \ell^q$ are not completely continuous for any $1 \leq p \leq q \leq \infty$: the sequence (e_n) is bounded in ℓ^p, but it has no convergent subsequence in ℓ^q, because $\|e_n - e_m\|_q \geq 1$ for all $n \neq m$.

*Remarks

• If A is completely continuous, then repeating the proof given in Proposition 1.24 we obtain that

$$x_n \rightharpoonup x \quad \Longrightarrow \quad Ax_n \rightarrow Ax. \tag{2.9}$$

---

[71]Hilbert [209] and Riesz [383].

- Conversely, property (2.9) implies the complete continuity if X is *reflexive*: we may repeat the proof given in 1.24. The reflexivity condition cannot be omitted: for example, the identity map of $X = \ell^1$ is not completely continuous (see Proposition 2.1 (d), p. 55), although (2.9) is satisfied (because the strong and weak convergences coincide here by Proposition 2.26, p. 84).

Now we prove a deeper result:

*Proposition 2.38 (Schauder)[72] *If X, Y are Banach spaces and $A \in L(X, Y)$ is completely continuous, then $A^* \in L(Y', X')$ is completely continuous.*

Proof Let Δ be a bounded set in Y': $\|\varphi\| \leq L$ for all $\varphi \in \Delta$. We have to show that

$$\{A^*\varphi \ : \ \varphi \in \Delta\} = \{\varphi \circ A \ : \ \varphi \in \Delta\}$$

is totally bounded in X'. Introducing the closed unit ball F of X, by the definition of the norm of X' this is equivalent to the complete boundedness of

$$\{\varphi \circ A|_F \ : \ \varphi \in \Delta\}$$

in $C_b(F)$, or to the complete boundedness of

$$\{\varphi|_{A(F)} \ : \ \varphi \in \Delta\}$$

in $C_b(A(F))$. Setting finally $K := \overline{A(F)}$, the last property is equivalent to the complete boundedness of

$$\{\varphi|_K \ : \ \varphi \in \Delta\} \tag{2.10}$$

in $C_b(K)$.

Since Y is complete and A is completely continuous, K is compact in Y. Furthermore, the system (2.10) is uniformly bounded and equicontinuous because

$$|\varphi(x)| \leq L\,\|A\| \quad \text{and} \quad |\varphi(x) - \varphi(y)| \leq L\,\|x - y\|$$

for all $\varphi \in \Delta$ and $x, y \in K$. Applying the classical Arzelà–Ascoli theorem,[73] we conclude that the system (2.10) is totally bounded. $\qquad\qquad\square$

---

[72]Schauder [414].

[73]See Proposition 8.7, p. 268. For its proof we will use only basic notions of topology.

2.10 * Fredholm–Riesz Theory

The fact that we restrict ourselves to continuous functions in this work, is inessential.

F. Riesz [383]

In the applications we often encounter operators of the form $I - K$ where K is completely continuous.[74] The purpose of this section is to clarify their structure.

Definition The vector space X is the *direct sum* of the subspaces N and R if

$$X = N + R \quad \text{and} \quad N \cap R = \{0\}.$$

We express this by the notation $X = N \oplus R$.

Remark If $X = N \oplus R$, then $\dim N = \dim X/R$, where X/R denotes the *quotient space* formed by the equivalence classes $y + R$, $y \in N$. Indeed, one may easily check that the linear map $y \mapsto y + R$ is a bijection between N and X/R.

In this section we denote by $N(A)$ and $R(A)$ the *null set* (or *kernel*) and *range* of a linear map A. By an *automorphism* of a normed space X we mean an isomorphism of X onto itself.

Theorem 2.39 (Riesz)[75] *Let X be a Banach space, $K \in L(X, X)$ a completely continuous operator and $T = I - K$. There exists a decomposition $X = N \oplus R$ such that*

- *N and R are T-invariant;*
- *N is finite-dimensional;*
- *R is closed, and the restriction $T|_R$ is an automorphism of R;*
- *there exists a constant C such that*

$$\|y\| + \|z\| \leq C \, \|y + z\|$$

for all $y \in N$ and $z \in R$.

Furthermore, there exists an integer $n \geq 0$ such that $N = N(T^n)$, $R = R(T^n)$, and

$$\{0\} = N(T^0) \subsetneq \cdots \subsetneq N(T^n) = N(T^{n+1}) = \cdots,$$
$$X = R(T^0) \supsetneq \cdots \supsetneq R(T^n) = R(T^{n+1}) = \cdots.$$

We proceed in several steps.

---

[74]For example in electrostatics: see Riesz and Sz.-Nagy [394], §81.
[75]Riesz [383].

Lemma 2.40 *For any fixed integer* $n \geq 0$

(a) $N(T^n)$ *is a finite-dimensional subspace;*
(b) $R(T^n)$ *is a closed subspace.*

Proof The case $n = 0$ is obvious: since $T^0 = I$, $N(I) = \{0\}$ and $R(I) = X$. The case $n \geq 2$ may be reduced to the case $n = 1$, because $T^n = I - K_n$ where

$$K_n = I - (I - K)^n$$

$$= nK - \binom{n}{2}K^2 + \binom{n}{3}K^3 - \cdots + (-1)^{n-1}K^n$$

is a *completely continuous* operator. Assume Henceforth That $n = 1$.

(a) We have $I = K$ on $N(T)$, i.e., the identity map of $N(T)$ is completely continuous. By a lemma of Riesz (p. 55) we conclude that $N(T)$ is finite-dimensional.
(b) We have to show that if

$$Tx_n \to y \quad \text{in} \quad X, \tag{2.11}$$

then $y \in R(T)$. We may assume that $y \neq 0$, and that $Tx_n \neq 0$ for all n. Since $\text{dist}(x_n, N(T)) > 0$ for each n, there exists a $z_n \in N(T)$ such that

$$\|x_n - z_n\| \leq 2 \text{ dist}(x_n, N(T)).$$

Changing x_n to $x_n - z_n$ we have

$$\|x_n\| \leq 2 \text{ dist}(x_n, N(T)), \tag{2.12}$$

and (2.11) remains valid.

Assume for the moment that the sequence (x_n) is bounded. Then there exists a subsequence (x_{n_k}) for which (Kx_{n_k}) is convergent, say $Kx_{n_k} \to z$. It follows that

$$x_{n_k} = Tx_{n_k} + Kx_{n_k} \to y + z,$$

and hence $Tx_{n_k} \to T(y + z)$. Using (2.11) we conclude that $y = T(y + z) \in R(T)$.

Assume on the contrary that (x_n) is not bounded, and choose a subsequence satisfying $\|x_{n_k}\| \to \infty$. Changing (x_n) to $(x_{n_k}/\|x_{n_k}\|)$ the properties (2.11), (2.12) remain valid with $y = 0$, and we also have

$$\|x_n\| = 1 \quad \text{for all} \quad n. \tag{2.13}$$

Repeating the previous reasoning we may get a convergent subsequence $x_{n_k} \to z$. Since $Tx_n \to 0$, hence $Tz = 0$, i.e., $z \in N(T)$. On the other hand, we infer from (2.13) that $\|z\| = 1$. Applying the estimate (2.12) for $n = n_k$, letting $k \to \infty$ we arrive at the impossible inequality $1 \leq 0$. □

Lemma 2.41

(a) *There exists an integer $n \geq 0$ such that*

$$\{0\} = N(T^0) \subsetneqq \cdots \subsetneqq N(T^n) = N(T^{n+1}) = \cdots .$$

(b) *The subspace $N(T^n)$ is T-invariant.*

Proof

(a) If $N(T^k) = N(T^{k+1})$ for some k, then $N(T^{k+1}) = N(T^{k+2})$, because

$$x \in N(T^{k+2}) \iff Tx \in N(T^{k+1}) = N(T^k) \iff x \in N(T^{k+1}).$$

It remains to prove the existence of such a k.

Assuming the contrary, using Proposition 2.1 (p. 55) we could construct a sequence (x_n) satisfying

$$x_n \in N(T^n) \quad \text{and} \quad \|x_n\| = \text{dist}(x_n, N(T^{n-1})) = 1$$

for all $n = 1, 2, \ldots$. Then (x_n) would be bounded, but (Kx_n) would not have any convergent subsequence because

$$\|Kx_n - Kx_m\| \geq 1 \quad \text{for all} \quad n > m.$$

Indeed,

$$Kx_n - Kx_m = x_n - y,$$

where

$$y = x_m - Tx_m + Tx_n \in N(T^{n-1}),$$

and hence

$$\|Kx_n - Kx_m\| \geq \text{dist}(x_n, N(T^{n-1})) = 1.$$

This contradicts the compactness of K.

(b) If $x \in N(T^n)$, then $Tx \in N(T^{n+1}) = N(T^n)$.

□

Lemma 2.42

(a) *There exists an integer $r \geq 0$ such that*

$$X = R(T^0) \supsetneq \cdots \supsetneq R(T^r) = R(T^{r+1}) = \cdots .$$

(b) *The subspace $R(T^r)$ is T-invariant.*
(c) *$T|_{R(T^r)}$ is an automorphism of the subspace $R(T^r)$.*

Proof

(a) If $R(T^k) = R(T^{k+1})$ for some k, then $R(T^{k+1}) = R(T^{k+2})$ because

$$R(T^{k+2}) = TR(T^{k+1}) = TR(T^k) = R(T^{k+1}).$$

It remains to prove the existence of such a k.

Assuming the contrary, using Proposition 2.1 again we could construct a sequence (x_n) satisfying

$$x_n \in R(T^n), \quad \|x_n\| = 2 \quad \text{and} \quad \text{dist}(x_n, R(T^{n+1})) > 1$$

for all $n = 0, 1, \ldots$. Then (x_n) would be bounded, but (Kx_n) would not have any convergent subsequence because

$$\|Kx_n - Kx_m\| > 1 \quad \text{for all} \quad n < m.$$

Indeed,

$$Kx_n - Kx_m = x_n - y,$$

where

$$y = x_m - Tx_m + Tx_n \in R(T^{n+1}),$$

and hence

$$\|Kx_n - Kx_m\| \geq \text{dist}(x_n, R(T^{n+1})) > 1.$$

This contradicts the compactness of K again.
(b) Observe that $TR(T^r) = R(T^{r+1}) = R(T^r)$.
(c) The restriction of T to $R(T^r)$ is onto because

$$TR(T^r) = R(T^{r+1}) = R(T^r).$$

It follows that $T^k|_{R(T^r)}$ is onto for every $k \geq 0$.

The restriction of T to $R(T^r)$ is also injective. Indeed, let $x \in R(T^r)$ satisfy $Tx = 0$, and consider the integer n of the preceding lemma. By the surjectivity there exists a $y \in R(T^r)$ such that $x = T^n y$. Then $0 = Tx = T^{n+1} y$, i.e., $y \in N(T^{n+1}) = N(T^n)$. Consequently, $x = T^n y = 0$.

The inverse of $T|_{R(T^r)}$ is continuous. For the proof we assume on the contrary that there exists a sequence (x_n) in $R(T^r)$, satisfying $Tx_n \to 0$, and $\|x_n\| = 1$ for all n. Since K is compact, there exists a convergent subsequence $Kx_{n_k} \to z$. Then $x_{n_k} = Tx_{n_k} + Kx_{n_k} \to z$. Here we have $z \in R(T^r)$ because $R(T^r)$ is closed,

$$\|z\| = \lim \|x_{n_k}\| = 1 \quad \text{and} \quad Tz = \lim Tx_{n_k} = 0.$$

This contradicts the injectivity of $T|_{R(T^r)}$. $\square$

The following lemma completes the proof of Theorem 2.39.

Lemma 2.43

(a) *The integers n and r of Lemmas 2.41 and 2.42 are equal.*
(b) *We have $X = R(T^n) \oplus N(T^n)$.*
(c) *There exists a constant C such that*

$$\|y\| + \|z\| \le C \|y + z\|$$

for all $y \in N(T^n)$ and $z \in R(T^n)$.

Proof

(a) If $T^{r+1}x = 0$, then $T^r x \in R(T^r)$ and $T(T^r x) = 0$, so that $T^r x = 0$ by the injectivity of $T|_{R(T^r)}$. Hence $N(T^{r+1}) \subset N(T^r)$, whence in fact $N(T^{r+1}) = N(T^r)$. This proves that $r \ge n$.

   If $T^n x \in R(T^n)$, then $T^{n+r}x \in R(T^{n+r}) = R(T^{n+r+1})$ by the preceding lemma, so that there exists $y \in X$ satisfying $T^{n+r+1}y = T^{n+r}x$. Then

$$x - Ty \in N(T^{n+r}) = N(T^n),$$

   whence $T^n x = T^{n+1}y \in R(T^{n+1})$. This implies $R(T^n) \subset R(T^{n+1})$, whence in fact $R(T^n) = R(T^{n+1})$. This proves that $n \ge r$.
(b) Since T^r is injective on $R(T^r)$, $R(T^r) \cap N(T^r) = \{0\}$. On the other hand, for any given $x \in X$ we have $T^r x \in R(T^r)$. Applying the lemma, there exists a unique $u \in R(T^r)$ satisfying $T^{2r}u = T^r x$. Then $y := T^r u \in R(T^r)$ and $z := x - T^r u \in N(T^r)$.
(c) Using the notations of (b) the linear map $T^r x \mapsto u$ is continuous by part (c) of the preceding lemma. By the continuity of T^r we infer that the formula $Px := y$ defines a *continuous* projection $P : X \to R(T^r)$.[76] Then the projection

---

[76]A linear map $P : X \to X$ is called a *projection* if $P^2 = P$.

$Q : X \to N(T^r)$ defined by $Qx := z$ is also continuous because $Q = I - P$. This yields the required estimate with $C := \|P\| + \|Q\|$.

□

As a first application of the theorem we study the spectrum of a completely continuous operator.

Definition The *resolvent set* $\rho(A)$ of an operator $A \in L(X, X)$ is the set of real numbers λ for which $A - \lambda I$ is invertible, i.e., there exists an operator $B \in L(X, X)$ satisfying

$$(A - \lambda I)B = B(A - \lambda I) = I.$$

The complement $\sigma(A) := \mathbb{R} \setminus \rho(A)$ is called the *spectrum* of A.[77]

Examples

- The spectrum contains the eigenvalues.
- If X is finite-dimensional, then the spectrum of A is exactly the set of eigenvalues.
- Using the openness[78] of the set of invertible operators in $L(X, X)$, one can show that $\sigma(A)$ is *closed* and $\sigma(A) \subset [-\|A\|, \|A\|]$.
- Consider the *right shift* of $X = \ell^2$ defined by the formula

$$S_r(x_1, x_2, \ldots) := (x_2, x_3, \ldots).$$

It can be shown that the set of its eigenvalues is $(-1, 1)$. Since $\|S_r\| = 1$, we conclude by using the previous remark that $\sigma(S_r) = [-1, 1]$.[79]

- Consider the *left shift* of $X = \ell^2$ defined by the formula

$$S_l(x_1, x_2, \ldots) := (0, x_1, x_2, \ldots)$$

We have $\|S_l\| = 1$ and $\sigma(S_l) = [-1, 1]$. But S_l has no eigenvalues.[80]

Proposition 2.44 (Riesz)[81] *Let K be a completely continuous operator on the Banach space X.*

(a) *If $\lambda \in \sigma(K)$ and $\lambda \neq 0$, then λ is an eigenvalue of K.*
(b) *The eigensubspaces of K are linearly independent.*
(c) *The spectrum of K is countable.*
(d) *If K has infinitely many eigenvalues, then their sequence tends to zero.*

---

[77]Hilbert [209].

[78]This is proved in most books on differential calculus as a preliminary step for the inverse function theorem.

[79]One can check that $(n^{-1}) \notin R(S_r - I)$ and $((-1)^n n^{-1}) \notin R(S_r + I)$.

[80]$S_l - \lambda I$ is not onto for any $\lambda \in [-1, 1]$ because $e_1 \notin R(S_l - \lambda I)$.

[81]Riesz [383].

Proof (a) We apply Theorem 2.39 for $T := I - \lambda^{-1}K$ instead of $I - K$. Since T is not an isomorphism on X, $R(T^n) \neq X$, i.e., $n \geq 1$. But then $N(T) \neq \{0\}$, so that λ is an eigenvalue of K.

(b) Assume on the contrary that there exist linearly dependent eigenvectors $x_1, \ldots, x_m$, belonging to pairwise different eigenvalues. Choose such a system with a minimal m, and consider a nontrivial linear combination

$$x := c_1 x_1 + \cdots + c_m x_m = 0.$$

Then we have $(A - \lambda_m I)x = 0$, i.e.,

$$c_1(\lambda_1 - \lambda_m)x_1 + \cdots + c_{m-1}(\lambda_{m-1} - \lambda_m)x_{m-1} = 0.$$

This contradicts the minimality of m.

(c) and (d) It suffices to show that for any fixed $\varepsilon > 0$ there are at most finitely many eigenvalues satisfying $|\lambda| > \varepsilon$. Assume on the contrary that there exists an infinite sequence (λ_n) of such eigenvalues. Let $M_0 := \{0\}$, and denote by M_n the vector sum of the eigensubspaces corresponding to $\lambda_1, \ldots, \lambda_n$, for $n = 1, 2, \ldots$. Then M_{n-1} is a *proper* subspace of M_n by property (b), and we have clearly

$$(\lambda_n I - K)M_n \subset M_{n-1}.$$

Applying Proposition 2.1 (p. 55) we may fix for each $n \geq 1$ a point $x_n \in M_n$ satisfying

$$\|x_n\| = \mathrm{dist}(x_n, M_{n-1}) = 1.$$

Since the sequence $(\lambda_n^{-1}x_n)$ is bounded, $(K(\lambda_n^{-1}x_n))$ has a convergent subsequence. But this is impossible because

$$\left\| K(\lambda_n^{-1}x_n) - K(\lambda_m^{-1}x_m) \right\| \geq 1$$

for all $n > m$. This follows from the choice of x_n because

$$K(\lambda_n^{-1}x_n) - K(\lambda_m^{-1}x_m) = x_n - y$$

where

$$y = \lambda_n^{-1}(\lambda_n I - K)x_n + \lambda_m^{-1}Kx_m \in M_{n-1}.$$

□

Now we investigate the equations

$$x - Kx = y \quad \text{and} \quad \varphi - K^*\varphi = \psi$$

where K is a compact operator. The following theorem is a far-reaching generalization of Proposition 1.31 (p. 44):

Theorem 2.45 (Fredholm Alternative)[82] *If K is a compact operator in a Banach space X, then*

(a) $R(I - K) = N(I - K^*)^\perp$;
(b) $R(I - K^*) = N(I - K)^\perp$;
(c) $\dim N(I - K^*) = \dim N(I - K)$;
(d) $N(I - K) = \{0\} \iff R(I - K) = X$.

Proof Let $T = I - K$, then $T^* = I - K^*$.

(a) We have the following equivalences for every $\varphi \in X'$:

$$\varphi \in R(T)^\perp \iff \langle \varphi, Tx \rangle = 0 \quad \text{for all} \quad x \in X$$
$$\iff \langle T^* \varphi, x \rangle = 0 \quad \text{for all} \quad x \in X$$
$$\iff \varphi \in N(T^*).$$

Since $R(T)$ is closed, applying Corollary 2.9 (p. 64) we obtain the required equality:

$$R(T) = \overline{R(T)} = R(T)^{\perp\perp} = N(T^*)^\perp.$$

(b) If $\varphi = T^* \psi \in R(T^*)$ and $x \in N(T)$, then

$$\langle \varphi, x \rangle = \langle T^* \psi, x \rangle = \langle \psi, Tx \rangle = \langle \psi, 0 \rangle = 0.$$

Hence $R(T^*) \subset N(T)^\perp$.

For the proof of the converse relation we fix a subspace Z of $N(T^n)$ such that $N(T^n) = N(T) \oplus Z$, and we set $Y := Z + R(T^n)$. Then the restriction $T|_Y : Y \to R(T)$ is an isomorphism by Theorem 2.39 (p. 103).

If $\varphi \in N(T)^\perp$, then $\varphi \circ (T|_Y)^{-1}$ is a continuous linear functional on $R(T)$. Applying the Helly–Hahn–Banach theorem (p. 65) it can be extended to a functional $\psi \in X'$. Then we have $T^* \psi = \varphi$ because

$$\langle T^* \psi, x \rangle = \langle \psi, Tx \rangle = \varphi(T|_Y)^{-1} Tx = \varphi(x)$$

for all $x \in X$. This proves the relation $N(T)^\perp \subset R(T^*)$.

---

[82]Fredholm [150, 151], Riesz [383], Hildebrandt [212], and Schauder [413].

(c) Let $T' = T|_{N(T^n)}$, and fix a subspace M of $N(T^n)$ satisfying

$$N(T^n) = R(T') \oplus M.$$

Then $X = R(T) \oplus M$ because $X = N(T^n) \oplus R(T^n)$ and $R(T^n) \subset R(T)$. Consequently, $\dim M = \dim X/R(T)$.

Let us observe that $\dim N(T') = \dim M$ because $N(T^n)$ is finite-dimensional, and that $N(T') = N(T)$ because $N(T) \subset N(T^n)$. It follows that

$$\dim N(T) = \dim M = \dim X/R(T). \qquad (2.14)$$

Notice that $N(T^*)$ is finite-dimensional because $T^* = I - K^*$ and K^* is completely continuous by Schauder's theorem (p. 102). Choose a basis $\varphi_1, \ldots, \varphi_m$ in $N(T^*)$, then choose $x_1, \ldots, x_m \in X$ satisfying $\varphi_i(x_j) = \delta_{ij}$. Let us admit temporarily that $X = R(T) \oplus M'$ with $M' = \mathrm{Vect}\{x_1, \ldots, x_m\}$. Then

$$\dim N(T^*) = m = \dim M' = \dim X/R(T), \qquad (2.15)$$

and the equality $\dim N(T) = \dim N(T^*)$ follows from (2.14) and (2.15).

It remains to prove the relations

$$X = R(T) + M' \quad \text{and} \quad R(T) \cap M' = \{0\}.$$

For any given $x \in X$ we consider the vector

$$y := \varphi_1(x)x_1 + \cdots + \varphi_m(x)x_m \in M'.$$

We have for each $i = 1, \ldots, m$ the equality

$$\varphi_i(x - y) = \varphi_i(x) - \sum_{j=1}^{m} \varphi_j(x)\varphi_i(x_j) = \varphi_i(x) - \sum_{j=1}^{m} \varphi_j(x)\delta_{ij} = 0,$$

so that $x - y \in N(T^*)^{\perp} = R(T)$. Hence $X = R(T) + M'$.

On the other hand, if $x \in R(T) \cap M'$, then

$$x = c_1 x_1 + \cdots + c_m x_m$$

with suitable coefficients c_i, and $\varphi_i(x) = 0$ for all $i = 1, \ldots, m$. Hence

$$0 = \varphi_i(x) = \sum_{j=1}^{m} c_j \varphi_i(x_j) = \sum_{j=1}^{m} c_j \delta_{ij} = c_i$$

for all i, i.e., $x = 0$.

(d) follows from (a) and (c).

$\square$

2.11 * The Complex Case

We list the modifications for complex normed spaces.

Section 2.1. In the definition of hyperplanes and in Lemmas 2.2 and 2.4 X is still considered to be a real vector or normed space. Lemma 2.3 remains valid in the complex case: in the last line of the proof we obtain that $\varphi(U)$ is inside the unit disk of the complex plane.

In the statement of Theorem 2.5 we change $\varphi(a)$ and $\varphi(b)$ to their real parts $\Re\varphi(a)$ and $\Re\varphi(b)$. The result follows from the real case because the correspondence $\psi := \Re\varphi$ is a bijection between complex and real linear functionals: its inverse is given by the formula[83]

$$\varphi(x) := \psi(x) - i\psi(ix). \tag{2.16}$$

Corollaries 2.9 and 2.10 remain valid.

Section 2.2. Theorem 2.11 and Corollary 2.13 remain valid by changing $\mathbb{R}$ to $\mathbb{C}$ in their statement. In the proof first we extend the real part of φ by using the real case theorem, and then we *complexify* the extended functional with the help of the above formula. This leads to a suitable extension because the complexification does not alter the norm. Indeed, it follows at once from the formula that $\|\psi\| \leq \|\varphi\|$. On the other hand, for each $x \in X$ there exists a complex number λ such that $|\lambda| = 1$ and $\lambda\varphi(x) = |\varphi(x)|$. Then

$$|\varphi(x)| = \varphi(\lambda x) = \psi(\lambda x) \leq \|\psi\| \cdot |\lambda x| = \|\psi\| \cdot |x|,$$

i.e., $\|\varphi\| \leq \|\psi\|$.

Section 2.3. All results and proofs remain valid if we define the sign of a complex number by the formulas $\operatorname{sign} 0 := 0$, and $\operatorname{sign} y := |y|/y$ for $y \neq 0$. The map $j : \ell^q \to (\ell^p)'$ is still linear. If we wish to get back the Riesz–Fréchet theorem for

[83]This complexification method was discovered by Murray [328], Bohnenblust–Sobczyk [48], and Soukhomlinov [430].

$p = 2$, then it is better to define j by the formula

$$(jy)(x) = \varphi_y(x) := \sum_{n=1}^{\infty} x_n \overline{y_n};$$

then $j : \ell^q \to (\ell^p)'$ is antilinear.

Section 2.4. The definition of ℓ^p, c_0, $C(I, \mathbb{C})$ and $L^p(I)$ is analogous, by using complex valued sequences and functions instead of real values.

Section 2.5. Only one change is needed: we write $\Re\varphi$ instead of φ in the proof of Proposition 2.22.

Section 2.6. No change is needed.

Section 2.7. We have to write $\Re\varphi$ instead of φ in the statement of Proposition 2.31 and in the proof of the implication (b) $\implies$ (c).

Sections 2.8–2.10. All results and proofs remain valid. The resolvent set $\rho(A)$ is now defined as the set of *complex* numbers λ for which $A - \lambda I$ is invertible, and the spectrum $\sigma(A)$ is its complement in $\mathbb{C}$.

2.12 Exercises

Exercise 2.1 Prove that c_0 is a closed subspace of ℓ^∞.

Exercise 2.2 We have seen that if $1 \le p < q \le \infty$, then $\ell^p \subset \ell^q$, and the identity map $i : \ell^p \to \ell^q$ is continuous.

(i) Investigate the validity of the following equalities:

$$\bigcap_{q>p} \ell^q = \ell^p \quad \text{and} \quad \bigcup_{p<q} \ell^p = \ell^q.$$

(ii) What happens if we change ℓ^∞ to c_0 in the above questions?

In the following exercises we denote by X^p the vector space X of continuous functions $f : [0, 1] \to \mathbb{R}$, endowed with the norm $\|\cdot\|_p$, $1 \le p \le \infty$.

Exercise 2.3 Indicate the convex sets in X among the following:

(i) the polynomials of degree k;
(ii) the polynomials of degree $\le k$;
(iii) the continuous functions x satisfying

$$\int_0^1 |x(t)| \, dt \le 1;$$

(iv) the continuous functions x satisfying

$$\int_0^1 |x(t)|^2 \, dt \leq 1.$$

Exercise 2.4 Do the sequences $(x_n), (y_n)$ defined by

$$x_n(t) = t^n - t^{n+1} \quad \text{and} \quad y_n(t) = t^n - t^{2n}$$

converge in $X^p, p \in [1, \infty]$?

Exercise 2.5 Is the linear functional $f(x) := x(1)$ continuous

(i) on X^∞;
(ii) on X^2?

Exercise 2.6 Is the nonlinear map $f(x) := x^2$ continuous

(i)   from X^∞ into X^∞;
(ii) from X^2 into X^2;
(iii) from X^∞ into X^2?

Exercise 2.7 Prove that the linear operators

$$Ax(t) := \frac{x(t) + x(1-t)}{2} \quad \text{and} \quad Bx(t) := \frac{x(t) - x(1-t)}{2}$$

are continuous projectors in X^p for all p, and compute their norms.

Exercise 2.8 Consider the linear functional

$$\varphi(x) := \int_0^1 \left(t - \frac{1}{2}\right)^3 x(t) \, dt$$

on $X^p, 1 \leq p \leq \infty$.

(i)   For which p is φ continuous?
(ii) Compute $\|\varphi\|$ when φ is continuous.
(iii) Is the norm $\|\varphi\|$ attained?

Exercise 2.9 Consider the set

$$M := \left\{ x \in X : \int_0^{1/2} x(t) \, dt - \int_{1/2}^1 x(t) \, dt = 1 \right\}.$$

(i)   Show that M is a non-empty convex set.
(ii) Show that M is closed in X^∞.
(iii) Show that M has no element of minimal norm in X^∞.
(iv) Reconsider the questions (ii), (iii) in X^1.

(v) Reconsider the questions (ii), (iii) in X^p for $1 < p < \infty$.
(vi) How are these results related to the theorems of this chapter?

Exercise 2.10 (Quotient Norm) Let L be a closed subspace of a normed space X. Consider the equivalence relation $x \sim y \Longleftrightarrow x - y \in L$ in X and let X/L be the quotient vector space. Show that

(i) the formula $\|\xi\|_{X/L} = \inf_{x \in \xi} \|x\|$ defines a norm in X/L;
(ii) if X a Banach space, then X/L is also a Banach space.

Exercise 2.11

(i) Prove that in a Banach space every decreasing sequence of closed balls has a non-empty interior.
(ii) Does it remain true in normed spaces as well?

Exercise 2.12

(i) Prove that in a reflexive space every decreasing sequence of non-empty bounded closed convex sets has a non-empty interior.
(ii) Does it remain true in non-reflexive spaces?

Exercise 2.13

(i) Prove that in finite-dimensional normed spaces every decreasing sequence of non-empty bounded closed sets has a non-empty interior.
(ii) Does it remain true in all normed spaces?

Exercise 2.14 Let X, Y be two normed spaces and $A \in L(X, Y)$.

(i) Prove that[84]

$$N(A^*) = R(A)^{\perp} \quad \text{and} \quad N(A) = R(A^*)^{\perp}.$$

(ii) Prove that[85] if there exists an $\alpha > 0$ satisfying $\|Ax\| \geq \alpha \|x\|$ for all $x \in X$, then $R(A) = Y$.

Exercise 2.15 (Banach Limit)[86] Set $e := (1, 1, 1, \ldots)$ and

$$M := \{(x_1, x_2 - x_1, x_3 - x_2, \ldots) \ : \ x = (x_1, x_2, \ldots) \in \ell^{\infty}\}.$$

Prove the following properties:

(i) M is a subspace ℓ^{∞};
(ii) $\text{dist}(e, M) = \|e\| = 1$;

---

[84]Compare with Exercise 1.23.
[85]Banach [24] proved much more in his *closed range theorem*, see also Yosida [488].
[86]Banach [24, p. 34]. See Mazur [318] for other interesting properties.

(iii) there exists an $L \in (\ell^\infty)'$ satisfying $\|L\| = Le = 1$, and $L = 0$ on M;
(iv) Lx does not change if we remove the first element of x;
 (v) $\liminf x_n \leq Lx \leq \limsup x_n$ for all $x = (x_1, x_2, \ldots) \in \ell^\infty$;
(vi) $Lx = \lim x_n$ for all convergent sequences $x = (x_1, x_2, \ldots)$.

Exercise 2.16 Let $f : \mathbb{R} \to \mathbb{R}$ be a continuous function satisfying for some numbers $p, q > 1$ the condition

$$|f(t)| \leq |t|^{p/q} \quad \text{for all} \quad t \in \mathbb{R}.$$

Set $F(x) = (f(x_1), f(x_2), \ldots)$ for every $x = (x_1, x_2, \ldots) \in \ell^p$. Show the following results:

 (i) $F(x) \in \ell^q$, and the map $F : \ell^p \to \ell^q$ is continuous;
(ii) if $x^k \rightharpoonup x$ in ℓ^p, then $A(x^k) \rightharpoonup A(x)$ in ℓ^q.

Exercise 2.17 A sequence (x_n) in a normed space X is called a *weak Cauchy sequence* if

$$\varphi(x_1), \varphi(x_2), \varphi(x_3), \ldots$$

is a Cauchy sequence in $\mathbb{R}$ for each $\varphi \in X'$.

  (i) Show that in finite-dimensional normed spaces every weak Cauchy sequence is convergent.
 (ii) Show that in Hilbert spaces every weak Cauchy sequence is weakly convergent.[87]
(iii) Does the conclusion of (ii) remain valid in reflexive spaces?
(iv) Does the conclusion of (ii) remain valid in $X = \ell^1$?
 (v) Does the conclusion of (ii) remain valid in $X = c_0$?
(vi) Does the conclusion of (ii) remain valid in $X = \ell^\infty$?

Exercise 2.18 Let X be an infinite-dimensional normed space. Prove that there exist non-continuous linear functionals on X.[88]

Exercise 2.19 The Hamel dimension of a Banach space cannot be countably infinite.

---

[87] We say that Hilbert spaces are *weakly sequentially complete*. On the other hand, the duals of infinite-dimensional normed spaces are never weakly complete: they contain weak Cauchy *nets* having no weak limits. See Grothendieck [174] and Schaefer [411].

[88] We may use a *Hamel basis*, i.e., a maximal linearly independent set.

Exercise 2.20

(i) Construct a family $\{N_t\}_{0<t<1}$ of sets of positive integers such that $N_1 \cap N_{t'}$ is finite for $t \neq t'$, and each N_t is infinite.
(ii) The Hamel dimension of an infinite-dimensional Banach space is at least $2^{\aleph_0}$.[89]

Exercise 2.21 We prove again that the Hamel dimension of an infinite-dimensional Banach space is at least $2^{\aleph_0}$.[90]

(i) The Hamel dimension of ℓ^∞ is $2^{\aleph_0}$.
(ii) For each infinite-dimensional Banach space X there exists a one-to-one linear map of ℓ^∞ into X.

Exercise 2.22 An infinite matrix $A := (a_{nk})_{n,k=0}^\infty$ of real numbers is called *convergence-preserving* if for each convergent real sequence $x_k \to \ell$ the formula

$$y_n := \sum_{k=0}^\infty a_{nk}x_k, \quad n = 0, 1, \ldots$$

defines a sequence satisfying $y_n \to \ell$.

Prove that A is convergence-preserving $\iff$ the following three conditions are satisfied[91]:

(i) $\sup_n \sum_{k=0}^\infty |a_{nk}| < \infty$;
(ii) $\sum_{k=0}^\infty a_{nk} \to 1$ as $n \to \infty$;
(iii) for each fixed $k = 0, 1, \ldots$, $a_{nk} \to 0$ as $n \to \infty$.

Express conditions (ii) and (iii) in terms of the matrix (a_{nk}).

---

[89]Lacey [276].
[90]See also Bauer and Brenner [31] and Tsing [459].
[91]Steinhaus–Toeplitz theorem.

Chapter 3
Locally Convex Spaces

There was far more imagination in the head of Archimedes than in that of Homer.

Voltaire

We have seen in the preceding chapters the usefulness of weak convergence. From a theoretical point of view, it would be more satisfying to find a norm associated with weak convergence. In finite dimensions every norm is suitable because the weak and strong convergences are the same. In infinite dimensions the situation is quite different. For example, we have the following

*Proposition 3.1 *In infinite-dimensional Hilbert spaces weak convergence is never metrizable.*[1]

Proof Fix an orthonormal sequence $e_1, e_2, \ldots$, and consider the set

$$A := \{e_m + me_n \ : \ n > m \geq 1\}.$$

Let us determine the set $\tilde{A}$ of limits of weakly convergent sequences in A. If a sequence $(x_k) = (e_{m_k} + m_k e_{n_k}) \subset A$ converges weakly to some $x \in H$, then it is bounded, and hence the sequence (m_k) of integers is bounded. We may take a subsequence, still converging weakly to x, for which the (m_k) sequence is constant: $m_k = m$ every k. If some element $e_m + me_n$ appears infinitely many times in (x_k), then $x = e_m + me_n \in A$. Otherwise we have $n_k \to \infty$ and $x = \lim e_m + me_{n_k} = e_m$. Hence

$$\tilde{A} \subset A \cup \{e_m \ : \ m = 1, 2, \ldots\}.$$

Since $(e_m + me_k)$ converges weakly to e_m for each fixed m, we have equality here.

If weak convergence were metrizable, then $\tilde{A}$ would be the closure of A and hence closed. However, $(e_m)_{m=1}^{\infty} \subset \tilde{A}$ and $e_m \rightharpoonup 0 \notin \tilde{A}$. $\square$

---

[1] von Neumann [336].

© Springer-Verlag London 2016
V. Komornik, *Lectures on Functional Analysis and the Lebesgue Integral*,
Universitext, DOI 10.1007/978-1-4471-6811-9_3

Similar non-metrizable convergence notions are often encountered in analysis. We may try at least to *topologize* them.[2] This attempt leads in the present case to an important generalization of normed spaces, called *locally convex spaces*. Since these spaces are often non-metrizable, in this section we sometimes use *nets* instead of sequences.

3.1   Families of Seminorms

We generalize the normed spaces.

Definition A *seminorm* on a vector space X is a function $p : X \to \mathbb{R}$ satisfying for all $x, y \in X$ and $\lambda \in \mathbb{R}$ the following conditions:

- $p(x) \geq 0$,
- $p(\lambda x) = |\lambda|\, p(x)$,
- $p(x + y) \leq p(x) + p(y)$.

Examples

- Every norm is a seminorm.
- If φ is a linear functional on X, then $|\varphi|$ is a seminorm.
- More generally, if $A : X \to Y$ is a linear map and q is a seminorm in Y, then $q \circ A$ is a seminorm in X.
- If p is a seminorm and $\lambda \geq 0$, then λp is a seminorm.
- If $p_1, \ldots, p_n$ are seminorms, then $p_1 + \cdots + p_n$ is a seminorm.

Definition By a *ball of center a* in a vector space X we mean a set of the form

$$B_{p,r}(a) = B_p(a; r) := \{x \in X \ : \ p(x - a) < r\}$$

where p is a seminorm in X and $r > 0$.

Remark It is clear that every ball is convex.

Consider a non-empty family $\mathcal{P}$ of seminorms in a vector space X. Let us denote by $\overline{\mathcal{P}}$ the set of seminorms q in X for which there exist finitely many seminorms $p_1, \ldots, p_n \in \mathcal{P}$ and a positive number N satisfying

$$q \leq N(p_1 + \cdots p_n).$$

---

[2]Even this may fail: see the last result of this book: Corollary 10.12, p. 362.

Furthermore, we denote by $\mathcal{T}_{\mathcal{P}}$ the family of sets $U \subset X$ having the following property: for every $a \in U$ there exist $q \in \overline{\mathcal{P}}$ and $r > 0$ such that

$$B_{q,r}(a) \subset U.$$

The following proposition is straightforward:

Proposition 3.2

(a) $\mathcal{T}_{\mathcal{P}}$ *is a topology on* X.
   Henceforth we consider this topology.
(b) *The topology* $\mathcal{T}_{\mathcal{P}}$ *is Hausdorff (or separated)* $\iff$ *for each non-zero point* $x \in X$ *there exists a* $p \in \mathcal{P}$ *such that* $p(x) \neq 0$.
(c) *For any sequence or net,* $x_n \to x$ $\iff$ $p(x_n - x) \to 0$ *for all* $p \in \mathcal{P}$.
(d) *Addition and multiplication by scalars, i.e, the operations*

$$X \times X \ni (x, y) \mapsto x + y \in X \quad and \quad \mathbb{R} \times \ni (\lambda, x) \mapsto \lambda x \in X$$

   are continuous.
(e) $\overline{\mathcal{P}}$ *contains precisely the continuous seminorms.*
(f) *A linear functional* φ *on* X *is continuous* $\iff$ $|\varphi| \in \overline{\mathcal{P}}$.
(g) *A ball* $B_{q,r}(a)$ *is open* $\iff$ $q \in \overline{\mathcal{P}}$.

Definition By a *locally convex space* we mean a vector space X equipped with a topology $\mathcal{T}_{\mathcal{P}}$ associated with a family $\mathcal{P}$ of seminorms.[3]

Examples

- If $\mathcal{P}$ has a single element, and this is a norm, then our definition reduces to that of normed spaces.
- Given a non-empty set K we denote by $\mathcal{F}(K)$ the vector space of the functions $f : K \to \mathbb{R}$. Considering the family of seminorms

$$p_t(f) := |f(t)|, \quad f \in \mathcal{F}(K)$$

where t runs over the elements of K, $\mathcal{F}(K)$ becomes a separated locally convex space, and the corresponding convergence is *pointwise convergence:*

$$f_n \to f \text{ in } \mathcal{F}(K) \iff f_n(t) \to f(t) \text{ for every } t \in K.$$

We will soon show that $\mathcal{F}(K)$ is not always normable.

---

[3] von Neumann [233]. The terminology will be explained by Proposition 3.25, p. 145.

Let us generalize the bounded sets of normed spaces:

Definition In a locally convex space X associated with a family $\mathcal{P}$ of seminorms a set A is *bounded* if every seminorm $p \in \mathcal{P}$ is bounded on A.

Remarks

- If A is bounded, then every continuous seminorm $p \in \overline{\mathcal{P}}$ is bounded on A.
- Since a *continuous* seminorm is bounded on every compact set, compact sets of locally convex spaces are bounded. It follows that in a separated locally convex space every compact set is bounded and closed. We recall[4] that the converse is false in *every* infinite-dimensional *normed* space.

Our last remark stresses the interest of the following result:

*Proposition 3.3 *Consider the spaces $\mathcal{F}(K)$.*

(a) *For the sets in $\mathcal{F}(K)$ we have compact $\iff$ bounded and closed.*
(b) *If K is infinite, then $\mathcal{F}(K)$ is not normable.*
(c) *If K is uncountable, then $\mathcal{F}(K)$ is not even metrizable.*

Proof

(a) Since $\mathcal{F}(K)$ is a separated locally convex space, it remains to show that if C is bounded and closed in $\mathcal{F}(K)$, then it is compact.

Since C is bounded in $\mathcal{F}(K)$, the sets $C(t) := \{f(t) : f \in C\} \subset \mathbb{R}$ are bounded for all $t \in K$. Choose a compact interval $F_t \supset C(t)$ for each t. The product space $F := \prod_{t \in K} F_t$ is compact by Tychonoff's theorem. Let us observe that topologically $\mathcal{F}(K)$ is the product space $\prod_{t \in K} X_t$ where $X_t = \mathbb{R}$ for every $t \in K$. Hence F is a compact subset of $\mathcal{F}(K)$. We complete the proof by observing that C is a closed subset of F, and hence compact.

(b) In view of (a) it suffices to recall that the closed balls are bounded and closed, but not compact in infinite-dimensional normed spaces.[5]

Let us also give a direct proof: we show that $\mathcal{F}(K)$ has no continuous *norms*. Indeed, if q is a continuous *seminorm* on $\mathcal{F}(K)$, then there exist $t_1, \ldots, t_n \in K$ and a number $N > 0$ such that

$$q(f) \leq N(|f(t_1)| + \cdots + |f(t_n)|)$$

for all $f \in \mathcal{F}(K)$. Since K is infinite, there exists a non-zero function $f \in \mathcal{F}(K)$ for which $f(t_1) = \cdots = f(t_n) = 0$. Then $q(f) = 0$, i.e., q is not a norm.

(c) If the topology of $\mathcal{F}(K)$ is metrizable by some metric d, then for each $n = 1, 2, \ldots$ there exist points $t_{n,1}, \ldots, t_{n,k_n} \in K$ and a number $N_n > 0$ such that

$$N_n(|f(t_{n,1})| + \cdots + |f(t_{n,k_n})|) < 1 \implies d(f, 0) < \frac{1}{n}, \quad n = 1, 2, \ldots$$

---

[4] Proposition 2.1, p. 55.
[5] Proposition 2.1, p. 55.

for all $f \in \mathcal{F}(K)$. If K were uncountable, then there would exist a point $t' \in K$ differing from all points t_{n,k_n}, and then the non-zero function

$$f(t) := \begin{cases} 1 & \text{if } t = t'; \\ 0 & \text{if } t \neq t' \end{cases}$$

would satisfy $d(f, 0) = 0$, contradicting the metric property of d.

$\square$

Remark If the seminorm family is countable: $\mathcal{P} = \{p_1, p_2, \ldots\}$, and the corresponding topology is separated, then it is also metrizable by the metric

$$d(x, y) := \sum_{p \in \mathcal{P}} \frac{1}{2^n} \cdot \frac{p_n(x - y)}{1 + p_n(x - y)}.$$

We end this section with a characterization of normable locally convex spaces:

*Proposition 3.4 (Kolmogorov)[6] *For a separated locally convex space X the following properties are equivalent:*

(a) *X is normable;*
(b) *there exists a bounded neighborhood of 0;*
(c) *there exists a non-empty bounded open set.*

Proof The implications (a) $\Longrightarrow$ (b) $\Longrightarrow$ (c) are obvious.

(c) $\Longrightarrow$ (b). If V is a non-empty bounded open set and $a \in V$, then $V - a$ is a bounded neighborhood of 0.

(b) $\Longrightarrow$ (a). Let U be a bounded neighborhood of 0. Fix an open ball $B_{p,r}(0) \subset U$. If q is a continuous seminorm, then, since U is bounded, there exists a sufficiently large number R such that $U \subset B_{q,R}(0)$. Hence $B_{p,r}(0) \subset B_{q,R}(0)$ and therefore $q \leq Cp$ with $C := R/r$. This shows that p alone defines the topology of X. Since X is separated, p is a norm.

$\square$

3.2 Separation and Extension Theorems

One of the main reasons for the usefulness of locally convex spaces is that the Helly–Hahn–Banach type theorems remain valid. We start with the geometrical results. If X is a locally convex space, then we denote by X' the *vector space*[7] of continuous linear functionals $X \to \mathbb{R}$.

---

[6] Kolmogorov [253].

[7] We will define later (in Sect. 3.5, p. 135) a locally convex topology on X'.

Theorem 3.5 *Let A and B be two disjoint non-empty convex sets in a locally convex space X.*

(a) *If A is open, and B is a subspace, then there exists a closed hyperplane H such that*

$$B \subset H \quad and \quad A \cap H = \varnothing.$$

(b) *If A is open, then there exist $\varphi \in X'$ and $c \in \mathbb{R}$ such that*

$$\varphi(a) < c \leq \varphi(b) \quad for\ all \quad a \in A \quad and \quad b \in B.$$

(c) *(Tukey–Klee)*[8] *If A is closed and B is compact, then there exist $\varphi \in X'$ and $c_1, c_2 \in \mathbb{R}$ such that*

$$\varphi(a) \leq c_1 < c_2 \leq \varphi(b) \quad for\ all \quad a \in A \quad and \quad b \in B.$$

See Figs. 2.1, 2.2, and 2.3 again, pp. 59–60.

Proof

(a) We may repeat the proof of Theorem 2.5 (a) (p. 61) with one small modification: in the proof of Lemma 2.4 (b) (p. 58) we take an open ball $B_{p,r}(a)$ instead of $B_r(a)$. Then we get $|\varphi| < 1$ on $B_{p,r}(0)$, whence $|\varphi| \leq r^{-1}p$. This implies the continuity of φ.
(b) The proof of Theorem 2.5 (b) remains valid.
(c) We modify the proof of Theorem 2.5 (c) as follows. Since A is closed, for each $b \in B$ we can find an open ball $B_b := B_{p_b,r_b}(b)$ of center b, disjoint from A. A finite number of them covers the compact set B, say

$$B \subset \bigcup_{j=1}^{n} B_{b_j}.$$

Introduce the open ball $U := B_{p,r}(0)$ with

$$p := p_{b_1} + \cdots + p_{b_n}, \quad and \quad r := 2^{-1} \min\{r_1, \ldots, r_n\}.$$

Then $A + U$ and $B + U$ are disjoint non-empty convex open sets satisfying $A \subset A + U$ and $B \subset B + U$.

---

[8]Tukey [460], Klee [250].

Applying (b) there exist $\varphi \in X'$ and $c \in \mathbb{R}$ such that

$$\varphi(a') < c \leq \varphi(b') \quad \text{for all} \quad a' \in A + U \quad \text{and} \quad b' \in B + U.$$

Hence

$$\varphi(a) + \sup_U |\varphi| \leq c \leq \varphi(b) - \sup_U |\varphi| \quad \text{for all} \quad a \in A \quad \text{and} \quad b \in B.$$

Since φ is non-zero, $s := \sup_U |\varphi| > 0$, and the required inequalities follow with $c_1 = c - s, c_2 = c + s$. $\qquad \square$

The extension theorem 2.11 (p. 65) takes the following form:

Theorem 3.6 *Let X be a locally convex space. If $\varphi : M \to \mathbb{R}$ is a continuous linear functional on a subspace $M \subset X$, then φ may be extended to a continuous linear functional $\Phi : X \to \mathbb{R}$.*

Proof We may assume that $\varphi \not\equiv 0$. Fix $a \in M$ with $\varphi(a) = 1$, and then a continuous seminorm p with $|\varphi| < 1$ on $M \cap B_{p,1}(0)$. Repeating the proof of Theorem 2.11 we obtain a linear extension Φ of φ to X, satisfying $\Phi^{-1}(0) \cap B_{p,1}(a) = \varnothing$. We conclude that Φ is continuous. $\qquad \square$

Let X be a locally convex space. Similarly to the case of normed spaces, we define the *orthogonal complements* of $D \subset X$ and $\Delta \subset X'$ by the formulas

$$D^{\perp} = \{\varphi \in X' : \varphi(x) = 0 \quad \text{for all} \quad x \in D\}$$

and

$$\Delta^{\perp} = \{x \in X : \varphi(x) = 0 \quad \text{for all} \quad x \in \Delta\}.$$

Using the preceding theorem we may repeat the proof of Corollary 2.9 (p. 64); we get the following

Corollary 3.7 *Let X be a locally convex space, $D \subset X$, and $M \subset X$ a subspace.*

(a) *We have $\overline{\text{Vect}(D)} = (D^{\perp})^{\perp}$.*
(b) *If $D^{\perp} = \{0\}$, then D generates X.*
(c) *If $M^{\perp} = \{0\}$, then M is dense in X.*

In *separated* locally convex spaces Corollary 2.10 (p. 65) and its proof remain valid:

Corollary 3.8 *Let X be a separated locally convex space.*

(a) *For any two distinct points $a, b \in X$ there exists a $\varphi \in X'$ such that $\varphi(a) \neq \varphi(b)$.*

(b) *If* $x_1, \ldots, x_n \in X$ *are linearly independent vectors, then there exist linear functionals* $\varphi_1, \ldots, \varphi_n \in X'$ *such that*

$$\varphi_i(x_j) = \delta_{ij} \quad \text{for all} \quad i, j = 1, \ldots, n.$$

Consequently, $\dim X' \geq \dim X$.

Remark Every finite-dimensional *separated* locally convex space is normable. Indeed, choose a basis $\varphi_1, \ldots, \varphi_m$ in X'. Then the formula

$$\|x\| := |\varphi_1(x)| + \cdots + |\varphi_m(x)|$$

defines a continuous *norm* by the above corollary, and every continuous seminorm p satisfies the inequality $p \leq c\|\cdot\|$ with $c := \max\{p(x) : \|x\| = 1\}$.[9] Hence this norm induces the topology of X.

3.3   Krein–Milman Theorem

Every bounded convex polygon is the convex hull of its *vertices*; see Fig. 3.1. This was generalized by Minkowski for every non-empty bounded closed convex set in $\mathbb{R}^N$ by a suitable modification of the notion of vertex. His result was further extended by Krein and Milman for every separated locally convex space. This section is devoted to this result.

Definition A point x of a convex set C in a vector space is called *extremal* if $C \setminus \{x\}$ is convex.

It is clear that in locally convex spaces the extremal point of a convex set must lie on its boundary. For example, on Fig. 3.2 all boundary points are extremal, while on Figs. 3.3 and 3.4 only the vertices are extremal.

Examples Let us denote by $E = E_X$ the set of extremal points of the closed unit ball $B = B_X$ of a normed space X. We recall that $E \subset S$ where $S = S_X$ denotes the *unit sphere*, i.e., the boundary of B.

- If X is a Euclidean space, then $E = S$.
- If $X = \ell^p$ $(1 < p < \infty)$, then we still have $E = S$.
- More generally, $E = S \iff X$ is strictly convex.
- If $X = \ell^1$, then $E = \{\lambda e_k : |\lambda| = 1, k = 1, 2, \ldots\}$.
- If $X = \ell^\infty$, then $E = \{x = (x_n) : |x_n| = 1 \quad \text{for all} \quad n\}$.
- If $X = c_0$, then $E = \varnothing$.

---

[9] As the maximum of a continuous function on a compact set, c is finite.

Fig. 3.1 Vertices of a convex
polygon

Fig. 3.2 Extremal points of a
disk

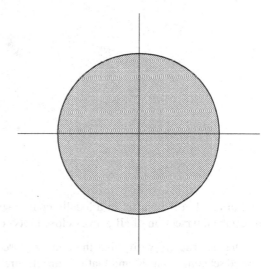

Fig. 3.3 "Unit ball" of
$(\mathbb{R}^2, \|\cdot\|_1)$

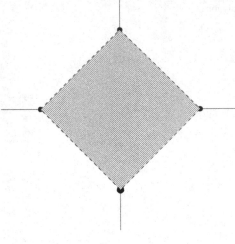

Fig. 3.4 "Unit ball" of
$(\mathbb{R}^2, \|\cdot\|_\infty)$

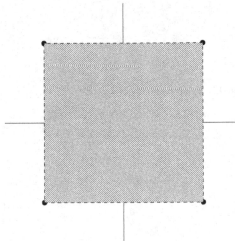

Definition Let E be a set in a locally convex space. By its *convex closed hull* we mean the intersection of all convex closed sets containing E.

One can readily verify that the convex closed hull of E is the smallest convex closed set containing E, and that it is the closure of its *convex hull,* i.e., of the set of all convex linear combinations of the elements of E.

> **Theorem 3.9 (Krein–Milman)**[10] *Let C be a non-empty convex* compact *set in a* separated *locally convex space X. Then C is the convex closed hull of its extremal points.*

For the proof we generalize the extremal points. We consider segments of the form

$$[a, b] := \{ta + (1 - t)b \; : \; 0 \le t \le 1\} \quad \text{with} \quad a \ne b.$$

Definition Let C be a non-empty set in a vector space. A subset $E \subset C$ is called a *side* of C if for every segment $[a, b] \subset C$, $[a, b] \cap E$ is one of the following four sets: $\varnothing, \{a\}, \{b\}$ and $[a, b]$.

Remark One can check the following properties:

(a) C is a side of itself;
(b) the intersection of any family of sides is a side;
(c) if E is a side of C and F is a side of E, then F is a side of C;
(d) if a linear functional φ has a maximum on C, then

$$E := \left\{ z \in C \; : \; \varphi(z) = \max_C \varphi \right\}$$

  is a side of C;
(e) the one-point sides $\{x\}$ of a *convex* set correspond exactly to its extremal points x.

Proof of Theorem 3.9 We proceed in two steps.

First step. We show that C has at least one extremal point. By the above properties (a) and (b) the family of compact sides of C satisfies the conditions of Zorn's lemma, and hence C has at least one minimal compact side E. In view of property (e) it remains to show that E cannot contain more than one point.
Assume that E contains two distinct points $x \ne y$. Applying Corollary 3.8 we fix $\varphi \in X'$ satisfying $\varphi(x) < \varphi(y)$. Since φ is continuous,

$$F := \left\{ z \in E \; : \; \varphi(z) = \max_E \varphi \right\}$$

is a well defined compact set, and it is a side of C by (d) and (c). Since $x \notin F$, F is a proper side of E, contradicting the minimality of E.

---

[10]Minkowski [325] (p. 160), Krein–Milman [269]. See Phelps [359] for further improvements and generalizations.

Second step. We already know that the convex closed hull K of the extremal points of C is a *non-empty* convex compact subset of C. Assume on the contrary that there exists a point $x \in C \setminus K$, and applying Theorem 3.5 (c) choose $\varphi \in X'$ satisfying

$$\max_K \varphi < \varphi(x).$$

Then the convex compact set

$$E := \left\{ z \in C \ : \ \varphi(z) = \max_C \varphi \right\}$$

is a side of C, disjoint from K. Applying the first step, E has an extremal point y. Then y is also an extremal point of C by properties (c) and (e). But this is impossible because the extremal points of C belong to the set K which is disjoint from E.

$\square$

3.4   * Weak Topology. Farkas–Minkowski Lemma

Given a locally convex space X, we denote by $\sigma(X, X')$ the locally convex topology defined by the seminorms $|\varphi|$ where φ runs over X'. By Proposition 3.2 (p. 121) we have

$$x_n \to x \text{ in } \sigma(X, X') \quad \Longleftrightarrow \quad \varphi(x_n) \to \varphi(x) \text{ for all } \varphi \in X' \qquad (3.1)$$

for every sequence or net in X. This motivates the following terminology:

Definition $\sigma(X, X')$ is called the *weak topology* of X.[11] The corresponding space $(X, \sigma(X, X'))$ is also denoted briefly by X_σ.

Proposition 3.10 *Let X be a locally convex space.*

(a) *The weak topology of X is coarser than its original topology.*
(b) *The same linear functionals are continuous for both topologies.*
(c) *The closed convex sets are the same for both topologies.*
(d) *The two topologies are separated at the same time.*

Proof (a) follows from (3.1), (b) follows from (a) and (3.1), (c) follows from Theorem 3.5 (c) similarly to the proof of Proposition 2.22 (e), and (d) follows from Corollary 3.8 (a). $\square$

---

[11] von Neumann [336].

In general the weak topology is not normable, and not even metrizable:

*Proposition 3.11

(a) *The weak topology of infinite-dimensional locally convex spaces is not normable.*
(b) *The weak topology of infinite-dimensional normed spaces is not metrizable.*[12]

Remarks

- The theorem of choice (p. 90) is not completely satisfactory in non-metrizable cases because the convergent sequences do not characterize the topology. We return to this question later.[13]
- The basic properties of the weak convergence (Propositions 1.17 and 2.22, pp. 30 and 80) and the characterizations of continuous linear maps (Propositions 1.22 and 2.24, pp. 35 and 82) and their proofs remain valid for nets instead of sequences.

Proof

(a) We show that there is no continuous *norm* on X. Indeed, if q is a continuous *seminorm* on X, then there exist functionals $\varphi_1, \ldots, \varphi_n \in X'$ and a positive number N such that

$$q(x) \leq N \sum_{i=1}^{n} |\varphi_i(x)| \quad \text{for all} \quad x \in X.$$

Since X is infinite-dimensional, there exists a point $x \neq 0$ such that $\varphi_1(x) = \cdots = \varphi_n(x) = 0$. Then $q(x) = 0$, so that q is not a norm.

(b) Assume that the weak topology of a normed space X may be defined by a metric d; then X_σ is separated. For each $n = 1, 2, \ldots$ we fix finitely many functionals $\varphi_{n1}, \ldots, \varphi_{nk_n} \in X'$ such that

$$\bigcap_{j=1}^{k_n} \{x \in X : |\varphi_{nj}(x)| < 1\} \subset \left\{x \in X : d(x,0) < \frac{1}{n}\right\}.$$

For each $\varphi \in X'$ there exists an n such that

$$d(x,0) < \frac{1}{n} \implies |\varphi(x)| < 1.$$

Consequently

$$\varphi_{n1}(x) = \cdots = \varphi_{nk_n}(x) = 0 \implies |\varphi(x)| < 1,$$

---

[12]Wehausen [479].
[13]See Theorem 3.21, p. 140.

and hence, changing x to tx and letting $t \to \infty$,

$$\varphi_{n1}(x) = \cdots = \varphi_{nk_n}(x) = 0 \Longrightarrow \varphi(x) = 0.$$

Applying a well-known lemma from linear algebra[14] this implies that φ is a linear combination of $\varphi_{n1}, \ldots, \varphi_{nk_n}$.

The finite-dimensional (and thus closed) subspaces

$$F_n := \mathrm{Vect}\{\varphi_{n1}, \ldots, \varphi_{nk_n}\}$$

cover X'. By Baire's lemma (p. 32) at least one of them, say F_n, has interior points. Then we have $F_n = X'$ and hence $\dim X' < \infty$. Applying Corollary 2.10 (p. 65) we conclude that $\dim X < \infty$.   □

Let us recall a proof of the lemma:

Lemma 3.12 *Let* $\varphi_1, \ldots, \varphi_n$ *and* φ *be linear functionals on a vector space X. Assume that*

$$x \in X \quad and \quad \varphi_1(x) = \cdots = \varphi_n(x) = 0 \quad \Longrightarrow \quad \varphi(x) = 0.$$

Then φ is a linear combination of $\varphi_1, \ldots, \varphi_n$.

Proof Consider the subspace

$$M := \{(\varphi_1(x), \ldots, \varphi_n(x)) \in \mathbb{R}^n : x \in X\}$$

of $\mathbb{R}^n$. By our assumption the formula

$$(\varphi_1(x), \ldots, \varphi_n(x)) \mapsto \varphi(x)$$

defines a linear functional $\psi : M \to \mathbb{R}$. Introducing the usual scalar product of $\mathbb{R}^n$ and considering the orthogonal projection P onto M, $\psi \circ P$ is a continuous linear functional on $\mathbb{R}^n$, and hence it can be represented by some vector $(c_1, \ldots, c_n) \in \mathbb{R}^n$:

$$\psi(Py) = c_1 y_1 + \cdots + c_n y_n$$

for all $y = (y_1, \ldots, y_n) \in \mathbb{R}^n$. In particular, we have

$$\varphi(x) = c_1 \varphi_1(x) + \cdots + c_n \varphi_n(x)$$

for all $x \in X$.   □

------

[14] See Lemma 3.12 below.

We recall from Proposition 2.24 (p. 82) that in a normed space every weakly convergent sequence is bounded. This also follows from our next result[15]:

Proposition 3.13 *If is a X normed space, then X and X_σ have the same bounded sets.*

Proof If A is bounded in X, then $\varphi(A)$ is bounded for every $\varphi \in X'$ by the characterization of continuity (p. 100), so that A is bounded for X_σ by definition.

For the proof of the converse we consider the linear isometry $J : X \to X''$ of Corollary 2.21 (p. 79). If A is bounded for X_σ, then $J(A)$ is pointwise bounded because

$$\{(Jx)(\varphi) \ : \ x \in A\} = \{\varphi(x) \ : \ x \in A\} \subset \mathbb{R}$$

is bounded for all $\varphi \in X'$. Applying the Banach–Steinhaus theorem (p. 81) we obtain that $J(A)$ is bounded in X''. Since J is an isometry, this is equivalent to the boundedness of A in X.   □

We end this section by proving a famous variant of Lemma 3.12, of fundamental importance in convex analysis and linear programming.[16] We denote the usual scalar product of $\mathbb{R}^n$ by (x, y).

Proposition 3.14 (Farkas–Minkowski)[17] *Given $a, a_1, \ldots, a_k \in \mathbb{R}^n$, the inequality $(a, x) \leq 0$ is a logical consequence of the system of inequalities $(a_1, x) \leq 0, \ldots, (a_k, x) \leq 0 \Longleftrightarrow a$ is a nonnegative linear combination of $a_1, \ldots, a_k$.*

In the following elementary proof we avoid the use of topology. For this we give an algebraic proof of the following lemma where we denote by K the convex cone generated by $a_1, \ldots, a_k$, i.e., the set of linear combinations of these vectors with nonnegative coefficients.

Lemma 3.15 *The distance $d(a, K)$ is attained by some point $b \in K$ for each fixed $a \in \mathbb{R}^n$.*

Remarks

- The point b is clearly unique but we will not need this here.
- The lemma implies that K is closed but we will not need this explicitly either.

Using the lemma we can quickly prove the nontrivial part of the proposition: if $(a, x) \leq 0$ is a logical consequence of the system

$$(a_1, x) \leq 0, \ldots, (a_k, x) \leq 0,$$

then $a \in K$.

---

[15] The proposition holds in all locally convex spaces: see, e.g., Reed–Simon [367], Theorem V. 23.
[16] See, e.g., Dantzig [94], Rockafellar [398], Vajda [462].
[17] Minkowski [323] (pp. 39–45), Farkas [135]. We follow Komornik [258].

First we observe that

$$(a_j, a - b) \le 0, \quad j = 1, \ldots, k \tag{3.2}$$

and

$$(-b, a - b) \le 0. \tag{3.3}$$

For otherwise we would have for every sufficiently small $t \in (0, 1)$ the following relations:

$$\begin{aligned}
|a - (b + ta_j)|^2 &= |(a - b) - ta_j|^2 \\
&= |a - b|^2 - t\left(2(a_j, a - b) - t|a_j|^2\right) \\
&< |a - b|^2,
\end{aligned}$$

and

$$\begin{aligned}
|a - (b - tb)|^2 &= |(a - b) + tb|^2 \\
&= |a - b|^2 - t\left(2(-b, a - b) - t|b|^2\right) \\
&< |a - b|^2.
\end{aligned}$$

This would contradict the choice of b because

$$b + ta_j \in K \quad \text{and} \quad b - tb = (1 - t)b \in K.$$

By our assumption (3.2) implies $(a, a - b) \le 0$. Combining this with (3.3) we obtain $(a - b, a - b) \le 0$. Hence $a = b$, and therefore $a \in K$.

Proof of the lemma The case $k = 1$ is obvious. Let $k \ge 2$, and assume by induction that for each $j = 1, \ldots, k$, the convex cone K_j generated by the vectors

$$a_1, \ldots, a_{j-1}, a_{j+1}, \ldots, a_k$$

has a closest point b_j from a. Now we distinguish three cases.

(a) If $a \in K$, then we may choose $b := a$.
(b) If $a \in \mathrm{Vect}\{a_1, \ldots, a_k\} \setminus K$, then let b be at a minimal distance from a among $b_1 \ldots, b_k$. We show that $|a - b| \le |a - c|$ for all $c \in K$.
    The segment $[a, c]$ meets one of the sides K_i of the cone K. More precisely, let

$$a = \alpha_1 a_1 + \cdots + \alpha_k a_k \quad \text{and} \quad c = \gamma_1 a_1 + \cdots + \gamma_k a_k$$

with $\gamma_1, \ldots, \gamma_k \geq 0$, and set

$$t := \min \left\{ \gamma_j / (\gamma_j - \alpha_j) \ : \ \alpha_j < 0 \right\}.$$

(There is at least one such j, because $a \notin K$.) Then $0 \leq t < 1$, and the minimum is attained for some i. Consequently,

$$t\alpha_j + (1 - t)\gamma_j \geq 0 \quad \text{for every} \quad j,$$

and

$$t\alpha_i + (1 - t)\gamma_i = 0.$$

Now $ta + (1 - t)c \in K_i$, so that

$$|a - b| \leq |a - b_i| \leq |a - (ta + (1 - t)c)| = (1 - t)|a - c| \leq |a - c|.$$

(c) If $a \notin L := \text{Vect}\{a_1, \ldots, a_k\}$, then we apply the above results to the orthogonal projection a' of a onto L: there exists a b at a minimal distance from a' in K. Since

$$|a - b|^2 = |a - a'|^2 + |a' - b|^2 \leq |a - a'|^2 + |a' - c|^2 = |a - c|^2$$

for all $c \in K$, b is also at a minimal distance from a in K.

$\square$

3.5 * Weak Star Topology: Theorems of Banach–Alaoglu and Goldstein

Until now the dual X' of a locally convex space X was not endowed with any topology. Now we introduce in X' the locally convex topology $\sigma(X', X)$ defined by the seminorms

$$\varphi \mapsto |\varphi(x)|$$

where x runs over X.

Definition The topology $\sigma(X', X)$ is called the *weak star topology* of X'.[18] The space $(X', \sigma(X', X))$ is also denoted briefly by $X'_{\sigma*}$. The corresponding *weak star convergence* is denoted by $\varphi_n \overset{*}{\rightharpoonup} \varphi$.

It follows from the definitions that

$$\varphi_n \overset{*}{\rightharpoonup} \varphi \quad \Longleftrightarrow \quad \varphi_n(x) \to \varphi(x) \quad \text{for all} \quad x \in X$$

for both sequences and nets.

Before giving some examples, we formulate the dual of Lemma 2.25 (p. 83); its proof is a simple adaptation of the proof of Lemma 1.20 (p. 33).

Lemma 3.16 *Let (φ_k) be a bounded sequence or net in the dual X' of some normed space X.*

(a) *For any given $\varphi \in X'$ the set*

$$\{x \in X \,:\, \varphi_k(x) \to \varphi(x)\}$$

is a closed subspace of X.

(b) *The set*

$$\{x \in X \,:\, (\varphi_k(x)) \quad converges\ in \quad \mathbb{R}\}$$

is a closed subspace of X.

Examples (Compare with the examples on pages 83 and 86)

- Let $(\varphi_n) \in \ell^1$, and let $k \mapsto (\varphi_n^k)$ be a *bounded* sequence or net in ℓ^1. Lemmas 2.16 (p. 73) and 3.16 yield the following characterization of weak star convergence in $(c_0)' = \ell^1$:

$$(\varphi_n^k) \overset{*}{\rightharpoonup} (\varphi_n) \quad \Longleftrightarrow \quad \varphi_n^k \to \varphi_n \quad \text{for each} \quad n.$$

For example, $e_n \overset{*}{\rightharpoonup} 0$ in $(c_0)' = \ell^1$.
- We obtain the same characterization for the weak star convergence of bounded sequences or nets in $(\ell^1)' = \ell^\infty$. For example,

$$e_1 + \cdots + e_n \overset{*}{\rightharpoonup} a = (1, 1, \ldots).$$

Using the weak star topology we may also complete Corollaries 1.6 and 2.9 (pp. 15 and 64) on the characterization of generated closed subspaces. Similarly to the preceding chapter we define the *orthogonal complements* of $D \subset X$ and $\Delta \subset X'$

---

[18]Banach [22].

by the formulas

$$D^{\perp} := \{\varphi \in X' : \varphi(x) = 0 \quad \text{for all} \quad x \in D\}$$

and

$$\Delta^{\perp} := \{x \in X : \varphi(x) = 0 \quad \text{for all} \quad \varphi \in \Delta\}.$$

Let us establish the basic properties of the weak star topology. For simplicity we consider only separated spaces.

Proposition 3.17 *Let X be a* separated *locally convex space.*

(a) *The weak star topology of X' is separated.*
(b) *The formula $(Jx)(\varphi) := \varphi(x)$ defines a linear bijection between X and $(X'_{\sigma*})'$.*
(c) *If $\Delta \subset X'$, then $(\Delta^{\perp})^{\perp}$ is the* weak star closed *subspace generated by Δ.*

Proof

(a) This follows from the definition.
(b) The continuity of the linear functionals $Jx : X'_{\sigma*} \to \mathbb{R}$ follows from the definition of the weak star topology. The linearity of J is obvious, its injectivity follows from Corollary 3.8 (p. 125). For the proof of the surjectivity fix an arbitrary functional $\Phi \in (X'_{\sigma*})'$. By the definition of its continuity there exist $x_1, \ldots, x_n \in X$ and a number $\varepsilon > 0$ satisfying

$$\varphi \in X' \quad \text{and} \quad |\varphi(x_1)| < \varepsilon, \ldots, |\varphi(x_n)| < \varepsilon \implies |\Phi(\varphi)| < 1.$$

We may thus apply Lemma 3.12 (p. 132) to $Jx_1, \ldots, Jx_n, \Phi \in X'$: we get

$$\Phi = c_1 Jx_1 + \cdots + c_n Jx_n = J(c_1 x_1 + \cdots + c_n x_n)$$

with suitable numbers $c_1, \ldots, c_n$.
(c) Let us denote temporarily by M the weak star closure of $\text{Vect}(\Delta)$. The inclusion $M \subset (\Delta^{\perp})^{\perp}$ is obtained easily, as in Corollary 1.6 (p. 15). For the converse we fix $\varphi \in X' \setminus M$ arbitrarily. We have to show that $\varphi \notin (\Delta^{\perp})^{\perp}$.

Applying Theorem 3.5 (c) (p. 124) and using property (b) above, there exist $x \in X$ and numbers $c_1 < c_2$ such that $\psi(x) \leq c_1$ for all $\psi \in M$, and $\varphi(x) \geq c_2$. Since $\{\psi(x) : \psi \in M\}$ is a subspace of $\mathbb{R}$, hence $\psi(x) = 0$ for all $\psi \in M$, and therefore $\varphi(x) > 0$. Hence $x \in \Delta^{\perp}$ and $\varphi \notin (\Delta^{\perp})^{\perp}$.

$\square$

In the rest of this section we consider only normed spaces.

Remark For a *normed* space X we may define three natural topologies on X': the usual norm topology, which we will denote here by $\beta(X', X)$, the weak star topology $\sigma(X', X)$ and the weak topology $\sigma(X', X'')$. Since X may be identified with a subspace of X'' via the map $J : X \to X''$ of Corollary 2.21 (p. 79), the weak star topology is coarser than the weak topology. We thus have the inclusions

$$\sigma(X', X) \subset \sigma(X', X'') \subset \beta(X', X).$$

They all coincide in finite dimensions, but they usually differ in infinite dimensions.[19]

Proposition 3.18 *If X is a* Banach *space, then the same sets are bounded in the three topologies.*

Proof In a coarser locally convex topology we have fewer (or the same) continuous seminorms, and hence the same (or more) sets are bounded.

It remains to show that a weak star bounded set $\Delta \subset X'$ is also norm bounded. This follows by applying the Banach–Steinhaus theorem because the $X'_{\sigma*}$-boundedness of Δ is equivalent by definition to its pointwise boundedness on X. $\square$

Example It follows from the proposition that in the dual X' of a *Banach space* every weak star convergent sequence is bounded.

This may fail for non-complete normed spaces X. Consider for example the subspace X of ℓ^2 formed by the sequences having at most finitely many non-zero elements. The formula

$$\varphi_n(x) := n x_n$$

defines a sequence $(\varphi_n) \subset X'$ for which $\varphi_n \overset{*}{\rightharpoonup} 0$ and $\|\varphi_n\| \to \infty$.

Next we establish a new variant of Theorems 1.21 and 2.30 (pp. 33 and 90):

Proposition 3.19 (Theorem of choice)[20] *If X is a* separable *normed space, then every bounded sequence $(\varphi_k) \subset X'$ has a weak star convergent subsequence.*

Proof Fix a dense sequence (x_n) in X. Applying Cantor's diagonal method, similarly to the proofs of Theorems 1.21 and 2.30 we obtain a subsequence (ψ_k) of (φ_k) such that the numerical sequences $k \mapsto \psi_k(x_n)$ converge for each fixed n.

Since (φ_k) is bounded and (x_n) is dense in X, by Lemma 3.16 the numerical sequence $k \mapsto \psi_k(x)$ converges for each $x \in X$, and[21] the formula

$$\varphi(x) := \lim \psi_k(x)$$

---

[19]More precisely, $\sigma(X', X'')$ is strictly coarser than $\beta(X', X)$, and one can show that $\sigma(X', X) = \sigma(X', X'') \iff X$ is reflexive.

[20]Banach [22].

[21]Similarly to the proof of Theorem 2.30.

defines a continuous linear functional $\varphi \in X'$. Then $\psi_k \overset{*}{\rightharpoonup} \varphi$ by the *definition* of weak star convergence. $\qquad\qquad\qquad\qquad\qquad\qquad\qquad\qquad\qquad\qquad\qquad\qquad\square$

Example The separability condition cannot be omitted. For example, the sequence of functionals defined by the formula

$$\varphi_k(x) := x_k, \quad x = (x_n) \in \ell^\infty, \quad k = 1, 2, \ldots$$

belongs to the closed unit sphere of $(\ell^\infty)'$, and it has no weak star convergent subsequence.

Indeed, for any given subsequence (φ_{k_m}) we may consider a vector $x = (x_n) \in \ell^\infty$ satisfying $x_{k_m} = (-1)^m$ for all m. Then the numbers $\varphi_{k_m}(x) = (-1)^m$ form a divergent sequence, so that $(\varphi_{k_m}(x))$ is not weak star convergent.

However, we may remove the separability assumption by considering nets instead of sequences: part (b) of the following theorem implies that every bounded net has a weak star convergent subnet in X'. This compactness property is perhaps the most important and useful feature of the weak star topology, because it can be used to obtain existence theorems.[22]

Theorem 3.20 *Let X be a normed space, and denote by B, B', B'' the closed unit balls of X, X', X''.*

(a) *(Banach–Alaoglu)*[23] *B' is compact in X' with respect to the weak star topology $\sigma(X', X)$.*

(b) *(Goldstine)*[24] *$J(B)$ is dense in B'', and $J(X)$ is dense in X'' with respect to the weak star topology $\sigma(X'', X')$.*

Proof

(a) As a topological space, $X'_{\sigma*}$ is a subspace of $\mathcal{F}(X)$. In view of Proposition 3.3 (p. 122) it is sufficient to show that B' is bounded and closed in $\mathcal{F}(X)$.

Since $|\varphi(x)| \leq \|x\|$ for all $\varphi \in B'$, B' is pointwise bounded on X, and hence bounded in $\mathcal{F}(X)$.

Now consider a net (φ_n) in B', converging to some φ in $\mathcal{F}(X)$.[25] We have to show that $\varphi \in B'$.

---

[22] See, e.g., Lions [304] for many applications.

[23] Banach [24], Alaoglu [3].

[24] Goldstine [171].

[25] We could avoid the use of nets, but the proof becomes less transparent: see, e.g., Rudin [406] or Brezis [65].

For any given $x, y \in X$ and $\lambda \in \mathbb{R}$, letting $n \to \infty$ in the relations

$$\varphi_n(x + y) = \varphi_n(x) + \varphi_n(y), \quad \varphi_n(\lambda x) = \lambda \varphi_n(x) \quad \text{and} \quad |\varphi_n(x)| \le \|x\|$$

we obtain that φ is linear, and that $|\varphi(x)| \le \|x\|$ for all x. In other words, $\varphi \in B'$.

(b) For the first result we show that if $\Phi \in X''$ does not belong to the $\sigma(X'', X')$-closure K of $J(B) \subset X''$, then $\|\Phi\| > 1$. Since K is a non-empty closed convex set for this topology, by Theorem 3.5 (p. 124) there exist $\varphi \in X'$ and $c_1, c_2 \in \mathbb{R}$ satisfying

$$\varphi(x) \le c_1 < c_2 \le \Phi(\varphi)$$

for all $x \in B$. Hence $\|\varphi\| < \Phi(\varphi)$, and therefore $\|\Phi\| > 1$.

The second result follows from the first one by homogeneity.   □

Example Combining the Banach–Alaoglu and Krein–Milman theorems (p. 129) we obtain that the closed unit ball of every dual space has an extremal point. Since this property is not true for c_0 (see pp. 126), c_0 is *not* isomorphic to X' for any normed space X.

Remark We mention the following equivalences[26]:

- X is separable $\Longleftrightarrow$ the restriction of $\sigma(X', X)$ to B' is metrizable;
- X' is separable $\Longleftrightarrow$ the restriction of $\sigma(X, X')$ to B is metrizable.

Using the first direct implication we may also deduce Proposition 3.19 from the Banach–Alaoglu theorem.

3.6 * Reflexive Spaces: Theorems of Kakutani and Eberlein–Šmulian

Using the weak *topology* instead of weak *convergence*, we may complete the results of Sects. 2.6–2.7 by giving new characterizations of reflexivity:

Theorem 3.21 *For a normed space X the following properties are equivalent:*

(a) *X is reflexive;*
(b) *every bounded sequence has a weakly convergent subsequence;*
(c) *the closed unit ball of X is weakly compact.*[27]

---

[26]See, e.g., Dunford–Schwartz [117]. The direct implications $\Longrightarrow$ are due to Banach [24].

[27]Banach [24], Bourbaki [64], Kakutani [239], Šmulian [424, 425], Eberlein [118]. See also Dunford–Schwartz [117], Whitley [486], Rolewicz [400].

For the proof of the implication (b) $\Longrightarrow$ (c) we will use the following simple lemma:

Lemma 3.22 *For any given finite-dimensional subspace $F \subset X''$ there exist vectors $\varphi_1, \ldots, \varphi_n \in X'$ of norm one such that*

$$\max_{1 \leq m \leq n} |\Phi(\varphi_m)| \geq \frac{1}{2} \|\Phi\| \quad \text{for all} \quad \Phi \in F.$$

Proof Since the unit sphere S of F is compact, there exist $\Phi_1, \ldots, \Phi_n \in S$ such that

$$\min_{1 \leq m \leq n} \|\Phi - \Phi_m\| \leq \frac{1}{4}$$

for all $\Phi \in S$. Fix $\varphi_m \in X'$ of norm one with $|\Phi_m(\varphi_m)| \geq \frac{3}{4}$ for each m. Then for each $\Phi \in S$, choosing m such that $\|\Phi - \Phi_m\| \leq \frac{1}{4}$, we have the estimate

$$|\Phi(\varphi_m)| \geq |\Phi_m(\varphi_m)| - |(\Phi - \Phi_m)(\varphi_m)| \geq \frac{3}{4} - \frac{1}{4} = \frac{1}{2}.$$

The lemma hence follows by homogeneity. $\qquad\square$

Proof of the theorem (a) $\Longrightarrow$ (b) This is Theorem 2.30, p. 90.

(b) $\Longrightarrow$ (c)[28] We use the notations of Theorem 3.20. The weak compactness of B is equivalent by definition to the weak star compactness of $J(B)$. Since $J(B) \subset B''$ and B'' is weak star compact by Theorem 3.20 (b), it is sufficient to show that $J(B)$ is weak star closed.

Since $J : X \to X''$ is a linear isometry, the weak closedness of B implies that the weak star closure $\overline{J(B)}$ of $J(B)$ satisfies

$$\overline{J(B)} \cap J(X) = J(B).$$

The weak star closedness of $J(B)$ will thus follow if we prove that $\overline{J(B)} \subset J(X)$.

Fix $\Phi_0 \in \overline{J(B)}$ arbitrarily. We are going to construct a sequence (n_k) of positive integers, a sequence $(\varphi_n) \subset X'$ of norm one functionals, and a sequence of points $(x_k) \subset B$ satisfying the following two conditions for $k = 1, 2, \ldots$:

$$\max_{1 \leq n \leq n_k} |\Phi(\varphi_n)| \geq \frac{1}{2} \|\Phi\| \quad \text{for all} \quad \Phi \in \text{Vect}\{\Phi_0, Jx_1, \ldots, Jx_{k-1}\}; \tag{3.4}$$

$$\max_{1 \leq n \leq n_k} |(\Phi_0 - Jx_k)(\varphi_n)| < \frac{1}{k}. \tag{3.5}$$

---

[28] We follow Whitley [486].

For $k = 1$, (3.4) is satisfied with $n_1 = 1$ if we choose a functional $\varphi_1 \in X'$ of norm one, satisfying $|\Phi(\varphi_1)| \geq \frac{1}{2} \|\Phi\|$. Then we may choose $x_1 \in B$ satisfying (3.5) by applying the definition of $\Phi_0 \in \overline{J(B)}$.

If the sequences are defined until some n_{k-1}, $\varphi_{n_{k-1}}$ and a_{k-1}, then applying Lemma 3.22 we may choose $n_k > n_{k-1}$ and functionals $\varphi_n \in X'$ of norm one for $n_{k-1} < n \leq n_k$ so as to satisfy (3.4), and then we may choose $x_k \in B$ satisfying (3.5) by applying the definition of $\Phi_0 \in \overline{J(B)}$.

There exists a weakly convergent subsequence $x_{k_\ell} \rightharpoonup x \in B$ by our assumption. Then we deduce from (3.4) by continuity that

$$\max_{1 \leq n < \infty} |(\Phi_0 - Jx)(\varphi_n)| \geq \frac{1}{2} \|\Phi_0 - Jx\| .$$

It remains to prove that $(\Phi_0 - Jx)(\varphi_n) = 0$ for all n. Indeed, then we will deduce from the last inequality that $\|\Phi_0 - Jx\| = 0$, i.e., $\Phi_0 = Jx \in J(X)$.

For any fixed index n we deduce from (3.5) that

$$|(\Phi_0 - Jx)(\varphi_n)| = |(\Phi_0 - Jx_{k_\ell})(\varphi_n) + \varphi_n(x_{k_\ell} - x)| < \frac{1}{k_\ell} + |\varphi_n(x_{k_\ell} - x)|$$

for all $\ell = 1, 2, \ldots$. Letting $\ell \to \infty$ we conclude $(\Phi_0 - Jx)(\varphi_n) = 0$ as required.

(c) $\Longrightarrow$ (a) If $J(B)$ is weak star compact, then it is also closed in B'' for this topology. Since $J(B)$ is also dense in B'' with respect to this topology by Goldstein's theorem, we must have $J(B) = B''$. Hence $J(X) = X''$, i.e., X is reflexive.

$\square$

*Remarks Let X be a reflexive space.

- According to property (c) the theorem of choice 2.30 (p. 90) holds for nets as well.
- In the weak topology of X we have the equivalence[29]

$$\text{compact} \Longleftrightarrow \text{bounded and closed.}$$

Indeed, the implication $\Longrightarrow$ holds in every separated locally convex space. For the converse let A be a weakly bounded and weakly closed set in X. Then A is also norm bounded (Proposition 3.13, p. 133), and therefore a subset of some closed ball K. Since B is weakly compact, the same holds for K by homogeneity, and then for its weakly closed subset A as well.

- Using the previous remark and applying the Tukey–Klee theorem (p. 124) for X_σ we obtain a new proof of Proposition 2.31 (p. 91) on the separation of disjoint, non-empty, convex, *bounded and closed* sets in *reflexive* spaces.

---

[29] We recall once again that this is false in every infinite-dimensional norm topology.

- Using the same remark and applying the Krein–Milman theorem (p. 129) for X_σ we obtain that in a *reflexive* space every non-empty, convex, *bounded and closed* set is the convex hull of its extremal points.

We end this section by establishing two further properties of reflexive spaces:

Proposition 3.23 (Pettis)[30] *Let X be a Banach space.*

(a) *If X is reflexive, then its closed subspaces are also reflexive.*
(b) *X is reflexive $\iff$ X' is reflexive.*

Proof Let Y be a closed subspace of X and denote the closed unit balls of X, Y, X' by B_X, B_Y, $B_{X'}$.

(a) Let Y be a closed subspace of X. In view of the preceding theorem, it is sufficient to prove that every bounded sequence $(y_n) \subset Y$ has a weakly convergent subsequence in Y.

   Since X is reflexive, (y_n) has a weakly convergent subsequence (y_{n_k}) in X, i.e, there exists an $a \in X$ such that $\varphi(y_{n_k}) \to \varphi(a)$ for all $\varphi \in X'$.

   Since the closed subspace Y is also weakly closed in X, we have $a \in Y$. Furthermore, since each $\psi \in Y'$ may be extended to a functional $\varphi \in X'$ by the Helly–Hahn–Banach theorem, it follows that $\psi(y_{n_k}) \to \psi(a)$ for all $\psi \in Y'$. In other words, $y_{n_k} \rightharpoonup a$ in Y.

(b) The closed unit ball B' of X' is $\sigma(X', X)$-compact by the Banach–Alaoglu theorem. If X is reflexive, then the topologies $\sigma(X', X)$ and $\sigma(X', X'')$ coincide, so that B' is also $\sigma(X', X'')$-compact. Applying the preceding theorem we conclude that X' is reflexive.

   If X' is reflexive, then X'' is reflexive by the just proved result. Using the linear isometry $J : X \to X''$ of Proposition 2.28 (p. 87), $J(X)$ is reflexive, as a complete and therefore closed subspace of X''. Since X and $J(X)$ are isomorphic, we conclude that X is reflexive.

$\square$

Examples

- We have proved in Sect. 2.6 (p. 87) separately that none of c_0, ℓ^1 and ℓ^∞ is reflexive. Since[31] $(c_0)' = \ell^1$ and $(\ell^1)' = \ell^\infty$, these results follow from one another by property (b) above.
- Since c_0 is a closed subspace of ℓ^∞ the non-reflexivity of c_0 directly implies the non-reflexivity of ℓ^∞.

---

[30]Pettis [357]. See Dunford and Schwartz [117] for more direct proofs.
[31]See Proposition 2.15, p. 73.

3.7 * Topological Vector Spaces

At first sight the following notion is more natural than that of locally convex spaces:

Definition By a *topological vector space* we mean a vector space endowed with a topology $\mathcal{T}$ for which the operations

$$X \times X \ni (x, y) \mapsto x + y \in X \quad \text{and} \quad \mathbb{R} \times \ni (\lambda, x) \mapsto \lambda x \in X$$

are continuous.

Remark It follows from the definition that the topology $\mathcal{T}$ is invariant for translations and multiplications by scalars: if A is an open, closed or compact set, then $A + x$ and λA are also open, closed or compact for all $x \in X$ and $\lambda \in \mathbb{R}$.

Every locally convex space is a topological vector space by Proposition 3.2 (d) (p. 121).

The following elementary inequality will allow us to give interesting examples of *non*-locally convex topological vector spaces.

Lemma 3.24 *If x, y are nonnegative real numbers and $0 < p \le 1$, then*

$$(x + y)^p \le x^p + y^p.$$

Proof Consider in $\mathbb{R}^2$ the norm $\|\cdot\|_{1/p}$ and apply the triangle inequality for the vectors $a := (x^p, 0)$ and $b := (0, y^p)$:

$$(x + y)^p = \|a + b\|_{1/p} \le \|a\|_{1/p} + \|b\|_{1/p} = x^p + y^p.$$

$\square$

Example Given $0 < p \le 1$ we denote by ℓ^p the set of real sequences $x = (x_n)$ satisfying $\sum |x_n|^p < \infty$. By the preceding lemma this is a vector space, and the formula

$$d_p(x, y) := \sum_{n=1}^{\infty} |x_n - y_n|^p$$

defines a metric on ℓ^p. For the corresponding topology ℓ^p is a topological vector space.[32] For $p = 1$ we obtain the already known Banach space ℓ^1.

---

[32]We will prove a more general theorem later in Proposition 10.5, p. 348.

Now we may explain the terminology *locally convex:*

Proposition 3.25 (Kolmogorov)[33] *A topological vector space is locally convex* $\Longleftrightarrow$ *every 0-neighborhood contains a* convex *0-neighborhood.*

Proof Every locally convex space has this property because the balls $B_{p,r}(0)$ are convex. Conversely, assume that a topological vector space X has this property, and consider an arbitrary 0-neighborhood V. It suffices to find a continuous seminorm p satisfying $B_{p,1}(0) \subset V$.

Let $U \subset V$ be a convex 0-neighborhood, then $-U$ and thus $W := -U \cap U$ is also a convex 0-neighborhood. One can readily verify that the formula[34]

$$p(x) := \inf \{t > 0 \ : \ x \in tW\}$$

defines a seminorm on X, satisfying

$$B_{p,1}(0) \subset W \subset \overline{B_{p,1}(0)}.$$

In particular, $B_{p,1}(0) \subset V$.

We show that p is continuous. For any given $a \in X$ and $r > 0$, $a + rW$ is a neighborhood of a. If $b \in a + rW$, then $r^{-1}(b - a) \in \overline{B_{p,1}(0)}$. Consequently,

$$|p(b) - p(a)| \le p(b - a) \le r.$$

$\square$

Example If $0 < p < 1$, then ℓ^p is not locally convex because the unit ball

$$B_1(0) := \{x \in \ell^p \ : \ d_p(0, x) < 1\}$$

contains no convex 0-neighborhood. Indeed, if K is a convex 0-neighborhood, then there exists a sufficiently small $r > 0$ such that $B_{2r}(0) \subset K$. Then the relations

$$r^{1/p} e_n \in \overline{B_r(0)} \subset B_{2r}(0) \subset K$$

hold for all $n = 1, 2, \ldots$, and hence

$$z_n := r^{1/p} \frac{e_1 + \cdots + e_n}{n} \in K$$

by the convexity of K. Since

$$d_p(0, z_n) = rn^{1-p} \to \infty,$$

K cannot belong to $B_1(0)$.

---

[33]Kolmogorov [253].
[34]Minkowski [325], pp. 131–132.

Remarks The non-locally convex topological vector spaces may have surprising pathological properties:

- There exist infinite-dimensional separated topological vector spaces X, in which $\varnothing$ and X are the only convex open sets.[35] In these spaces there are no *closed* hyperplanes because $X' = \{0\}$.
- Some separated topological vector spaces contain non-empty convex compact sets having no extremal points.[36]

3.8 Exercises

Exercise 3.1 Let B be a set in a normed space X. Prove that the following conditions are equivalent:

(i) B is bounded;
(ii) for every neighborhood V of 0 there exists an $r > 0$ such that $r'B \subset V$ for all $r' \in (0, r)$;
(iii) for every sequence $(x_n) \subset B$ and for every real sequence $r_n \to 0$ we have $r_n x_n \to 0$ in X.

Exercise 3.2 Prove that the conditions (i) and (ii) of the preceding exercise are equivalent in every topological vector space.

Exercise 3.3 We recall that the formula

$$\varphi_y(x) := \sum_{n=1}^{\infty} x_n y_n$$

defines a functional $\varphi_y \in c_0'$ for each $y = (y_n) \in \ell^1$, and that the linear map $y \mapsto \varphi_y$ is a bijection between ℓ^1 and c_0'.

(i) Prove that this result remains valid if we change c_0' to c', where c is the subspace of ℓ^∞ formed by the convergent sequences.
(ii) In view of (i) we may define two weak star topologies on ℓ^1. Are they the same?

Exercise 3.4 Prove the equivalences mentioned in the last remark of Sect. 3.5, p. 140.

Exercise 3.5 We recall that the formula

$$\varphi_y(x) := \sum_{n=1}^{\infty} x_n y_n$$

---

[35] We will encounter some examples at the end of Sects. 10.2 and 10.3, pp. 350 and 355.
[36] Roberts [395, 396]. See also the footnote on p. 349.

defines a functional $\varphi_y \in (\ell^1)'$ for each $y = (y_n) \in \ell^\infty$, and that the linear map $y \mapsto \varphi_y$ is a bijection between ℓ^∞ and $(\ell^1)'$.

(i) Prove that this result remains valid if we change $(\ell^1)'$ to $(\ell^p)'$ with $0 < p < 1$.
(ii) In view of (a) we may define a weak star topology on ℓ^∞ for each $0 < p \le 1$. Do they coincide?

Exercise 3.6 Does the Krein–Milman theorem remain valid in ℓ^p for $0 < p < 1$?

Exercise 3.7 Let us denote by ℓ^0 the vector space of the sequences $x = (x_n) \subset \mathbb{R}$ having at most finitely many non-zero elements. For $x, y \in \ell^0$ we denote by $d_0(x, y)$ the number of indices n such that $x_n \ne y_n$.

(i) Show that d_0 is a metric on ℓ^0.
(ii) Show that $d_0(x, y) = \lim_{p \to 0} d_p(x, y)$ for all $x, y \in \ell^0$.
(iii) Identify the topology associated with the metric $d_0(x, y)$.
(iv) Prove that every linear functional is continuous on ℓ^0.
(v) Is ℓ^0 a topological vector space?

Exercise 3.8 Given $1 \le p < \infty$, we recall from Exercise 2.2 (p. 113) that

$$\ell^p_w := \cap_{q>p} \ell^q$$

is strictly bigger than ℓ^p.

The family of norms $\|\cdot\|_q$ $(p < q \le \infty)$ defines a locally convex topology on ℓ^p_w. Is it normable?

Exercise 3.9 Generalize Exercise 2.2 for $0 \le p < q \le \infty$.

Exercise 3.10 Let $0 \le p < 1$.

(i) Prove that

$$\ell^p_w := \cap_{q>p} \ell^q$$

is a topological vector space for the family of metrics d_q, $q \in (p, 1]$.
(ii) Is it locally convex?

Part II
The Lebesgue Integral

Integration (in geometrical form) goes back to Archimedes [6], but he had practically no followers for almost two millennia. The Newton–Leibniz formula revolutionized the discipline in the seventeenth century, and led to the solution of a great number of geometrical and mechanical problems. A solid theoretical foundation became indispensable, especially after the publication of Fourier's work on heat propagation in [148].

Riemann [371] extended Cauchy's integral [80] to a class of not necessarily continuous functions. Subsequently much research was devoted to the construction of more general integrals and to the simplification of their manipulation. Following the works of Harnack [192, 194], Hankel [190], du Bois-Reymond [52], Jordan [230], Stolz [437] and Cantor [74], Peano [353] introduced the *finitely additive measures*, based on finite covers by intervals or rectangles.

Borel [59] discovered that *countable* covers lead to better, *σ-additive measures*. Baire [16, 17] enlarged the class of continuous functions by the repeated operation of pointwise limits of function sequences. Motivated by the works of Borel and Baire, Lebesgue [287, 288] defined a very general integral. He obtained a much wider class of integrable functions, and at the same time simpler limit theorems than before. He also greatly extended the validity of the Newton–Leibniz formula.

The extraordinary strength of the Lebesgue integral was demonstrated by subsequent important discoveries of Vitali, Beppo Levi, Fatou, Riesz, Fischer, Fréchet, Fubini (1905–1910) and others. These works also led to the development of *Functional Analysis*. The Lebesgue integral later allowed Kolmogorov to give a solid foundation of probability theory [252] and Sobolev to introduce new function spaces for the successful investigation of partial differential equations [426, 427].

F. Riesz gave nice historical accounts in two papers [390, 391]; for more complete surveys we refer to [61, 115, 198, 360–362].

More than a half-century after its publication, the monograph of Riesz and Sz.-Nagy [394] contains still perhaps the most elegant presentation of this theory. We follow this approach, with some minor subsequent improvements. Further results and exercises may be found in the following works: [68, 92, 188, 270, 351, 403, 406, 409, 451].

Part II
The Lebesgue Integral

Chapter 4
* Monotone Functions

I see it, but I don't believe it!

Letter of G. Cantor to Dedekind

No one shall expel us from the Paradise that Cantor has created for us.

D. Hilbert

In this chapter the letter I denotes a non-degenerate interval (having more than one point).

4.1 Continuity: Countable Sets

A *monotone* function $f : I \to \mathbb{R}$ has one-sided limits in each interior point a, and f is continuous at $a \iff$ they are equal. (See Fig. 4.1.)

What can we say about the set of points of continuity? In order to answer this question we recall the following notion:

Definition A set A is *countable*[1] if there exists a sequence (a_n) containing each element of A (at least once).[2]

Remarks

- The finite sets are countable.[3]
- If A is an *infinite* countable set, then there exists a sequence $(a_n) \subset A$ containing each element of A exactly once.[4]
- *The image of a countable set is also countable.* More precisely, if $g : A \to B$ is a *surjective* function and A is countable, then B is also countable.
- If $g : A \to B$ is an *injective* function and B is countable, then A is also countable.

---

[1]Cantor [71].

[2]Vilenkin's books [467, 468] give a very pleasant introduction to infinite sets.

[3]The sequence (a_n) may contain points outside A.

[4]We may take a suitable subsequence of the sequence in the definition.

© Springer-Verlag London 2016
V. Komornik, *Lectures on Functional Analysis and the Lebesgue Integral*,
Universitext, DOI 10.1007/978-1-4471-6811-9_4

Fig. 4.1 Graph of a
monotone function

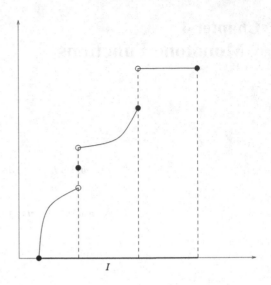

I

Examples

- The number sets $\mathbb{N}$, $\mathbb{Z}$ and $\mathbb{Q}$ are countable.
- A set $\mathcal{P}$ of pairwise disjoint non-degenerate intervals is always countable. Indeed, selecting a rational number in each interval we get an injective map $g : \mathcal{P} \to \mathbb{Q}$.

The last example motivates the following terminology:

Definition A set system or set sequence is *disjoint* if its elements are pairwise disjoint.

Let us state the basic properties of countable sets. The last result contains a famous theorem of Cantor[5]: the set $\mathbb{R}$ of real numbers is *uncountable*.

Proposition 4.1

(a) *A subset of a countable set is also countable.*
(b) *The union of countably many countable sets is also countable.*
(c) *The* non-degenerate *intervals are uncountable.*

Proof

(a) If $B \subset A$, then the formula $f(x) := x$ defines an injective function $f : B \to A$. Since A is countable, B is countable, too.
(b) Let (A_n) be a countable set sequence. Fix for each n a sequence $a_{n1}, a_{n2}, \ldots$ containing the elements of A_n. If $p_1, p_2, \ldots$ is the sequence of prime numbers, then the following formula defines a sequence (a_n) containing the elements of

---

[5]Cantor [70], pp. 117–118.

$\cup A_n$:

$$a_m := \begin{cases} a_{nk} & \text{if } m = (p_n)^k \text{ for some } n \text{ and } k, \\ 0 & \text{otherwise.} \end{cases}$$

(c) We show that no sequence (a_n) contains all points of a non-degenerate interval I. First we choose a non-degenerate *compact* subinterval $I_1 \subset I$ such that $a_1 \notin I_1$. Then we choose a non-degenerate compact subinterval $I_1 \subset I$ such that $a_2 \notin I_2$. Continuing by induction we obtain a non-increasing sequence of non-degenerate compact intervals

$$I \supset I_1 \supset I_2 \supset \cdots$$

such that $a_n \notin I_n$ for every n. By Cantor's intersection theorem these intervals have a common point x. Then $x \in I$ and x does not belong to the sequence (a_n).

$\square$

Now we return to the study of monotone functions.

Proposition 4.2

(a) *The set of discontinuity of a monotone function is countable.*
(b) *Every countable set of real numbers is the set of discontinuity of a suitable monotone function.*

Proof

(a) Multiplying our monotone function $f : I \to \mathbb{R}$ by -1 if necessary, we may assume that it is non-decreasing. Let A denote the set of interior points a of I where f is not continuous. Since f is non-decreasing, the non-degenerate open intervals

$$(f(a-0), f(a+0)), \quad a \in A$$

are pairwise disjoint. By a preceding remark this implies that A is countable. The set of discontinuity of f has at most two more points (the endpoints of I), hence it is also countable.

(b) For the empty set we may choose any constant function. Otherwise, denoting by (a_n) the (finite or infinite) sequence of the points of the given countable set, the sum of the uniformly convergent series

$$f(x) := \sum_{\{n \,:\, a_n < x\}} 2^{-n}$$

is a suitable function $f : \mathbb{R} \to \mathbb{R}$.

$\square$

4.2 Differentiability: Null Sets

In this section we investigate the differentiability of monotone functions. The following notion will be very useful:

Definition A set A of real numbers is a *null set*[6] if for each fixed $\varepsilon > 0$ it may be covered by a set of intervals of total length $\le \varepsilon$:

$$A \subset \bigcup I_k \quad \text{and} \quad \sum |I_k| \le \varepsilon.$$

Here and in the sequel we denote by $|I|$ the length of an interval I.

Remarks

- A set of intervals of finite total length L is necessarily countable. Indeed, it contains less then nL intervals of length $\ge 1/n$ for each $n = 1, 2, \ldots$, and the union of countably many finite sets is countable.
- If A is a null set, then there exists an interval sequence (J_m) of finite total length such that each point of A is covered infinitely many times. Indeed, we may cover A for each $n = 1, 2, \ldots$ by an interval set (I_{nk}) of total length $< 2^{-n}\varepsilon$.[7] We conclude by arranging all the intervals I_{nk} into a sequence (J_m).

  Conversely, the existence of such a sequence (J_m) implies that A is a null set. Indeed, for any fixed $\varepsilon > 0$ there exists a large integer N such that

$$\sum_{m>N} |J_m| < \varepsilon,$$

and the intervals $J_{m+1}, J_{m+2}, \ldots$ still cover A.

Examples

- (Harnack)[8] Every countable set $\{a_n\}$ of real numbers is a null set: for each $\varepsilon > 0$: it is covered by the intervals

$$(a_n - \varepsilon 3^{-n}, a_n + \varepsilon 3^{-n})$$

of total length ε.
- (*Cantor's ternary set*)[9] There exist uncountable null sets. Let us remove from the unit segment $[0, 1]$ its middle third, i.e., the open interval $(1/3, 2/3)$. There

---

[6]Hankel [190] (p. 86), Ascoli [11], Smith [423] (p. 150), du Bois-Reymond [52], Harnack [194].

[7]By slightly enlarging them we may assume that all the intervals are open.

[8]Harnack [194].

[9]Smith [423], Cantor [72] (p. 207). Many analogous sets appear "naturally" in combinatorial number theory, see, e.g., Erdős–Joó–Komornik [127], Komornik–Loreti [260], de Vries–Komornik [101], Komornik–Kong–Li [259], de Vries–Komornik–Loreti [102].

Fig. 4.2 The sets C_n

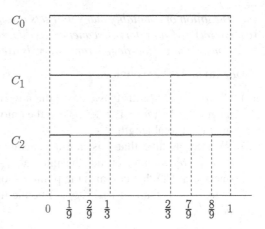

remain two disjoint segments $[0, 1/3]$ and $[2/3, 1]$ of total length $2/3$. Next remove from each of them their middle thirds: there remain four disjoint segments of total length $(2/3)^2$; see Fig. 4.2. Continuing by induction, after n steps we obtain a set C_n, which is the union of 2^n disjoint compact segments of length 3^{-n} each. The intersection C of this decreasing set sequence is a compact set, called *Cantor's ternary set*.

It is a null set. Indeed, for each $\varepsilon > 0$ there is a large integer n such that $(2/3)^n < \varepsilon$; then the 2^n disjoint segments of C_n form a finite cover of C with total length $= (2/3)^n < \varepsilon$.

By construction C is formed by the real numbers x that may be written in base 3 in the form

$$x = \sum_{i=1}^{\infty} \frac{c_i}{3^i}$$

with $(c_i) \subset \{0, 2\}$, i.e., without using the digit $c_i = 1$. Since all sequences $(c_i) \subset \{0, 2\}$ occur here, the formula

$$\sum_{i=1}^{\infty} \frac{c_i}{3^i} \mapsto \sum_{i=1}^{\infty} \frac{c_i}{2^{i+1}}$$

defines a map of C onto $[0, 1]$. The latter set is uncountable, hence C is also uncountable.

- It follows from our next proposition that $\mathbb{R}$ is *not* a null set.

Let us resume the basic properties of null sets.

Proposition 4.3

(a) *The empty set is a null set.*
(b) *The subsets of a null set are null sets.*

(c) *The union of countably many null sets is a null set.*
(d) *(Borel)*[10] *If an interval sequence* (I_k) *covers an interval* I *then* $|I| \leq \sum |I_k|$. *Consequently, non-degenerate intervals are* not *null sets.*

Proof (a) and (b) are obvious.

(c) Given $\varepsilon > 0$ arbitrarily, we cover the null set A_n by an interval set (I_{nk}) of total length $\leq \varepsilon 2^{-n}$, $n = 1, 2, \ldots$. Then the union of all these intervals form a cover of $\cup A_n$ of total length $\leq \varepsilon$.
(d) We may assume that I is non-degenerate. First we consider the case where $I = [a, b]$ is compact and the intervals I_k are open. Let (a_1, b_1) be the first interval in (I_k) that contains the point a. Continuing by induction, if $b_n \leq b$ for some $n \geq 1$, then let (a_{n+1}, b_{n+1}) be the first interval in (I_k) that contains the point b_n.

The construction stops after a finite number of steps because $b_N > b$ for some N. For otherwise the bounded sequence (b_n) would converge to some $x \leq b$, and we would have $x \in I_\ell$ for some ℓ. Since I_ℓ is open, there would exist an index m such that $b_n \in I_\ell$ for all $n \geq m$. By construction this would mean that the intervals (a_n, b_n) would precede I_ℓ in the sequence (I_k) for all $n > m$. But this is absurd because $b_1 < b_2 < \cdots$ by construction, so that the intervals (a_n, b_n) are pairwise distinct.

It follows that

$$|I| = b - a < b_N - a_1 = \sum_{i=2}^{N}(b_i - b_{i-1}) + b_1 - a_1 \leq \sum_{i=1}^{N}(b_i - a_i) \leq \sum |I_k|.$$

In the general case we fix a number $\alpha > 1$, a compact subinterval $J \subset I$ of length $|I|/\alpha$, and for each n an open interval $J_n \supset I_n$ of length $\alpha|I_n|$. The sequence (J_n) covers J, so that $\sum |J_n| \geq |J|$ by the first part of the proof. In other words we have $\alpha \sum |I_n| \geq |I|/\alpha$, and we conclude by letting $\alpha \to 1$. □

Let us introduce a convenient terminology:

Definition A property holds *almost everywhere*[11] (shortly *a.e.*) if it holds outside a null set.

[10]Borel [59]. His proof was based on a construction of Heine [200] (p. 188).
[11]Lebesgue [293] (p. 7).

We may now state a deep theorem:

Theorem 4.4

(a) *(Lebesgue)[12] Every monotone function $f : I \to \mathbb{R}$ is a.e. differentiable.*
(b) *For each null set A there exists a non-decreasing, continuous function $f : \mathbb{R} \to \mathbb{R}$ that is non-differentiable at the points of A.*

Part (a) of this theorem will be proved in the next two sections.

*Proof of part (b) Choose a sequence (J_m) of open intervals, of finite total length, and covering each point of A infinitely many times. Denoting the length of the interval $J_m \cap (-\infty, x)$ by $f_m(x)$, the formula $f := \sum f_m$ defines a non-decreasing function $f : \mathbb{R} \to \mathbb{R}$.

Since the series is uniformly convergent and each f_m is continuous, f is also continuous.

We complete the proof by establishing the relation

$$\lim_{h \searrow a} \frac{f(a+h) - f(a)}{h} = \infty$$

for each $a \in A$.

Fix an arbitrarily large number N, and then choose a sufficiently small number $\delta > 0$ such that at least N intervals J_m contain $[a, a + \delta]$, say $J_{m_1}, \ldots, J_{m_N}$. Then

$$f(a+h) - f(a) \geq \sum_{k=1}^{N} f_{m_k}(a+h) - f_{m_k}(a) = Nh$$

for all $0 < h < \delta$. $\qquad\square$

4.3 Jump Functions

Since every interval is the union of countably many compact intervals, it is sufficient to prove Lebesgue's theorem for compact intervals $I = [a, b]$.

In this section we follow an approach of Lipiński and Rubel[13] to prove some special cases of the theorem.

---

[12]Lebesgue [290], pp. 128–129. He considered only the case of continuous functions. Before him Weierstrass conjectured the existence of continuous and monotone, but nowhere differentiable functions; see Hawkins [198], p. 47.
[13]Lipiński [307], Rubel [401].

Fig. 4.3 Meaning of E_C

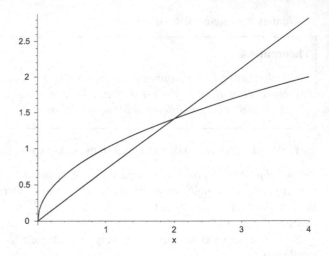

We start with a lemma:

Lemma 4.5 *Let* $f : [a, b] \to \mathbb{R}$ *be a non-decreasing function. For each* $C > 0$ *we denote by* E_C *the set of points* $a < x < b$ *for which there exist numbers* $s = s_x$ *and* $t = t_x$ *satisfying* $s < x < t$ *and*

$$f(t) - f(s) > C(t - s). \tag{4.1}$$

Then E_C *is the union of countably many intervals* (a_n, b_n) *of total length* $\leq$ $4C^{-1}(f(b) - f(a))$.

Remark The set E_C contains all points at which f has a derivative $> C$, but it may contain other points as well. For example, consider the function $f(x) := \sqrt{x}$ in the interval $[0, 4]$. For $C = 1/\sqrt{2}$ we have

$$\{f' > C\} = (0, 1/2) \quad \text{and} \quad E_C = (0, 2).$$

(See Fig. 4.3: for $0 < x < 2$ we may choose $s_x = 0$ and $t_x = (x + 2)/2$.)

Proof The set E_C is open by definition, hence it is the union of disjoint open intervals (a_n, b_n). We also observe that if $x \in (a_n, b_n)$, then $(s_x, t_x) \subset (a_n, b_n)$ by definition.

Fix for each n a compact subinterval $[a'_n, b'_n] \subset (a_n, b_n)$ of length

$$b'_n - a'_n = (b_n - a_n)/2. \tag{4.2}$$

It is covered by the intervals

$$(s_x, t_x), \quad x \in [a'_n, b'_n].$$

Since $[a'_n, b'_n]$ is compact, there exists a finite subcover $(s_1, t_1), \ldots, (s_N, t_N)$. Choose a finite subcover with N as small as possible. Then no point of $\cup(s_k, t_k)$ is covered more than twice, because if three intervals have a common point, then one of them belongs to the union of the other two. Consequently, using (4.1) and the relations $(s_k, t_k) \subset (a_n, b_n)$, we have

$$b'_n - a'_n \le \sum_{k=1}^{N}(t_k - s_k) \le C^{-1}\sum_{k=1}^{N}(f(t_k) - f(s_k)) \le 2C^{-1}(f(b_n) - f(a_n)).$$

Using (4.2) this yields the required inequality:

$$\sum(b_n - a_n) \le 4C^{-1}\sum(f(b_n) - f(a_n)) \le 4C^{-1}(f(b) - f(a)).$$

$\square$

As a first application of this lemma, we prove that a non-decreasing function cannot have an infinite derivative at many points. More precisely, we have the

Lemma 4.6 *If* $f : [a, b] \to \mathbb{R}$ *is a non-decreasing function, then*

$$Df(x) := \limsup_{y \to x} \frac{f(y) - f(x)}{y - x} < \infty \quad a.e. \ in \ [a, b].$$

Proof If $Df(x) = \infty$, then $x \in E_C$ for every $C > 0$, so that the set of these points may be covered by a set of intervals of total length $\le 4(f(b) - f(a))/C$. We conclude by letting $C \to \infty$. $\square$

As a second application we prove Lebesgue's theorem in a special case.

Definition By a *jump function* we mean a function $f : I \to \mathbb{R}$ of the form $f = \sum f_k$ where $(a_k) \subset I$ is a given sequence of points, $\sum S_k$ is a nonnegative convergent numerical sequence, and

$$f_k(x) = 0 \quad \text{if} \quad x < a_k,$$
$$f_k(x) = S_k \quad \text{if} \quad x > a_k,$$
$$0 \le f_k(a_k) \le S_k.$$

Every jump function is non-decreasing.

Proposition 4.7 *If* $f : I \to \mathbb{R}$ *is a jump function, then* $f' = 0$ *a.e.*

Proof We may assume that $I = [a, b]$ is compact. It suffices to show that $Df \le C$ a.e. for every fixed $C > 0$.

Fix an arbitrary $\varepsilon > 0$,[14] then choose a large N such that

$$\sum_{k=N+1}^{\infty} S_k < \varepsilon.$$

Then the function

$$h := \sum_{k=N+1}^{\infty} f_k$$

is non-decreasing, and $h(b) - h(a) < \varepsilon$. By Lemma 4.5 we have $Dh \leq C$ outside a set of intervals of total length $< 4\varepsilon/C$.

Observe that the function

$$f - h = \sum_{k=1}^{N} f_k$$

has zero derivative everywhere, except $a_1, \ldots, a_N$. Hence $Df \leq C$ outside a set of intervals of total length $< 4\varepsilon/C$. We conclude by letting $\varepsilon \to 0$.   □

Using jump functions we may isolate the discontinuous part of non-decreasing functions:

Proposition 4.8 *Every* bounded *non-decreasing function $f : I \to \mathbb{R}$ is the sum of a* continuous *non-decreasing function and a jump function.*

Proof Since f is bounded, extending f by constants we may assume that $I = \mathbb{R}$. Let (a_k) be the (finite or infinite) sequence of discontinuities of f, and set $S_k = f(a_k+0) - f(a_k-0)$. The series $\sum S_k$ is convergent because f is bounded. Introduce the functions f_k as in the definition of the jump functions, and set $f_k(a_k) := f(a_k) - f(a_k - 0)$. Then $h := \sum f_k$ is a jump function by definition, while $g := f - h$ is non-decreasing and continuous.[15]   □

---

[14]The following proof is due to Á. Császár; see Sz.-Nagy [448].

[15]See Exercise 4.3 at the end of this chapter, p. 165.

Fig. 4.4 Dini derivatives

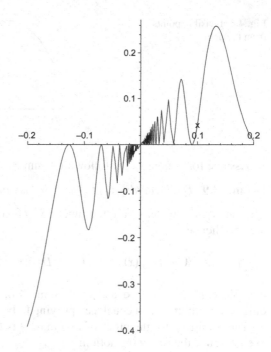

4.4 Proof of Lebesgue's Theorem

In view of Propositions 4.7 and 4.8 it is sufficient to consider a non-decreasing and *continuous* function $f : [a, b] \rightarrow \mathbb{R}$, defined on a compact interval. In this section we present an elementary proof due to F. Riesz.[16]

We introduce the *Dini derivatives*[17]:

$$D_-f(x) := \limsup_{\substack{y<x \\ y\to x}} \frac{f(y) - f(x)}{y - x}, \quad D_+f(x) := \limsup_{\substack{y>x \\ y\to x}} \frac{f(y) - f(x)}{y - x},$$

$$d_-f(x) := \liminf_{\substack{y<x \\ y\to x}} \frac{f(y) - f(x)}{y - x}, \quad d_+f(x) := \liminf_{\substack{y>x \\ y\to x}} \frac{f(y) - f(x)}{y - x}.$$

Since f is non-decreasing, they are all nonnegative.

Example For $f(x) := x + x\sin(1/x)$ we have

$$D_-f(0) = D_+f(0) = 1 \quad \text{and} \quad d_-f(0) = d_+f(0) = 0;$$

see Fig. 4.4.

---

[16]Riesz [386, 387]. The proof may be adapted to the discontinuous case: see Riesz and Sz.-Nagy [394], Sz.-Nagy [448]. See also other elementary proofs of Austin [14] and Botsko [63].
[17]Dini [109] (Sect. 145).

Fig. 4.5 Invisible points
from the right

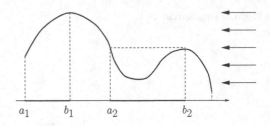

Assume for a moment the following lemma:

Lemma 4.9 *The inequality $D_+f \leq d_-f$ holds almost everywhere.*

Then applying this lemma to the function $-f(-x)$ we have also $D_-f(x) \leq d_+f(x)$
a.e., and hence

$$0 \leq D_+f(x) \leq d_-f(x) \leq D_-f(x) \leq d_+f(x) \leq D_+f(x)$$

a.e. Since $D_+f(x) < \infty$ a.e. by Lemma 4.6, we conclude that the four Dini
derivatives are finite and equal a.e., proving Lebesgue's theorem.

The main tool for the proof of Lemma 4.9 is the "Rising sun lemma" of Riesz.
We introduce the following notion:

Definition Let $g : [a, b] \rightarrow \mathbb{R}$ be a continuous function on a compact interval.
The point $a < x < b$ is *invisible (from the right)* if there exists a $y > x$ such that
$g(y) > g(x)$. (See Fig. 4.5.)

Lemma 4.10 ("Rising sun lemma")[18] *The invisible points (from the right) form a
union of disjoint open intervals (a_k, b_k), and $g(a_k) \leq g(b_k)$ for every k.*[19]

Proof The set of invisible points is open by the continuity of g, hence a union of
disjoint open intervals (a_k, b_k).

Assume on the contrary that $g(a_k) > g(b_k)$ for some k. Fix a number $g(a_k) >
c > g(b_k)$ and set

$$x := \sup\{a_k \leq t \leq b_k : g(t) \geq c\}.$$

By the continuity of g we have $g(x) = c$ and thus $a_k < x < b_k$. Since x is invisible,
there exists a $y > x$ such that $g(y) > g(x) = c$. Since $g < c$ on $(x, b_k]$ by the choice of
x, we have $y > b_k$. But this contradicts the visibility of b_k because $g(y) > c > g(b_k)$.
□

Proof of Lemma 4.9 It suffices to show that for any fixed *rational* numbers $c_1 < c_2$,

$$E := \{x \in (a, b) : d_-f(x) < c_1 < c_2 < D_+f(x)\}$$

---

[18]Riesz [386, 387]. See the correspondence of Riesz in [443, 444] for the history of this result.
[19]It is easy to see that we even have $g(a_k) = g(b_k)$ if $a_k \neq a$.

is a null set. Indeed, then their (countable) union is also a null set, and $d_-f(x) \geq D_+f(x)$ outside them.

We are going to show that for any fixed open subinterval (a', b') of (a, b), we may cover $E \cap (a', b')$ by a (countable) set of open intervals of total length $< (c_1/c_2)(b' - a')$. Iterating this procedure we will get that $E = E \cap (a, b)$ may be covered for each $n = 1, 2, \ldots$ by a set of open intervals of total length $< (c_1/c_2)^n(b - a)$. Since $c_1/c_2 < 1$, letting $n \to \infty$ we will conclude that E is a null set.

If $x \in E \cap (a', b')$, then

$$\frac{f(y) - f(x)}{y - x} < c_1$$

for some $a' < y < x$, i.e.,

$$f(y) - c_1 y > f(x) - c_1 x.$$

In other words, x is *invisible from the left*[20] for the function

$$g(t) := f(t) - c_1 t, \quad t \in [a', b'].$$

Applying Lemma 4.10 for the function $t \mapsto g(-t)$, $E \cap (a', b')$ may be covered by a countable set of disjoint open intervals (a_k, b_k) such that $g(a_k) \geq g(b_k)$, i.e.,

$$f(b_k) - f(a_k) \leq c_1(b_k - a_k)$$

for every k.

Now consider one of these intervals (a_k, b_k). If $x \in E \cap (a_k, b_k)$, then

$$\frac{f(y) - f(x)}{y - x} > c_2$$

for some $x < y < b_k$, i.e.,

$$f(y) - c_2 y > f(x) - c_2 x.$$

In other words, x is invisible from the right for the function

$$g(t) := f(t) - c_2 t, \quad t \in [a_k, b_k].$$

---

[20]We say that x is *invisible from the left* for a function g if $-x$ is invisible from the right for the function $t \mapsto g(-t)$.

Applying Lemma 4.10, $E \cap (a_k, b_k)$ may be covered by a countable set of disjoint open intervals (a_{km}, b_{km}) such that $g(a_{km}) \leq g(b_{km})$, i.e.,

$$f(b_{km}) - f(a_{km}) \geq c_2(b_{km} - a_{km})$$

for every m.

Consequently, the intervals (a_{km}, b_{km}) cover $E \cap (a', b')$, and

$$\sum_{k,m}(b_{km} - a_{km}) \leq \frac{1}{c_2} \sum_{k,m} f(b_{km}) - f(a_{km})$$

$$\leq \frac{1}{c_2} \sum_{k}\left(f(b_k) - f(a_k)\right)$$

$$\leq \frac{c_1}{c_2} \sum_{k}(b_k - a_k)$$

$$\leq \frac{c_1}{c_2}(b' - a').$$

$\square$

4.5   Functions of Bounded Variation

The difference of two monotone functions is not necessarily monotone. However, it follows from Proposition 4.2 and Theorem 4.4 (pp. 153 and 157) that these functions still also have at most countably many discontinuities, and they are differentiable a.e. In this section we briefly discuss these functions.

Definition A function $f : I \to \mathbb{R}$ is *of bounded variation*[21] if there exists a number A such that

$$\sum_{i=1}^{n} |f(x_i) - f(x_{i-1})| \leq A$$

for every finite set of points $x_0 < \cdots < x_n$ in I. The smallest such number A is called the *total variation* of f.

Remarks

• Every function of bounded variation is bounded.

---

[21]Jordan [229]. He introduced this notion in order to give an elegant formulation of Dirichlet's theorem on the convergence of Fourier series.

- In the case of a bounded interval I, f has a bounded variation $\Longleftrightarrow$ it is *rectifiable*, i.e, if its graph has a finite arc length.
- Every monotone and bounded function has a bounded variation.
- The functions of bounded variation form a vector space.

Our last remarks imply that the difference of two monotone and bounded functions has a bounded variation. The converse also holds:

Proposition 4.11 (Jordan)[22] *Every function of bounded variation is the difference of two non-decreasing and bounded functions.*

Proof If $f : I \to \mathbb{R}$ has bounded variation, then its restriction to any subinterval also has bounded variation. Let us denote by $g(x)$ the total variation of f on $I \cap (-\infty, x)$ for each $x \in I$. Then $0 \le g \le T$, where T denotes the total variation of f, so that g is a bounded function.

If $y \in I$ and $x < y$, then $g(x) + |f(y) - f(x)| \le g(y)$ by the definition of the total variation. It follows that g is non-decreasing, and then that $g - f$ is also non-decreasing because

$$(g - f)(y) - (g - f)(x) = \big(g(y) - g(x)\big) - \big(f(y) - f(x)\big)$$
$$\ge \big(g(y) - g(x)\big) - |f(y) - f(x)|$$
$$\ge 0.$$

Since f and g are bounded, $h := g - f$ is bounded, too. Therefore the decomposition $f = g - h$ has the required properties. $\qquad\qquad\square$

Remark It follows from the theorems of Jordan and Lebesgue that if $f : I \to \mathbb{R}$ has bounded variation and $\bar{I} = [a, b]$, then f has a finite left limit at every $a < x \le b$, a finite right limit at every $a \le x < b$, and (applying Lebesgue's theorem) that f is a.e. differentiable.

4.6 Exercises

Exercise 4.1 Given an arbitrary null set D, does there exist a monotone function $f : \mathbb{R} \to \mathbb{R}$ that is non-differentiable exactly at the points of D?

Exercise 4.2 If C denotes Cantor's ternary set, then $C - C = [0, 1]$.

Exercise 4.3 Prove that the function g in the proof of Proposition 4.8 (p. 160) is non-decreasing and continuous.

---

[22] Jordan [229].

In the remaining exercises we consider *bounded closed* intervals.

Exercise 4.4 (Lebesgue's criterium)[23] Let $f : [a, b] \to \mathbb{R}$, then

$$f \quad \text{is Riemann integrable} \quad \Longleftrightarrow f \quad \text{is bounded, and continuous a.e.}$$

Exercise 4.5 Let $f, g : [a, b] \to \mathbb{R}$ have bounded variations.

(i) fg, max $\{f, g\}$ and min $\{f, g\}$ also have bounded variations.
(ii) $|f|$ has bounded variation.
(iii) If moreover, inf $|g| > 0$, then f/g also has bounded variation.

Exercise 4.6 If $f : [a, b] \to \mathbb{R}$ is continuous, then f and $|f|$ have bounded variations at the same time. Is the continuity assumption necessary?

Exercise 4.7 For which values of α, β does $f(x) := x^\alpha \sin \dfrac{1}{x^\beta}$ have bounded variation on $[0, 1]$?

Exercise 4.8 If $f : [a, b] \to \mathbb{R}$ has bounded variation, then f has finite left and right limits everywhere, and f has at most countably many discontinuities.

Exercise 4.9

(i) If $f : [a, b] \to \mathbb{R}$ is Lipschitz continuous, then it has bounded variation.
(ii) Construct a Hölder continuous function $f : [a, b] \to \mathbb{R}$ which is not of bounded variation.

Exercise 4.10 Write the following functions as the difference of two non-decreasing functions:

(i) $f(x) = \text{sign} \, x$ in $[-1, 1]$;
(ii) $f(x) = \sin x$ in $[0, 2\pi]$.

Exercise 4.11 (Helly's selection theorem)[24] Let $f_n : [a, b] \to \mathbb{R}$, $n = 1, 2, \ldots$ be a uniformly bounded sequence of functions of bounded variation. Assume that their total variations are bounded by some constant. Prove the existence of an everywhere convergent subsequence by proving the statements below.

(i) We may assume that all functions f_n are non-decreasing. Henceforth we consider this special case.
(ii) There exists a subsequence $(f_n^1) \subset (f_n)$ converging in a, b and in all rational points of (a, b). Write

$$\psi(x) := \lim f_n^1(x), \quad x \in E := \{a, b\} \cup ((a, b) \cap \mathbb{Q}).$$

---

[23]Lebesgue [288], p. 29.
[24]Helly [204]. This is a weak compactness theorem in the space of functions of bounded variation. We follow Natanson [332].

(iii) ψ extends to a non-decreasing function $\psi : [a, b] \to \mathbb{R}$.

(iv) $f_n^1(x) \to \psi(x)$ at all points $x \in (a, b)$ where ψ is continuous.

(v) There exists a second subsequence $(f_n^2) \subset (f_n^1)$ which also converges at the points of discontinuity of ψ.

Chapter 5
The Lebesgue Integral in $\mathbb{R}$

I turn with terror and horror from this lamentable scourge of continuous functions with no derivatives!—Letter of Hermite to Stieltjes, 1893

In former times when one invented a new function it was for a practical purpose; today one invents them purposely to show up defects in the reasoning of our fathers and one will deduce from them only that.—H. Poincaré

The Riemann integral has the drawback that many important functions are not integrable and the limiting processes are complicated:

Examples

- (*Dirichlet function*)[1] The function

$$f(x) := \begin{cases} 1 & \text{if } x \text{ is rational;} \\ 0 & \text{if } x \text{ is irrational} \end{cases}$$

is not Riemann integrable. However, since $f = 0$ a.e., it is tempting to define $\int f \, dx := 0$.
- Let us enumerate the rational numbers into a sequence (r_n). Then the functions

$$f_n(x) := \begin{cases} 1 & \text{if } x = r_1, \ldots, r_n; \\ 0 & \text{otherwise} \end{cases}$$

are Riemann integrable, $\int f_n \, dx = 0$ for all n, and $f_n \to f$ a.e. We would like to conclude that $\int f_n \, dx \to \int f \, dx$, but the last integral is not defined.

---

[1]Dirichlet [112, pp. 131–132].

© Springer-Verlag London 2016
V. Komornik, *Lectures on Functional Analysis and the Lebesgue Integral*,
Universitext, DOI 10.1007/978-1-4471-6811-9_5

- The formula $\|g\| := \int |g| \, dx$ defines a natural norm in the vector space of Riemann integrable functions. For this norm the above sequence (f_n) satisfies the Cauchy criterion, but it is not convergent.

The Lebesgue integral eliminates these difficulties: much more functions are integrable and they are easier to manipulate. One key of this theory is that we do not distinguish between two functions if they are equal outside some null set:

Definition The functions $f_1 : D_1 \to \mathbb{R}$ and $f_2 : D_2 \to \mathbb{R}$ are *equal almost everywhere (a.e.)* if

$$D_1 \setminus D_2, \quad D_2 \setminus D_1 \quad \text{and} \quad \{x \in D_1 \cap D_2 : f_1(x) \neq f_2(x)\}$$

are null sets.

This is an equivalence relation that is compatible with the usual algebraic operations: if $f_1 = g_1$ and $f_2 = g_2$ a.e., then

$$|f_1| = |g_1| \quad \text{a.e.,}$$
$$f_1 \pm f_2 = g_1 \pm g_2 \quad \text{a.e.,}$$
$$f_1 f_2 = g_1 g_2 \quad \text{a.e.,}$$
$$\min\{f_1, f_2\} = \min\{g_1, g_2\} \quad \text{a.e.,}$$
$$\max\{f_1, f_2\} = \max\{g_1, g_2\} \quad \text{a.e.}$$

If, moreover, $f_2 \neq 0$ a.e., then $f_1/f_2 = g_1/g_2$ a.e.

Finally, if $f_n \to f$ a.e., and $f_n = g_n$ a.e. for every n, then $g_n \to f$ a.e.[2]

In view of these properties we often identify two functions if they are equal almost everywhere.[3] Hence we often write $f = g, f \geq g, f > g$ instead of $f = g$ a.e, $f \geq g$ a.e., $f > g$ a.e., and a sequence (f_n) is called simply nonnegative, non-decreasing or non-increasing if it is nonnegative a.e., non-decreasing a.e. or non-increasing a.e.

5.1 Step Functions

Definition $\varphi : \mathbb{R} \to \mathbb{R}$ is a *step function* if there exist finitely many points

$$-\infty < x_0 < \cdots < x_n < \infty$$

---

[2] It is essential here that we use *countable* covers in the definition of null sets.

[3] To be precise, we should use equivalence classes of functions but we follow the traditional, looser terminology.

Fig. 5.1 Step function

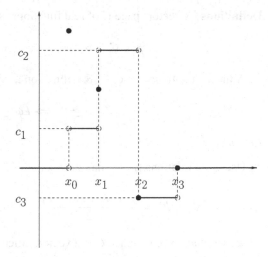

and real numbers $c_1, \ldots, c_n$ such that a.e.,

$$\varphi(x) = \begin{cases} 0 & \text{if } x < x_0, \\ c_1 & \text{if } x_0 < x < x_1, \\ \cdots \\ c_n & \text{if } x_{n-1} < x < x_n, \\ 0 & \text{if } x_n < x. \end{cases}$$

See Fig. 5.1. The class of step functions is denoted by C_0.

Remarks

- We may always add to the definition a finite number of arbitrary points x_i. Consequently, for finitely many given step functions we may always assume that they are defined by the same points x_i.
- Once the points x_i are given, the corresponding numbers c_i are uniquely determined because the non-degenerate intervals in the definition of $\varphi(x)$ are not null sets.

Definition By the *integral* of a step function we mean the number

$$\int \varphi \, dx := \sum_{i=1}^{n} c_k(x_k - x_{k-1}).$$

In order to show the correctness of this definition we introduce two useful notions:

Definitions A vector space C of real functions is a *vector lattice* if

$$\varphi, \psi \in C \Longrightarrow \max\{\varphi, \psi\}, \ \min\{\varphi, \psi\} \in C.$$

A linear functional $L : C \to \mathbb{R}$ defined on a vector lattice C is *positive* if

$$\varphi \geq 0 \Longrightarrow L\varphi \geq 0.$$

Remarks

- Using the relations $|\varphi| = \max\{\varphi, -\varphi\}$ and

$$\max\{\varphi, \psi\} = \frac{\varphi + \psi + |\varphi - \psi|}{2}, \quad \min\{\varphi, \psi\} = \frac{\varphi + \psi - |\varphi - \psi|}{2}$$

  we see that a vector space C is a vector lattice $\Longleftrightarrow$

$$\varphi \in C \Longrightarrow |\varphi| \in C.$$

- Every positive linear functional is *monotone*, i.e.,

$$\varphi \geq \psi \Longrightarrow L\varphi \geq L\psi.$$

Using the remark following the definition of step functions the next result can be shown easily:

Proposition 5.1

(a) C_0 *is a vector lattice.*
(b) *The integral of a step function does not depend on the particular choice of the points x_i.*
(c) *The integral of step functions is a positive linear functional on C_0.*

The following two "innocent-looking" lemmas are due to Riesz. Almost the whole theory of Lebesgue integral will follow from them.

The first one is a simple variant of a classical theorem of Dini[4]:

Lemma 5.2 *If a sequence (φ_n) of step functions satisfies[5] $\varphi_n(x) \searrow 0$ a.e., then $\int \varphi_n \, dx \to 0$.*

Proof Fix a compact interval $[a, b]$ and a number $M > 0$ such that $\varphi_1 = 0$ outside $[a, b]$, and $\varphi_1 < M$ on $[a, b]$. Changing the functions φ_n on some null set if necessary, we may assume that they all vanish outside $[a, b]$.

---

[4]See Proposition 8.24 below, p. 292.
[5]The notation means that the sequence is non-increasing and converges to zero for almost every x.

Fix an arbitrarily small number $\varepsilon > 0$. Outside a suitable null set E all functions φ_n are continuous, and the sequence tends to zero. Let us cover E by a countable open interval system $\{I\}$ of total length $< \varepsilon/(2M)$.

If $x_0 \notin E$, then $\varphi_n(x_0) \to 0$, so that

$$\varphi_{n_0}(x_0) < \frac{\varepsilon}{2(b-a)}$$

for a suitable index n_0. Since φ_{n_0} continuous at x_0, we have

$$\varphi_{n_0}(x) < \frac{\varepsilon}{2(b-a)}$$

at each point x of an open interval $J = J(x_0)$ containing x_0. Finally, by the non-increasingness of (φ_n) we have

$$\varphi_n(x) < \frac{\varepsilon}{2(b-a)}$$

for all $x \in J$ and $n \geq n_0$.

The *compact* interval $[a, b]$ may be covered by finitely many of the intervals I and J. Let us denote by N the largest index n_0 among the chosen intervals J, and by A the union of these intervals J. Then

$$\varphi_n(x) < \frac{\varepsilon}{2(b-a)}$$

for all $x \in A$ and $n \geq N$. Consequently, the integral of the step function $\varphi_n \chi_A$, where χ_A denotes the *characteristic function*[6] of the set A, is at most $\varepsilon/2$.

The remainder of $[a, b]$ is a union of closed intervals, covered by the chosen intervals I. Since the total length of the latter is less than $\varepsilon/(2M)$, the integral of $\varphi_n(1 - \chi_A)$ is at most $\varepsilon/2$.

Adding the two equalities we obtain that

$$0 \leq \int \varphi_n \, dx \leq \varepsilon$$

for all $n \geq N$. □

Our next result will be greatly extended later.[7]

Lemma 5.3 *Let (φ_n) be an a.e. non-decreasing sequence of step functions. If the sequence of their integrals is bounded from above, then (φ_n) has a finite limit a.e.*

---

[6]de la Vallée Poussin [465, p. 440]: $\chi_A := 1$ on A, and $\chi_A := 0$ outside A.

[7]See the Beppo Levi theorem, p. 178.

Remark In view of Proposition 5.1 the sequence of integrals $\int \varphi_n \, dx$ is non-decreasing, and hence convergent.

Proof Changing φ_n to $\varphi_n - \varphi_1$ if necessary we may assume that the functions φ_n are nonnegative. We have to show that the points x satisfying $\varphi_n(x) \to \infty$ form a null set E_0.

Let $\int f_n \, dx \leq A$ for all n. For any fixed $\varepsilon > 0$ let us denote by $E_{\varepsilon,n}$ the set of points x satisfying $\varphi_n(x) > A/\varepsilon$, for $n = 1, 2, \dots$. Since $E_\varepsilon := \cup E_{\varepsilon,n}$ contains E_0, it is sufficient to cover E_ε with a countable interval system of total length $\leq \varepsilon$.

The set sequence $(E_{\varepsilon,n})$ is non-decreasing by the analogous property of (φ_n). Consequently, $E_{\varepsilon,1}$ and each difference set $E_{\varepsilon,n} \setminus E_{\varepsilon,n-1}$ is the union of finitely many disjoint intervals, say

$$E_{\varepsilon,1} = \bigcup_{k=1}^{K_1} I_{1k}$$

and

$$E_{\varepsilon,n} \setminus E_{\varepsilon,n-1} = \bigcup_{k=1}^{K_n} I_{nk}, \quad n = 2, 3, \dots.$$

The set of all these intervals covers E_ε. Furthermore, their total length is at most ε, because

$$\sum_{n=1}^{m} \sum_{k=1}^{K_n} \frac{A}{\varepsilon} |I_{nk}| \leq \int_{E_{\varepsilon,1}} \varphi_1 \, dx + \sum_{n=2}^{m} \int_{E_{\varepsilon,n} \setminus E_{\varepsilon,n-1}} \varphi_n \, dx \leq \int \varphi_m \, dx \leq A$$

for each m. $\qquad\qquad\qquad\qquad\qquad\qquad\qquad\qquad\qquad\qquad\qquad\qquad\qquad\square$

5.2 Integrable Functions

We enlarge the class of integrable functions in two steps. The first one is based on Lemmas 5.2 and 5.3:

Definition We denote by C_1 the set of limit functions f of sequences (φ_n) satisfying the assumptions of Lemma 5.3, and we define the *integral* of these functions by the formula

$$\int f \, dx := \lim \int \varphi_n \, dx.$$

The following lemma will imply the correctness of this definition[8]:

Lemma 5.4 *If two sequences of step functions* (φ_n), (ψ_n) *satisfy the relations* $\varphi_n \nearrow f$, $\psi_n \nearrow g$ *and* $f \leq g$ *a.e., then*

$$\lim \int \varphi_n \, dx \leq \lim \int \psi_n \, dx.$$

Proof It suffices to prove for any fixed m the inequality

$$\int \varphi_m \, dx \leq \lim_{n \to \infty} \int \psi_n \, dx,$$

or the equivalent inequality

$$\lim_{n \to \infty} \int \varphi_m - \psi_n \, dx \leq 0.$$

Hence the lemma will follow by letting $m \to \infty$.

We prove the stronger relation

$$\lim_{n \to \infty} \int (\varphi_m - \psi_n)^+ \, dx \leq 0,$$

where

$$(\varphi_m - \psi_n)^+ := \max \{\varphi_m - \psi_n, 0\}$$

denotes the *positive part* of the function $\varphi_m - \psi_n$. For this it suffices to observe that the sequence $n \mapsto (\varphi_m - \psi_n)^+$ satisfies the conditions of Lemma 5.2.   □

Now we collect the properties of the integral on C^1:

Proposition 5.5

(a) *The integral does not depend on the particular choice of the sequence* (φ_n).
(b) *If* $f \in C_0$, *then the two definitions of the integral give the same value.*
(c) *If* $f \in C_1$ *and* $f = g$ *a.e., then* $g \in C_1$ *and* $\int f \, dx = \int g \, dx$.
(d) *If* $f, g \in C_1$ *and* $f \leq g$ *a.e., then*

$$\int f \, dx \leq \int g \, dx.$$

(e) *If* $f, g \in C_1$, *then* $\max \{f, g\} \in C_1$ *and* $\min \{f, g\} \in C_1$.

---

[8]For clarity, in this section we do not omit the notation a.e. for the equalities and inequalities.

(f) *If $f, g \in C_1$, then $f + g \in C_1$, and*

$$\int (f + g)\, dx = \int f\, dx + \int g\, dx.$$

(g) *If $f \in C_1$ and c is a nonnegative real number, then $cf \in C_1$ and*

$$\int cf\, dx = c \int f\, dx.$$

Proof

(a) We apply the preceding lemma with $f = g$.
(b) Let $\varphi_n = f$ for every n.
(c) If $\int f\, dx$ is defined by the sequence (φ_n), then we also have $\varphi_n \to g$ a.e.
(d) This is a reformulation of the preceding lemma.

   (e), (f) and (g) If $\int f\, dx$ and $\int g\, dx$ are defined by the sequences (φ_n) and (ψ_n), then the sequences given by the formulas

$$\max\{\varphi_n, \psi_n\}, \quad \min\{\varphi_n, \psi_n\}, \quad \varphi_n + \psi_n, \quad c\varphi_n$$

satisfy the conditions of Lemma 5.3, and they converge a.e. to $\max\{f, g\}$, $\min\{f, g\}, f + g$ and cf, respectively. The equalities in (f) and (g) follow from the similar equalities for step functions.

$\square$

Next we extend the integral from C_1 to the vector space spanned by C_1:

Definition A function f is *integrable* if $f = f_1 - f_2$ a.e. with suitable functions $f_1, f_2 \in C_1$. Its *integral* is defined by the formula

$$\int f\, dx := \int f_1\, dx - \int f_2\, dx.$$

We often write $\int f(x)\, dx$ instead of $\int f\, dx$.
   The set of integrable functions is denoted by C_2.

Proposition 5.6

(a) C_2 *is a vector lattice.*
(b) *The integral of a function $f \in C_2$ does not depend on the particular choice of the decomposition $f = f_1 - f_2$.*
(c) *If $f \in C_2$ and $f = g$ a.e., then $g \in C_2$, and $\int f\, dx = \int g\, dx$.*
(d) *The integral is a positive linear functional on C_2.*

Proof

(a) Let $f = f_1 - f_2$ and $g = g_1 - g_2$ a.e., where $f_1, f_2, g_1, g_2 \in C_1$, and let c be
a nonnegative number. Applying the preceding proposition, it follows from the
a.e. equalities

$$f + g = (f_1 + g_1) - (f_2 + g_2),$$
$$cf = cf_1 - cf_2,$$
$$-cf = cf_2 - cf_1,$$
$$|f| = \max\{f_1, f_2\} - \min\{f_1, f_2\}$$

that $f + g$, cf, $-cf$ and $|f|$ are integrable.

(b) If $f = f_1 - f_2 = g_1 - g_2$ a.e. with $f_1, f_2, g_1, g_2 \in C_1$, then $f_1 + g_2 = f_2 + g_1$ a.e.,
and hence

$$\int f_1 \, dx + \int g_2 \, dx = \int f_1 + g_2 \, dx = \int f_2 + g_1 \, dx = \int f_2 \, dx + \int g_1 \, dx$$

by Proposition 5.5. Consequently,

$$\int f_1 \, dx - \int f_2 \, dx = \int g_1 \, dx - \int g_2 \, dx.$$

(c) If $f = f_1 - f_2$ a.e., where $f_1, f_2 \in C_1$, then we also have $g = f_1 - f_2$ a.e.

(d) This follows from Proposition 5.5 (d), (f), (g) and from the definition of the
integral.

$\square$

5.3 The Beppo Levi Theorem

In the preceding section we started with a positive linear functional defined on a
vector lattice, and we extended it to a positive linear functional defined on a larger
vector lattice. It is tempting to reiterate this process in order to obtain new integrable
functions. It is surprising and remarkable that this step is useless:

Theorem 5.7 (Beppo Levi)[9] *Let (f_n) be a non-decreasing sequence of integrable functions. If their integrals are bounded from above, then (f_n) converges a.e. to an integrable function f, and*

$$\int f_n \, dx \to \int f \, dx. \tag{5.1}$$

Remark The use of a.e. convergence is essential here. Using *everywhere* convergent sequences the process could be iterated indefinitely by a celebrated theorem of Baire.[10]

We prove the theorem in two steps.

Proof in case $(f_n) \subset C_1$ Let

$$\int f_n \, dx \le A$$

for all n. Fix for each n a non-decreasing sequence (φ_{nk}) of step functions, converging a.e. to f_n. Then the formula

$$\varphi_n := \sup_{i,k \le n} \varphi_{ik}$$

defines a non-decreasing sequence of step functions, satisfying

$$\int \varphi_n \, dx \le A$$

for all n, because $\varphi_{ik} \le f_i \le f_n$ for all $i, k \le n$, and therefore $\varphi_n \le f_n$. By Lemma 5.3 we have $\varphi_n \to f$ a.e. for some function $f \in C_1$, and

$$\int \varphi_n \, dx \to \int f \, dx. \tag{5.2}$$

Since $\varphi_{nk} \le \varphi_k$ whenever $k \ge n$, letting $k \to \infty$ we obtain $f_n \le f$ for each n. Integrating the inequalities $\varphi_n \le f_n \le f$ and applying (5.2) we obtain (5.1).   □

Remark We emphasize that in the above special case the limit function is not only integrable, but even belongs to C_1. This will be used in the proof of the general case below.

---

[9]Levi [301].
[10]Baire [16, 17].

To proceed we need the following lemma:

Lemma 5.8 *Given a nonnegative function $f \in C_2$ and a positive number $\varepsilon > 0$, there exist nonnegative functions $f_1, f_2 \in C_1$ such that $f = f_1 - f_2$ and $\int f_2 \, dx < \varepsilon$.*

Remark We cannot take $f_2 = 0$ if f is unbounded from below.

Proof Let $f = g_1 - g_2$ with $g_1, g_2 \in C_1$. Choose a sequence (φ_n) of step functions such that $\varphi_n \nearrow g_2$ a.e. Then

$$\int \varphi_n \, dx \to \int g_2 \, dx,$$

and hence

$$\int g_2 - \varphi_n \, dx < \varepsilon$$

if n is sufficiently large. Since $-\varphi_n \in C_0 \subset C_1$, the functions $f_1 := g_1 - \varphi_n$ and $f_2 := g_2 - \varphi_n$ belong to C_1. Furthermore, $f = f_1 - f_2$ and $\int f_2 \, dx < \varepsilon$. Finally, $f_2 = g_2 - \varphi_n \geq 0$, because the sequence (φ_n) is non-decreasing, and $f_1 = f + f_2 \geq 0$ as the sum of two nonnegative functions. ☐

Proof of Theorem 5.7 in the General Case Applying the preceding lemma to the differences $f_{n+1} - f_n$ we obtain nonnegative functions $g_n, h_n \in C_1$ satisfying the conditions

$$f_{n+1} - f_n = g_n - h_n \quad \text{and} \quad \int h_n \, dx < 2^{-n}, \quad n = 1, 2, \ldots.$$

Hence

$$\int h_1 + \cdots + h_n \, dx < 1$$

for all n. Applying the already proven part of the theorem, the series $\sum h_i$ converges a.e. to some function $h \in C_1$, and

$$\sum_{n=1}^{\infty} \int h_n \, dx = \int h \, dx.$$

Consequently, assuming again that

$$\int f_n \, dx \leq A$$

for all n, the following inequalities also hold:

$$\int g_1 + \cdots + g_{n-1}\, dx = \int f_n - f_1 + h_1 + \cdots + h_{n-1}\, dx < A + 1 - \int f_1\, dx.$$

Applying once again the already proven part of the theorem, the series $\sum g_i$ converges a.e. to some function $g \in C_1$, and

$$\sum_{n=1}^{\infty} \int g_n\, dx = \int g\, dx.$$

Taking the difference of the two series we conclude that

$$(g_1 + \cdots + g_{n-1}) - (h_1 + \cdots + h_{n-1}) = f_n - f_1$$

converges a.e. to $g - h \in C_2$, and

$$\int f_n - f_1\, dx \to \int g - h\, dx.$$

Consequently, f_n converges a.e. to $f := f_1 + g - h \in C_2$, and (5.1) holds.   □

Let us mention some important corollaries of the theorem:

Corollary 5.9

(a) *If a non-decreasing sequence (f_n) of integrable functions converges a.e. to some integrable function f, then*

$$\int f_n\, dx \to \int f\, dx.$$

(b) *If (f_n) is a sequence of integrable functions, and the numerical series*

$$\sum_{n=1}^{\infty} \int |f_n|\, dx$$

is convergent, then the function series $\sum f_n$ converges a.e. to some integrable function f, and we may integrate this series termwise:

$$\int f\, dx = \sum_{n=1}^{\infty} \int f_n\, dx.$$

(c) *If f is integrable and $\int |f|\, dx = 0$, then $f = 0$ a.e.*

Proof

(a) The number $A := \int f \, dx$ is a uniform upper bound of the integrals $\int f_n \, dx$.
(b) If the functions f_n are nonnegative, then the partial sums of $\sum f_n$ satisfy the conditions of the Beppo Levi theorem.

    In the general case we consider instead the series $\sum f_n^+$ and $\sum f_n^-$, where

$$f_n^+ := \max\{f_n, 0\} \quad \text{and} \quad f_n^- := \max\{-f_n, 0\}$$

denote the *positive* and *negative parts* of the functions f_n: then $f_n^+, f_n^- \geq 0$ and $f_n = f_n^+ - f_n^-$.
(c) Apply (b) with $f_n := f$ for all n.

$\square$

5.4   Theorems of Lebesgue, Fatou and Riesz–Fischer

If $f_n \to f$ a.e., then the Beppo Levi theorem gives a sufficient condition for the relation

$$\int f_n \, dx \to \int f \, dx.$$

Another important sufficient condition is the following[11]:

Theorem 5.10 (Lebesgue)[12] *Let (f_n) be a sequence of integrable functions with $f_n \to f$ a.e. If there exists an integrable function g such that $|f_n| \leq g$ a.e. for every n, then f is integrable, and*

$$\int f_n \, dx \to \int f \, dx. \tag{5.3}$$

The function g is called an *integrable majorant* of the sequence (f_n).

---

[11]This theorem greatly extended and at the same time simplified earlier results of Arzelà [7], [10, pp. 723–724] and Osgood [350, pp. 183–189] on the Riemann integral. An elementary proof of the latter was given by Lewin [302].
[12]Lebesgue [288] (for uniformly bounded sequences), [294] (general case, pp. 9–10). It is also called *dominated convergence theorem*.

Fig. 5.2 Non-dominated sequence

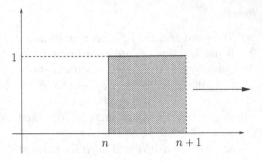

Fig. 5.3 Non-dominated sequence

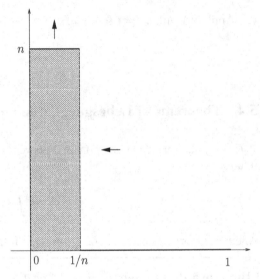

Examples The relation $f_n \rightarrow f$ a.e. alone does not imply the convergence of the integrals:

- If f_n is the characteristic function of the interval $[n, n + 1]$, then $f_n \rightarrow 0$ everywhere, but $\int f_n \, dx = 1$ for all n, and hence it does not converge to $\int 0 \, dx = 0$. See Fig. 5.2.
- Let $f_n(x) = n$, if $0 < x < n^{-1}$, and $f_n(x) = 0$ otherwise. Then $f_n \rightarrow 0$ everywhere, but $\int f_n \, dx = 1$ for all n, and hence it does not converge to $\int 0 \, dx = 0$. See Fig. 5.3.

Proof Let us introduce for each $n = 1, 2, \ldots$ the functions

$$g_n := \sup \{f_n, f_{n+1}, \ldots\}$$

and

$$g_{nm} := \sup \{f_n, f_{n+1}, \ldots, f_m\}, \quad m = n, n + 1, \ldots.$$

Since $|g_{nm}| \le g$ a.e. for all m, the functions g_{nm} are integrable, and their integrals are bounded from above by $\int g \, dx$. Since $g_{nm} \nearrow g_n$ a.e., the Beppo Levi theorem implies that g_n is integrable.

Observe that $g_n \searrow f$ a.e., and that $-\int g \, dx$ is a lower bound of the integrals of the functions g_n, because $|g_n| \le g$ a.e. for all m. Applying the Beppo Levi theorem to the sequence $(-g_n)$ we conclude that f is integrable, and

$$\int g_n \, dx \to \int f \, dx.$$

Similarly, the functions

$$h_n := \inf\{f_n, f_{n+1}, \ldots\}$$

satisfy $h_n \nearrow f$ a.e., and

$$\int h_n \, dx \to \int f \, dx.$$

Since $h_n \le f_n \le g_n$ a.e., and therefore

$$\int h_n \, dx \le \int f_n \, dx \le \int g_n \, dx,$$

(5.3) follows from the above two convergence relations.   □

We may also combine the assumptions of Beppo Levi and Lebesgue:

Theorem 5.11 (Fatou Lemma)[13]   *Let (f_n) be a sequence of nonnegative, integrable functions with $f_n \to f$ a.e. If the integrals $\int f_n \, dx$ are bounded from above, then f is integrable, and*

$$\int f \, dx \le \liminf \int f_n \, dx.$$

Remark The preceding examples show that we do not have equality in general. We will return to this question later.[14]

Proof Let us introduce again the functions

$$h_n := \inf\{f_n, f_{n+1}, \ldots\}, \quad n = 1, 2, \ldots.$$

---

[13]Fatou [136, p. 375].
[14]See Propositions 10.1 (c) and 10.6 (c), pp. 341 and 349.

Since the functions f_n are nonnegative, we may apply the Beppo Levi theorem to conclude that the functions h_n are integrable.

Since $0 \le h_n \le f_n$ a.e., we have

$$0 \le \int h_n \, dx \le \int f_n \, dx$$

for all n. Therefore we deduce from the assumptions of the theorem that the sequence of the integrals $\int h_n \, dx$ is bounded. Furthermore, since $h_n \nearrow f$ a.e., another application of the Beppo Levi theorem shows that f is integrable, and $\int h_n \, dx \to \int f \, dx$. $\qquad\square$

The integrable functions form a natural normed space[15]:

Definition Identifying two integrable functions if they are equal a.e., we obtain a vector space L^1 on which the formula

$$\|f\|_1 := \int |f| \, dx$$

defines a norm.[16]

A fundamental result is that the Cauchy convergence criterion holds in this space:

Theorem 5.12 (Riesz–Fischer)[17] L^1 *is a* Banach space.

The proof is based on the following lemma, important in itself:

Lemma 5.13 (Riesz)[18] *Given a Cauchy sequence (f_n) in L^1, there exists a subsequence (f_{n_k}) and two integrable functions f, g such that $|f_{n_k}| \le g$ for all k, and $f_{n_k} \to f$ a.e.*

Proof Choose a subsequence (f_{n_k}) satisfying

$$\int |f_n - f_{n_k}| \, dx \le 2^{-k} \quad \text{for all} \quad n \ge n_k, \quad k = 1, 2, \dots.$$

[15]The validity of the norm axioms is straightforward.

[16]More precisely, the elements of L^1 are equivalence classes of functions. As a vector space, $L^1 = C_2$. We write L^1 to emphasize that we have a normed space.

[17]Riesz [373, 374, 376] and Fischer [146] for the closely related L^2 spaces, Riesz [377, 379–381] for the more general L^p spaces. See also Chap. 9, p. 305.

[18]Riesz [378].

Since

$$\sum_{k=1}^{\infty} \int |f_{n_{k+1}} - f_{n_k}| \, dx \leq \sum_{k=1}^{\infty} 2^{-k} < \infty,$$

the function series

$$|f_{n_1}| + \sum_{k=1}^{\infty} |f_{n_{k+1}} - f_{n_k}| \quad \text{and} \quad f_{n_1} + \sum_{k=1}^{\infty} \left(f_{n_{k+1}} - f_{n_k}\right)$$

converge a.e. by Corollary 5.9 (p. 180) to two integrable functions g, f.

Applying the triangle inequality to the partial sums we obtain that $|f_{n_k}| \leq g$ for all k, and $f_{n_k} \to f$ a.e. □

Proof of Theorem 5.12 By the preceding lemma there exist a subsequence (f_{n_k}) and an integrable function f such that $f_{n_k} \to f$ a.e.

For any given $\varepsilon > 0$ choose a sufficiently large N such that

$$\int |f_m - f_n| \, dx < \varepsilon$$

for all $m, n \geq N$. Taking $n = n_k$ and letting $k \to \infty$, by applying the Fatou lemma we obtain that

$$\int |f_m - f| \, dx \leq \varepsilon$$

for all $m \geq N$. □

We end this section with two further applications of Lebesgue's theorem. The first one states the density of step functions in L^1:

Proposition 5.14 *For each $f \in L^1$ there exists a sequence (φ_n) of step functions such that $\int |f - \varphi_n| \, dx \to 0$.*

Remark Applying the preceding lemma and taking a subsequence we may also assume that $\varphi_n \to f$ a.e., and that there exists an integrable function k such that $|\varphi_n| \leq k$ for all n.

However, the proof below leads directly to such a sequence.

Proof Let $f = g - h$ with $g, h \in C_1$, and choose two sequences (ψ_n), (ϱ_n) of step functions such that $\psi_n \nearrow g$ and $\varrho_n \nearrow h$ a.e. Furthermore, set

$$k := \max\{g - \varrho_1, h - \psi_1\} \quad \text{and} \quad \varphi_n := \psi_n - \varrho_n, \quad n = 1, 2, \ldots.$$

Then k is integrable. Furthermore,

$$\varphi_n \to g - h = f \quad \text{a.e., and} \quad |\varphi_n| \leq k \quad \text{for all} \quad n,$$

because $\psi_n - \varrho_n \leq g - \varrho_1$ and $\varrho_n - \psi_n \leq h - \psi_1$. We conclude by applying Lebesgue's convergence theorem.   □

Finally we study integrals depending on a parameter:

Proposition 5.15 *Consider a function* $f : \mathbb{R} \times I \to \mathbb{R}$ *where* I *is an open interval. Assume that the functions* $x \mapsto f(x, t)$ *are integrable, and set*

$$F(t) := \int f(x, t) \, dx, \quad t \in I.$$

Let $t_0 \in I$.

(a) *Assume that*

- *the functions* $t \mapsto f(x, t)$ *are continuous at* t_0 *for a.e.* x;
- *the functions* $x \mapsto f(x, t)$ *have a uniform integrable majorant* g:

$$|f(x, t)| \leq g(x) \quad \text{for each} \quad t \in I.$$

Then F *is continuous at* t_0.

(b) *Assume that*

- *the functions* $t \mapsto f(x, t)$ *are differentiable at* t_0 *for a.e.* x;
- *the functions* $x \mapsto D_2 f(x, t)$ *have a uniform integrable majorant* g:

$$|D_2 f(x, t)| \leq g(x) \quad \text{for each} \quad t \in I;$$

Then F *is differentiable at* t_0, *and*

$$F'(t_0) = \int D_2 f(x, t_0) \, dx.$$

Proof

(a) It suffices to show that $F(t_n) \to F(t_0)$ for every sequence $t_n \to t_0$ in I. Setting $h_n(x) := f(x, t_n)$ and $h(x) := f(x, t_0)$, this is equivalent to the relation $\int h_n \, dx \to \int h \, dx$. It follows from our assumptions that the functions h_n are integrable, $h_n \to h$ a.e., and that $|h_n| \leq g$ a.e. for every n. We may therefore conclude by applying Lebesgue's theorem.

(b) Fix again a sequence $t_n \to t_0$ in I. Setting

$$h_n(x) := \frac{f(x, t_n) - f(x, t_0)}{t_n - t_0} \quad \text{and} \quad h(x) := D_2 f(x, t_0),$$

it follows from our assumptions that $h_n \to h$ a.e., and $|h_n| \leq g$ a.e. for every n. (We apply the Lagrange mean value theorem.) Applying Lebesgue's theorem we get $\int h_n \, dx \to \int h \, dx$, i.e.,

$$\frac{F(t_n) - F(t_0)}{t_n - t_0} \to \int D_2 f(x, t_0) \, dx.$$

□

Remarks

- Part (a) may be generalized for a metric space I in place of intervals.
- Part (b) may be generalized for higher-order derivatives (by induction) and to open subsets I of normed spaces in place of intervals.

5.5 * Measurable Functions and Sets

It is sometimes convenient to deal with infinite integrals. For this we introduce the following notion:

Definition A function $f : \mathbb{R} \to \overline{\mathbb{R}}$ is *measurable* if there exists a sequence (φ_n) of step functions such that $\varphi_n \to f$ a.e.

We emphasize that f may take infinite values.

Example Every continuous function $f : \mathbb{R} \to \mathbb{R}$ is measurable because the formula

$$\varphi_n(x) := \begin{cases} f\left(\frac{k}{n}\right) & \text{if } \frac{k}{n} \leq x < \frac{k+1}{n}, k = -n^2, \ldots, 0, \ldots, n^2, \\ 0 & \text{otherwise} \end{cases}$$

defines a sequence of step functions converging to f *everywhere*.

The following proposition clarifies the relationship between measurable and integrable functions, and it shows that all functions $f : \mathbb{R} \to \mathbb{R}$ usually encountered in analysis are measurable.[19]

Proposition 5.16

(a) *If f is measurable and $f = g$ a.e., then g is measurable.*
(b) *If $F : \mathbb{R}^N \to \mathbb{R}$ is continuous, $F(0) = 0$, and $f_1, \ldots, f_N$ are finite-valued measurable functions, then the composite function $h := F(f_1, \ldots, f_N)$ is measurable. In particular, if f and g are finite-valued measurable functions,*

---

[19]Even more is true: it is *impossible* to prove the existence of non-measurable functions without using the axiom of choice: see the remark on p. 192 below.

then

$$|f|, \quad f+g, \quad f-g, \quad fg, \quad \max\{f,g\} \quad and \quad \min\{f,g\}$$

are measurable.

(c) *If f is measurable and $f \neq 0$ a.e., then $1/f$ is measurable.*
(d) *Every integrable function f is measurable.*
(e) *If f is measurable, g is integrable, and $|f| \leq g$ a.e., then f is integrable.*
(f) *If (f_n) is a sequence of measurable functions and $f_n \to f$ a.e., then f is measurable.*

Remarks

- Since the constant functions are continuous and hence measurable, the assumption $F(0) = 0$ in (b) could be omitted. We made this assumption in order to keep the proposition valid in the much more general framework of Chap. 7 below.
- Property (b) may be generalized to the case where $f_1, \ldots, f_N$ also take infinite values, and $F : \overline{\mathbb{R}}^N \to \overline{\mathbb{R}}$ is continuous on the range of the vector-valued function $(f_1, \ldots, f_N)$.

Proof

(a) This follows from the definition.
(b) Fix for each f_k a sequence (φ_{kn}) of step functions converging to f_k a.e. Then the step functions

$$\varphi_n(x) := F(\varphi_{1n}(x), \ldots, \varphi_{Nn}(x))$$

converge to h a.e.
(c) Let (φ_n) be a sequence of step functions, converging to f a.e. Then the step functions

$$\psi_n(x) := \begin{cases} 0 & \text{if } \varphi_n(x) = 0, \\ 1/\varphi_n(x) & \text{otherwise} \end{cases}$$

converge to $1/f$ a.e.
(d) If f is integrable, then by Proposition 5.14 there exists a sequence (φ_n) of step functions satisfying $\int |f - \varphi_n|\, dx \to 0$. By Lemma 5.13 we may also assume (by taking a subsequence) that $\varphi_n \to f$ a.e.
(e) If (φ_n) is a sequence of step functions converging to f a.e., then the functions[20]

$$f_n := \mathrm{med}\{-g, \varphi_n, g\}$$

---

[20]See the definition of med $\{x, y, z\}$ on p. 9.

are integrable, and $f_n \to f$ a.e. Furthermore, $|f_n| \le g$ for all n. We conclude by applying Lebesgue's theorem.

(f) Fix a *strictly positive*, integrable function[21] $h : \mathbb{R} \to \mathbb{R}$, and set

$$g_n := \frac{h f_n}{h + |f_n|} \quad \text{and} \quad g := \frac{h f}{h + |f|}.$$

Then g_n is measurable and $|g_n| < h$, so that g_n is integrable. Since $g_n \to g$ a.e., by Lebesgue's theorem g is integrable, and then also measurable. Since $|g| \le h$, $\mathrm{sign} f = \mathrm{sign}\, g$ and hence $|f|g = f|g|$, then

$$f = \frac{hg}{h - |g|}$$

is also measurable.

$\square$

Now we are ready to generalize the integral. We recall that the positive and negative parts of a function f are defined by the formulas

$$f_+ := \max\{f, 0\}, \quad f_- := \max\{-f, 0\} = -\min\{f, 0\},$$

and that

$$f_+, f_- \ge 0, \quad f = f_+ - f_-, \quad |f| = f_+ + f_- \quad \text{and} \quad f_+ f_- = 0.$$

If f is measurable, then f_+ and f_- are also measurable.

Definition Let f be a measurable function.

- If $f \ge 0$ a.e. and non-integrable, then set $\int f \, dx = \infty$.
- If at least one of f_+ and f_- is integrable, then set

$$\int f \, dx = \int f_+ \, dx - \int f_- \, dx.$$

Remarks

- If neither f_+ nor f_- is integrable, then the right hand sum is undefined.
- If f is integrable, then f_+ and f_- are also integrable, and the above definition leads to the original integral of f by the linearity of the integral.
- We keep the adjective *integrable* for the case where the integral is finite.

---

[21] We emphasize that h has finite values a.e. We may take, for example, $h(x) = 1/(1 + x^2)$.

The usual integration rules remain valid:

Proposition 5.17

(a) *If $\int f\, dx$ exists and $f = g$ a.e., then $\int g\, dx$ also exists, and $\int f\, dx = \int g\, dx$.*
(b) *If $\int f\, dx$ exists and $c \in \mathbb{R}$, then $\int cf\, dx$ also exists, and[22]*

$$\int cf\, dx = c \int f\, dx.$$

(c) *If the integrals $\int f\, dx$, $\int g\, dx$ exist and $f \le g$ a.e., then*

$$\int f\, dx \le \int g\, dx.$$

(d) *If $\int f\, dx$, $\int g\, dx$ exist and the sum $\int f\, dx + \int g\, dx$ is defined, then $\int f + g\, dx$ exists, and*

$$\int f + g\, dx = \int f\, dx + \int g\, dx.$$

(e) *(Generalized Beppo Levi theorem) If the functions f_n are measurable, nonnegative, and $f_n \nearrow f$ a.e., then*

$$\int f_n\, dx \to \int f\, dx.$$

(f) *If the functions g_n are measurable and nonnegative, then*

$$\int \sum g_n\, dx = \sum \int g_n\, dx.$$

Proof

(a) and (b) are obvious.

(c) It is sufficient to consider the case where $\int g\, dx < \infty$ and $\int f\, dx > -\infty$, i.e., where g_+ and f_- are integrable. Then f_+ and g_- are integrable by Proposition 5.16 (e) because

$$0 \le f_+ \le g_+ \quad \text{and} \quad 0 \le g_- \le f_-$$

a.e. Hence f and g are integrable, and the required inequality follows from Proposition 5.6 (d) (p. 176).

---

[22] We adopt for $c = 0$ the convention $0 \cdot (\pm\infty) = (\pm\infty) \cdot 0 := 0$, useful in integral theory.

(d) If f and g are nonnegative a.e., then the equality follows from the definition of the generalized integral. In the general case we notice that the function

$$h := f_+ + g_+ - (f + g)_+ = f_- + g_- - (f + g)_-$$

is measurable and nonnegative a.e.; consequently,

$$\int (f + g)_+ \, dx + \int h \, dx = \int f_+ \, dx + \int g_+ \, dx$$

and

$$\int (f + g)_- \, dx + \int h \, dx = \int f_- \, dx + \int g_- \, dx$$

by our previous remark. If we show that in at least one of these rows all four integrals are finite, then we may conclude by taking the difference of the two rows.

If, for example, $\int f \, dx + \int g \, dx < \infty$ (the case $> -\infty$ is analogous), then f_+ and g_+ are integrable. Since

$$0 \leq h \leq f_+ + g_+ \quad \text{and} \quad 0 \leq (f + g)_+ \leq f_+ + g_+$$

a.e., it follows that h and $(f + g)_+$ are also integrable.

(e) The sequence of the integrals $\int f_n \, dx$ is non-decreasing by (c). If it is also bounded, then we may apply the Beppo Levi theorem. Otherwise we have $f_n \leq f$ a.e. (for every n) and $\int f_n \, dx \to \infty$; hence $\int f \, dx = \infty$, and therefore $\int f_n \, dx \to \infty = \int f \, dx$.

(f) We apply (e) with $f_n := g_1 + \cdots + g_n$.

$\square$

Next we generalize the length of intervals:

Definition A set A is *measurable* if its characteristic function is measurable; by its *Lebesgue measure* we mean the number $\mu(A) := \int \chi_A \, dx \in [0, \infty]$.

We introduce the following notion:

Definition A set system $\mathcal{M}$ is a *σ-ring*[23] if satisfies the following three conditions:

- $\varnothing \in \mathcal{M}$;
- if $A, B \in \mathcal{M}$, then $A \setminus B \in \mathcal{M}$;
- if (A_n) is a disjoint sequence in $\mathcal{M}$, then $\cup A_n \in \mathcal{M}$.

---

[23]Fréchet [158].

Remarks

- Here and in the sequel the letter σ refers to *countable unions*. If we use only finite unions in the definition, then we arrive at the notion of *rings*, to be considered later (p. 214).
- If $\mathcal{M}$ is a σ-ring, then $A := \cup A_n \in \mathcal{M}$ and $\cap A_n \in \mathcal{M}$ for every finite or infinite sequence $(A_n) \subset \mathcal{M}$. Indeed, in the infinite case the formulas

$$B_1 := A_1, \quad B_2 := A_2 \setminus A_1, \quad B_3 := A_3 \setminus (A_2 \cup A_1), \ldots$$

define a *disjoint* set sequence $(B_n) \subset \mathcal{M}$ with $A = \cup B_n$, so that

$$\cup A_n = \cup B_n \in \mathcal{M}.$$

The finite case may be reduced to the previous one by completing the sequence with empty sets. Finally, the formula

$$\cap A_n = A \setminus \cup (A \setminus A_n)$$

then shows that $\cap A_n \in \mathcal{M}$.

Let us list the basic properties of the Lebesgue measure:

Proposition 5.18

(a) *The measurable sets form a σ-ring, henceforth denoted by $\mathcal{M}$.*
(b) *The Lebesgue measure $\mu : \mathcal{M} \to \mathbb{R}$ is nonnegative, and*

$$\mu(\cup A_n) = \sum \mu(A_n)$$

for every finite or countable sequence (A_n) of pairwise disjoint measurable sets.
(c) *The null sets coincide with the measurable sets of zero Lebesgue measure.*
(d) *The Lebesgue measure is* complete *in the following sense: if $A \subset B$ and B is a set of zero Lebesgue measure, then A is also measurable (and has zero Lebesgue measure).*

Remark Using the axiom of choice, Vitali[24] proved that there exist non-measurable sets. Solovay[25] proved that the use of the axiom of choice cannot be avoided here. The application of the axiom of choice led to numerous paradoxical results.[26]

---

[24]Vitali [466]. See Exercise 5.5 below.
[25]Solovay [429].
[26]See Banach [21], Banach–Tarski [29], Hausdorff [195], von Neumann [335], Laczkovich [277, 278], Wagon [478].

Proof

(a) The zero function is integrable, so that $\varnothing \in \mathcal{M}$. If $A, B \in \mathcal{M}$, then $\chi_{A \setminus B} = \chi_A - \chi_A \chi_B$ is measurable by Proposition 5.16 (b), so that $A \setminus B \in \mathcal{M}$.

   If $(A_n) \subset \mathcal{M}$ is an infinite *disjoint* sequence and $A = \cup A_n$, then the finite sums $f_n := \chi_{A_1} + \cdots + \chi_{A_n}$ are measurable by Proposition 5.16 (b), and $f_n \to \chi_A$ everywhere. Applying Proposition 5.16 (f) we conclude that χ_A is measurable. Hence $A \in \mathcal{M}$.

(b) The properties $\mu(A) \geq 0$ and $\mu(\varnothing) = 0$ are obvious. If $(A_n) \subset \mathcal{M}$ is an infinite *disjoint* sequence and $A = \cup A_n$, then applying Proposition 5.16 (f) to the equality

$$\sum \chi_{A_n} = \chi_A$$

we obtain

$$\int \chi_A \, dx = \sum \int \chi_{A_n} \, dx,$$

i.e.,

$$\mu(A) = \sum \int \mu(A_n).$$

(c) If A is a null set, then $\chi_A = 0$ a.e., and then $\int \chi_A \, dx = 0$ by Proposition 5.6 (c) (p. 176). In other words, $\mu(A) = 0$.

   Conversely, if $\mu(A) = 0$, then $\int \chi_A \, dx = 0$ by definition. Applying Corollary 5.9 (c) (p. 180) this implies $\chi_A = 0$ a.e., i.e. A is a null set.

(d) This follows from (c) because the subsets of a null set are also null sets (p. 155).

$\square$

We end this chapter with a new characterization of measurable functions. Let us introduce for all $c \in \overline{\mathbb{R}}$ the *level sets* of a function f:

$$\{f > c\} := \{x \in \mathbb{R} : f(x) > c\},$$

$$\{f \geq c\} := \{x \in \mathbb{R} : f(x) \geq c\},$$

$$\{f < c\} := \{x \in \mathbb{R} : f(x) < c\},$$

$$\{f \leq c\} := \{x \in \mathbb{R} : f(x) \leq c\}.$$

Proposition 5.19 *A function f is measurable $\Longleftrightarrow$ its level sets*

$$\{f > c\}, \quad \{f < -c\}, \quad \{f \ge c\}, \quad \{f \le -c\}$$

are measurable for all $0 < c < \infty$.

Remark The measurability of $\mathbb{R}$ implies the measurability of *all* levels sets of all measurable functions $f : \mathbb{R} \to \mathbb{R}$. By considering only $0 < c < \infty$ the proposition will remain valid in the more general framework of Chap. 7 below.

Proof If f is measurable and $c > 0$, then the functions

$$\frac{\min\{f, c + n^{-1}\} - \min\{f, c\}}{n^{-1}} \quad \text{and} \quad \frac{\max\{f, -c + n^{-1}\} - \max\{f, -c\}}{n^{-1}}$$

are measurable for all $n = 1, 2, \ldots$ by Proposition 5.16 (b). Since these functions converge a.e. to the characteristic functions of $\{f > c\}$ and $\{f \le -c\}$, the latter sets are measurable. Since the function $-f$ is also measurable, the sets

$$\{f < -c\} = \{-f > c\} \quad \text{and} \quad \{f \ge c\} = \{-f \le -c\}$$

are also measurable.

Conversely, if the above sets are measurable, then the formula

$$f_n(x) := \operatorname{med}\left\{-n, \frac{[nf(x)]}{n}, n\right\}, \quad x \in \mathbb{R}, \quad n = 1, 2, \ldots,$$

where $[z]$ denotes the integer part of z, defines a sequence of measurable functions because each f_n is a finite linear combination of level sets of the given form. Since $f_n \to f$ a.e., the measurability of f follows. $\square$

5.6 Exercises

Exercise 5.1 The functions in C_1 are bounded from below by definition. Conversely, is it true that if $f \in C_2$ is bounded from below, then $f \in C_1$?

Exercise 5.2 What is the Lebesgue measure of the set of real numbers $x \in [0, 1]$ whose decimal expansion does not contain the digit 7?

Exercise 5.3 Let A be a set of finite measure $\mu(A) = \alpha > 0$ in $\mathbb{R}$. Prove the following:

(i) The function $x \mapsto \mu(A \cap (-\infty, x))$ is continuous on $\mathbb{R}$.
(ii) For each $0 < \beta < \alpha$ there exists a subset $B \subset A$ of measure β.

Exercise 5.4 A set of real numbers is a *Borel set* if it can be obtained from the open sets by taking countable unions, countable intersections and complements at most countably many times. Prove that they form the smallest σ-ring containing the open sets, the smallest σ-ring containing the closed sets, and that they have the power of continuum.

Exercise 5.5 (Vitali)[27] Consider in $\mathbb{R}$ the equivalence relation $x \sim y \Longleftrightarrow x - y$ is rational.

Prove that if a set contains exactly one point of each equivalence class, then it is not measurable.

Exercise 5.6

 (i) Every set of positive measure has the power of the continuum.
 (ii) The set of measurable sets has the same power as the set of all subsets of $\mathbb{R}$.
(iii) Every set $A \subset [0, 1]$ of positive measure contains two points whose distance is irrational.
(iv) Every set $A \subset [0, 1]$ of positive measure contains two points whose distance is rational.

Exercise 5.7 There exists a measurable set $A \subset [0, 1]$ such that

$$0 < \mu(A \cap V) < \mu(V)$$

for every non-empty open set of $V \subset [0, 1]$, where μ denotes the usual Lebesgue measure.

Exercise 5.8 Deduce Lebesgue's dominated convergence Theorem 5.10 from the Fatou lemma 5.11.

---

[27] Vitali [466].

Chapter 6
* Generalized Newton–Leibniz Formula

If Newton and Leibniz had thought that continuous functions need not have derivatives, and this is the general case, the differential calculus would not have been born.—É. Picard

One of the (if not *the*) most important theorems of classical analysis is the Newton–Leibniz formula:

$$\int_a^b f \, dx = F(b) - F(a),$$

allowing us to compute many integrals. The purpose of this chapter is to extend its validity to Lebesgue integrable functions.[1]

We consider in this chapter monotone functions defined on a *closed* interval of the extended real line $\overline{\mathbb{R}}$, where the latter is endowed with its usual compact topology. We thus allow the cases $a = -\infty$ and/or $b = \infty$ as well. We notice that all monotone functions $F : [a, b] \to \mathbb{R}$ are bounded.

In the preceding chapter we considered only integrals on the whole real line. Now we introduce the integrals on arbitrary intervals as follows:

Definition A function $f : D \to \mathbb{R}$ $(D \subset \mathbb{R})$ is *integrable on an interval I* if it is defined at a.e. point of I (i.e., $I \setminus D$ is a null set), and the function

$$g(x) := \begin{cases} f(x) & \text{if } x \in I \cap D, \\ 0 & \text{if } x \in \mathbb{R} \setminus (I \cap D) \end{cases}$$

---

[1] More complete results were obtained by Denjoy [99, 100] and Perron [356] by further generalizing the Lebesgue integral. Henstock [205, 206] and Kurzweil [274] showed later that these results may also be obtained by a suitable modification of the Riemann integral. See also Bartle [30].

© Springer-Verlag London 2016

V. Komornik, *Lectures on Functional Analysis and the Lebesgue Integral*,
Universitext, DOI 10.1007/978-1-4471-6811-9_6

is integrable. In this case the integral of f on I is defined by the formula

$$\int_I f \, dx := \int g \, dx.$$

Remarks

- An integrable function is integrable on every interval by Proposition 5.16 (b) and (e) (p. 187).
- Since the finite sets are null sets, for any function f and numbers and $a \le b$ the integrals

$$\int_{(a,b)} f \, dx, \quad \int_{(a,b]} f \, dx, \quad \int_{[a,b)} f \, dx, \quad \int_{[a,b]} f \, dx$$

exist or do not exist at the same time, and if they exist, they are equal. Hence we denote their common value simply by $\int_a^b f \, dx$.

6.1   Absolute Continuity

If f is integrable on $[a, b]$, then it is also integrable on every subinterval of $[a, b]$; we may therefore introduce its *indefinite integral* by the formula

$$F(y) := \int_a^y f \, dx, \quad a \le y \le b.$$

Let us investigate its properties.

Definition A function $F : I \to \mathbb{R}$, defined on an interval I, is *absolutely continuous*[2] if for every $\varepsilon > 0$ there exists a $\delta > 0$ such that

$$\sum |F(b_k) - F(a_k)| < \varepsilon$$

for every finite disjoint interval system $\{(a_k, b_k)\}$ of total length $< \delta$.

Remarks

- Every Lipschitz continuous function is absolutely continuous. On the other hand, the function $F(x) := \sqrt{x}$ is absolutely continuous on $[0, 1]$, but not Lipschitz continuous.

---

[2]Dini [110, p. 24], Harnack [193, p. 220], Lebesgue [290, pp. 128–129], Vitali [470]. We obtain an equivalent definition by using arbitrary intervals instead of open intervals.

Fig. 6.1 The Cantor function

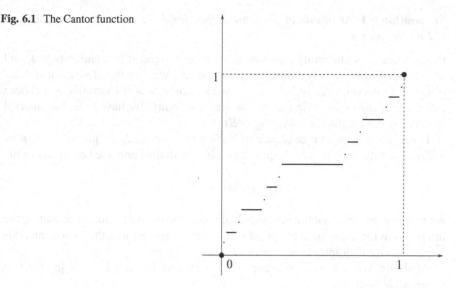

- (*Cantor function*)[3] Every absolutely continuous function is uniformly continuous. On the other hand, consider Cantor's ternary set C (p. 155), and define a function $F : C \to [0, 1]$ by the formula

$$\sum_{i=1}^{\infty} \frac{\varepsilon_i}{3^i} \mapsto \sum_{i=1}^{\infty} \frac{\varepsilon_i}{2^{i+1}}.$$

Then F is surjective, non-decreasing and continuous. (See Fig. 6.1.) By construction the set $[0, 1] \setminus C$ is a countable union of disjoint open intervals. If (a, b) is one of these intervals, then $F(a) = F(b)$ by the surjectivity of F. Set $F(x) := F(a)$ for $a < x < b$, then the extended function $F : [0, 1] \to [0, 1]$ is continuous on a compact set, hence uniformly continuous.

But F is not absolutely continuous. To see this we consider the sets C_n introduced during the construction of C. For each n, C_n is the union of 2^n disjoint intervals $[a_i, b_i]$ of length 3^{-n} each, hence of total length $(2/3)^n$. We have $\sum (F(b_i) - F(a_i)) = 1$ for every n by the definition of F, although the total length $(2/3)^n$ tends to zero as $n \to \infty$.

- If I is *bounded*, then every absolutely continuous function $f : I \to \mathbb{R}$ has bounded variation.[4] Applying Jordan's Proposition 4.11 and Lebesgue's Theorem 4.4 (pp. 157 and 165) it follows that every absolutely continuous function is a.e. differentiable.

---

[3]Cantor [73], Lebesgue [290], Vitali [470].
[4]The identity map of $\mathbb{R}$ shows that this is not necessarily true for unbounded intervals.

Proposition 6.1 *An absolutely continuous function* $F : I \to \mathbb{R}$ *sends every null set of I into a null set.*

Proof Since F is uniformly continuous, it can be extended by continuity to $\bar{I}$, and the extended function is still absolutely continuous. We may therefore assume that I is a *closed* interval. Fix a null set $E \subset I$ and a number $\varepsilon > 0$ arbitrarily, and choose $\delta > 0$ according to the definition of absolute continuity. We have to find an interval system of total length $\leq \varepsilon$, covering $F(E)$.

Let us cover E with a sequence of half-open intervals $I_k = [a_k, b_k) \subset I$, $k = 1, 2, \ldots$, of total length $< \delta$.[5] Replacing each I_k with the connected components of

$$I_k \setminus (I_1 \cup \cdots \cup I_{k-1})$$

we may also assume that the intervals I_k are pairwise disjoint. Moreover, uniting the intervals having a common endpoint we may even assume that the closed intervals $\bar{I}_k$ are pairwise disjoint.

Applying Weierstrass's theorem we may choose in each interval $[a_k, b_k]$ two points a'_k, b'_k such that

$$F(a'_k) \leq F(x) \leq F(b'_k) \quad \text{for all} \quad x \in [a_k, b_k].$$

Then the intervals $[F(a'_k), F(b'_k)]$ cover $F(E)$, and their total length is at most ε, because for each positive integer n we have

$$\sum_{k=1}^{n} |b'_k - a'_k| \leq \sum_{k=1}^{n} |b_k - a_k| < \delta,$$

whence

$$\sum_{k=1}^{n} \left| F(b'_k) - F(a'_k) \right| < \varepsilon$$

by the choice of δ. $\square$

Proposition 6.2 *If F is the indefinite integral of an integrable function $f : [a, b] \to \mathbb{R}$, then*[6]

(a) *F is absolutely continuous;*
(b) *F has bounded variation;*
(c) *$F' = f$ a.e.*

For the proof of (c) we temporarily admit the following

---

[5] We may assume that E does not contain the right endpoint of I.
[6] Lebesgue [290], Vitali [470].

Proposition 6.3 *(Fubini)*[7] *If a series $\sum G_n$ of nonnegative, non-decreasing functions converges a.e. on some interval I, then*

$$\left(\sum G_n\right)' = \sum G_n' \quad a.e. \ on \ \ I. \tag{6.1}$$

Proof of Proposition 6.2 (a)   Given any $\varepsilon > 0$, by Proposition 5.14 (p. 185) we may choose a step function φ satisfying

$$\int_a^b |f - \varphi| \ dx < \varepsilon/2.$$

Fix a number A such that $|\varphi| < A$.

Consider a finite number of pairwise disjoint intervals $(a_k, b_k) \subset [a, b]$, of total length $< \delta := \varepsilon/2A$. Then

$$\sum |F(b_k) - F(a_k)| = \sum \left| \int_{a_k}^{b_k} f \ dx \right|$$

$$\le \sum \int_{a_k}^{b_k} |f - \varphi| \ dx + \sum \int_{a_k}^{b_k} |\varphi| \ dx$$

$$\le \int_a^b |f - \varphi| \ dx + A \sum (b_k - a_k)$$

$$< \frac{\varepsilon}{2} + A\delta$$

$$= \varepsilon.$$

This proves the absolute continuity of F.

(b)   The nonnegative functions

$$f_+ := \max \{f, 0\} \quad \text{and} \quad f_- := \max \{-f, 0\}$$

are integrable, and $f = f_+ - f_-$. Their indefinite integrals are bounded, non-decreasing functions, hence their difference F has a bounded variation.

(c)   The proposition is obvious for step functions. If $f \in C_1$, then choose a non-decreasing sequence (f_n) of step functions, converging a.e. to f. Their indefinite integrals F_n satisfy $F_n' = f_n$ a.e. by our previous remark, and $F_n \to F$ by the definition of the integral.

Applying Proposition 6.3 with $G_n := F_{n+1} - F_n$ we obtain that $F_n' - F_1' \to F' - F_1'$ a.e., i.e., $f_n \to F'$ a.e. On the other hand, we have $f_n \to f$ a.e., so that $F' = f$ a.e.

---

[7]Fubini [165].

The general case follows because every integrable function is the difference of two functions of C_1.

$\square$

Proof of Proposition 6.3 Since every interval is a countable union of compact intervals, we may assume that $I = [a, b]$ is compact.

(a) We prove that the series $\sum G'_n$ converges a.e. Let $S_n = G_1 + \cdots + G_n$ and $S = \sum G_n$, then

$$S_n \to S \quad \text{on} \quad [a, b] \quad \text{everywhere.} \tag{6.2}$$

Since the functions S_n and S are non-decreasing, apart from a null set they are differentiable in $[a, b]$. The series $\sum G'_n(x)$, i.e., the sequence $(S'_n(x))$ converges at each differentiability point x. Indeed, by the non-decreasingness of G_n we have

$$\frac{S_n(x + h) - S_n(x)}{h} \le \frac{S_{n+1}(x + h) - S_{n+1}(x)}{h} \le \frac{S(x + h) - S(x)}{h}$$

for all h satisfying $x + h \in [a, b]$, and hence

$$S'_n(x) \le S'_{n+1}(x) \le S'(x) < \infty$$

for every n.

(b) For the proof of (6.1) it suffices to find a sequence $n_1 < n_2 < \cdots$ of indices such that

$$S' - S'_{n_k} \to 0 \quad \text{a.e.} \tag{6.3}$$

By (6.2) we may choose $n_1 < n_2 < \cdots$ satisfying $S(b) - S_{n_k}(b) < 2^{-k}$ for every k. Then the series

$$\sum (S(b) - S_{n_k}(b))$$

converges. Since

$$0 \le S(x) - S_{n_k}(x) \le S(b) - S_{n_k}(b)$$

for all $a \le x \le b$, it follows that the series $\sum (S - S_{n_k})$ converges on the whole interval $[a, b]$.

The last series is of the same type as $\sum G_n$. Applying the already proved property (a), we conclude that the series $\sum (S' - S'_{n_k})$ converges a.e. But then its general term tends to zero a.e., i.e., (6.3) holds.

□

Using Proposition 6.2 we may investigate the density of sets:

Definition A measurable set A set has *density d at a point* $x \in \mathbb{R}$ if

$$\frac{\mu(A \cap I_n)}{|I_n|} \to d \tag{6.4}$$

for every sequence (I_n) of non-degenerate intervals, containing x and satisfying $|I_n| \to 0$.

We always have $0 \le d \le 1$; for example a set has density one at each point of its interior. Much more is true:

Proposition 6.4 (Lebesgue)[8] *Every measurable set A set has density one at a.e. point of A.*

Proof Since density is a local property, we may assume that A is bounded. Then χ_A integrable, and its indefinite integral F satisfies $F' = \chi_A$ a.e. by Proposition 6.2 (p. 200).

The equality $F'(x) = \chi_A(x)$ means that (6.4) holds with $d = \chi_A(x)$ if x is an endpoint of each interval I_n. The general case follows from the identity

$$\frac{F(x+t) - F(x-s)}{t+s} = \frac{t}{t+s} \frac{F(x+t) - F(x)}{t} + \frac{s}{t+s} \frac{F(x) - F(x-s)}{s},$$

valid for all $t, s > 0$, and from the equality

$$\frac{t}{t+s} + \frac{s}{t+s} = 1.$$

□

6.2 Primitive Function

Proposition 6.2 motivates the following

Definition $F : [a, b] \to \mathbb{R}$ is a *primitive function* of $f : [a, b] \to \mathbb{R}$ if F is absolutely continuous, has bounded variation, and $F' = f$ a.e.

---

[8]Lebesgue [290, pp. 123–124]. See also Zajícek [491] for a direct proof using measure theory, and Riesz–Sz.-Nagy [394] for an extension to non-measurable sets A.

We have the following important generalization of the Newton–Leibniz formula:

Theorem 6.5 (Lebesgue–Vitali)[9] *Let $f : [a, b] \to \mathbb{R}$.*

(a) *f has a primitive function $\iff$ f is integrable.*
(b) *If F is a primitive function of f, then*

$$\int_a^b f \, dx = F(b) - F(a).$$

First we complement Lebesgue's differentiability theorem (p. 157):

Proposition 6.6

(a) *If $F : [a, b] \to \mathbb{R}$ has bounded variation, then F' is integrable.*
(b) *If $F : [a, b] \to \mathbb{R}$ is non-decreasing, then*[10]

$$\int_a^b F' \, dx \leq F(b) - F(a).$$

Examples In the absence of absolute continuity the last inequality may be strict.

- The simplest example is the discontinuous *sign* function:

$$\int_{-1}^1 \text{sign}' \, dx = 0 < 2 = \text{sign}(1) - \text{sign}(-1).$$

- The Cantor function $F : [0, 1] \to [0, 1]$ of the preceding section provides a more surprising example. We recall that F is continuous, non-decreasing and surjective. We also have $F'(x) = 0$ a.e. because F is constant on each interval of the complement of C by construction. Hence[11]

$$\int_0^1 F' \, dx = 0 < 1 = F(1) - F(0).$$

- There exist even continuous and *strictly* increasing functions F with $F' = 0$ a.e.[12]

---

[9]Lebesgue [290], Vitali [466]. The theorem greatly extended former results of Darboux [95, pp. 111–112] and Dini [109, Sect. 197]. Denjoy [98–100] obtained even more complete results; see, e.g., Natanson [332], Bartle [30].

[10]Lebesgue [290].

[11]Lebesgue [290], Vitali [466]. The graph of F is often called the *"Devil's staircase"*; see Fig. 6.1, p. 199. See a related, "natural" example in Komornik–Kong–Li [259].

[12]See, e.g., an example of F. Riesz in Sz.-Nagy [448].

Proof We may assume by Jordan's theorem (p. 165) that F is non-decreasing. Extending F as a constant to the left and to the right of its domain, we may also assume that $[a, b] = \mathbb{R}$. Finally, by Propositions 4.7 and 4.8 we may assume that F is continuous.

The formula

$$D_n(x) := n(F(x + n^{-1}) - F(x)), \quad n = 1, 2, \ldots$$

defines a sequence of nonnegative, continuous functions on $\mathbb{R}$. Their integrals form a bounded sequence on each compact interval $[-N, N]$ because by the continuity of F we have

$$\int_{-N}^{N} D_n \, dx = n \int_{N}^{N+n^{-1}} F \, dx - \int_{-N}^{-N+n^{-1}} F \, dx \to F(N) - F(-N)$$

as $n \to \infty$. Since $D_n \to F'$ a.e. on $[-N, N]$ by Lebesgue's theorem (p. 157), F' is integrable on $[-N, N]$ by the Fatou lemma (p. 183), and

$$\int_{-N}^{N} F' \, dx \leq F(N) - F(-N).$$

Since F is non-decreasing,

$$\int F' \chi_{[-N,N]} \, dx \leq F(\infty) - F(-\infty), \quad N = 1, 2, \ldots.$$

Finally, $F' \chi_{[-N,N]} \nearrow F'$ a.e., so that F' is integrable and

$$\int_{-\infty}^{\infty} F' \, dx \leq F(\infty) - F(-\infty)$$

by the Beppo Levi theorem.   □

Proof of Theorem 6.5 (a) If f is integrable, then its indefinite integral is a primitive function of f by Proposition 6.2. Conversely, if F is a primitive function of f, then $f = F'$ a.e., and f integrable by the preceding proposition.   □

For the proof of part (b) we need a lemma:

Lemma 6.7 *If $H : I \to \mathbb{R}$ is non-decreasing, absolutely continuous and $H' = 0$ a.e., then H is constant.*

Proof It is sufficient to consider the case where $I = [a, b]$ is compact. Let us denote by E the null set of the points $x \in [a, b]$ where the property $H'(x) = 0$ fails. By Proposition 6.1 its image $H(E)$ is also a null set.

We are going to show that the image of the complementary set $F := [a, b] \setminus E$ is a null set, too. Fix $\varepsilon > 0$ arbitrarily. Since $H' = 0$ on F, for each $x \in F$ there exists

$x < y < b$ such that

$$\frac{H(y) - H(x)}{y - x} < \varepsilon.$$

This means that x is invisible from the right with respect to the function $g(t) := \varepsilon t - H(t)$. Applying the "Rising Sun" lemma (p. 162), F has a countable cover by pairwise disjoint open intervals (a_k, b_k) satisfying $g(a_k) \leq g(b_k)$, i.e.,

$$H(b_k) - H(a_k) \leq \varepsilon(b_k - a_k).$$

Hence $H(F)$ may be covered by the system of intervals $[H(a_k), H(b_k)]$ of total length $\leq \varepsilon(b - a)$. Since ε can be chosen arbitrarily small, this proves that $H(F)$ is a null set.

We conclude from the preceding that the *interval* $H(I) = H(E) \cup H(F)$ is a null set, so that it is a one-point set. In other words, H is constant.   □

Proof of Theorem 6.5 (b) We have to show that if $F : [a, b] \to \mathbb{R}$ is absolutely continuous and has bounded variation, then

$$\int_a^b F' \, dx = F(b) - F(a).$$

Observing that in the Jordan decomposition $F = g - h$ of F (Proposition 4.11) the functions g, h are also absolutely continuous, we may assume that F is non-decreasing. By Proposition 6.6 $f := F'$ is integrable, and by Proposition 6.2 the indefinite integral G of f is absolutely continuous, and

$$\int_a^b F' \, dx = G(b) - G(a).$$

It suffices to show that $H := F - G$ is constant. This readily follows from Lemma 6.7 because H is absolutely continuous, and $H' = F' - G' = 0$ a.e.   □

Remark (Lebesgue Decomposition)[13] Let $F : [a, b] \to \mathbb{R}$ be a function of bounded variation, and denote by G the indefinite integral of F'. Then $H := F - G$ has bounded variation, and $H' = 0$ a.e. Functions having this property are called *singular*. Thus every function $F : [a, b] \to \mathbb{R}$ of bounded variation is the difference of an absolutely continuous and a singular function.

---

[13]Lebesgue [295, pp. 232–249].

6.3 Integration by Parts and Change of Variable

Proposition 6.8 *If f, g are integrable on $[a, b]$ and F, G are their primitive functions, then fG and Fg are also integrable on $[a, b]$, and*

$$\int_a^b fG \, dx + \int_a^b Fg \, dx = F(b)G(b) - F(a)G(a) =: [FG]_a^b.$$

Proof F and G are continuous functions on a compact interval, hence they are bounded by some constant M. It follows by applying Proposition 5.16 (b) and (e) (p. 187) that fG and Fg are integrable.

Furthermore, using for the subintervals $[\alpha, \beta]$ of $[a, b]$ the estimates

$$|F(\beta)G(\beta) - F(\alpha)G(\alpha)| = |(F(\beta) - F(\alpha))G(\beta) - F(\alpha)(G(\beta) - G(\alpha))|$$
$$\le M|F(\beta) - F(\alpha)| + M|G(\beta) - G(\alpha)|,$$

we conclude that FG is absolutely continuous and has bounded variation. Since

$$(FG)' = F'G + FG' = fG + Fg \quad \text{a.e.,}$$

applying Theorem 6.5 (p. 204) we conclude that

$$\int_a^b fG \, dx + \int_a^b Fg \, dx = \int_a^b fG + Fg \, dx = [FG]_a^b.$$

$\square$

Proposition 6.9 (de la Vallée-Poussin)[14] *Let $x : [\alpha, \beta] \to \mathbb{R}$ be an absolutely continuous, non-decreasing function. If f is integrable in $[x(\alpha), x(\beta)]$, then $(f \circ x)x'$ is integrable in $[\alpha, \beta]$, and*

$$\int_{x(\alpha)}^{x(\beta)} f(x) \, dx = \int_\alpha^\beta f(x(t))x'(t) \, dt. \tag{6.5}$$

Proof The statement is obvious if f is a step function. Since the general case may be reduced to the case of C^1 functions by using the decomposition $f = g - h$ with $g, h \in C_1$, it suffices to prove the proposition when $f \in C_1$.

Let $f \in C_1$, and choose a non-decreasing sequence (φ_n) of step functions, converging a.e. to f. Set

$$E := \{x \in [x(\alpha), x(\beta)] \ : \ \varphi_n(x) \not\to f(x)\} \tag{6.6}$$

---

[14]de la Vallée-Poussin [465, p. 467].

and

$$D := \{t \in [\alpha, \beta] \; : \; x(t) \in E \quad \text{and} \quad x'(t) > 0\}.$$

By assumption E is a null set. Assume temporarily that D is also a null set.
Since $x' \geq 0$, the sequence of measurable functions

$$t \mapsto \varphi_n(x(t))x'(t), \quad n = 1, 2, \ldots$$

is non-decreasing. Furthermore, we have

$$\varphi_n(x(t))x'(t) \to f(x(t))x'(t)$$

a.e. in $[\alpha, \beta]$ because the exceptional points belong either to D or to the non-differentiability set of x, both null sets. Finally, the corresponding integrals are uniformly bounded because using (6.5) for step functions we have

$$\int_\alpha^\beta \varphi_n(x(t))x'(t)\, dt = \int_{x(\alpha)}^{x(\beta)} \varphi_n(x)\, dx \to \int_{x(\alpha)}^{x(\beta)} f(x)\, dx.$$

Applying the Beppo Levi theorem we conclude that $(f \circ x)x'$ is integrable, and f satisfies (6.5).

It remains to prove that D is a null set in $[\alpha, \beta]$. For this we consider a system $\{I_k\}$ of open intervals, of finite total length, covering each point of E infinitely many times. Then

$$\sum_{k=1}^n \chi_{I_k}(x(t))x'(t), \quad n = 1, 2, \ldots$$

is a non-decreasing sequence of functions whose integrals are uniformly bounded because using (6.5) for step functions we have

$$0 \leq \int_\alpha^\beta \sum_{k=1}^n \chi_{I_k}(x(t))x'(t)\, dt = \sum_{k=1}^n \int_{x(\alpha)}^{x(\beta)} \chi_{I_k}(x)\, dx \leq \sum_{k=1}^\infty |I_k| < \infty.$$

The series converges a.e. by the Beppo Levi theorem. Since it tends to infinity for each $t \in D$, D is a null set.   □

Remark The formula (6.5) remains valid if f has an infinite integral. Considering the positive and negative parts of f, it suffices to study the case of nonnegative, measurable functions f. Choose a non-decreasing sequence (φ_n) of integrable functions, converging a.e. to f. Then we may repeat part (c) of the preceding proof by applying now the generalized Beppo Levi theorem, i.e., Proposition 5.17 (e) (p. 190).

6.4 Exercises

Exercise 6.1 Consider the Cantor function $F : [0, 1] \to [0, 1]$, and set $f(x) := x + F(x), x \in [0, 1]$. Prove the following[15]:

(i) f is a homeomorphism between the intervals $[0, 1]$ and $[0, 2]$;
(ii) f sends the null set C into a set of measure one;
(iii) there exists a subset of C whose image by f is non-measurable.

Exercise 6.2

(i) For each $\alpha \in [0, 1)$ there exists a perfect nowhere dense set $C_\alpha \subset [0, 1]$ of measure α.[16]
(ii) Construct a set $A \subset [0, 1]$ of measure one and of the first category.[17]
(iii) Construct a null set $B \subset [0, 1]$ of the second category.[18]

Exercise 6.3 If $f : [a, b] \to \mathbb{R}$ is continuous, then f and $|f|$ are absolutely continuous at the same time. Is the continuity assumption necessary?

Exercise 6.4 Given an integrable function $f : [a, b] \to \mathbb{R}$, $x \in (a, b)$ is a *Lebesgue point* if

$$\lim_{h \to 0} \frac{1}{2h} \int_{x-h}^{x+h} f(t) \, dt = f(x).$$

(i) If f is continuous at x, then x is a Lebesgue point.
(ii) If f has different finite left and right limits at x, then x is not a Lebesgue point.
(iii) Almost every x is a Lebesgue point.

---

[15]See Gelbaum–Olmsted [167, 168] for other interesting properties.

[16]A *perfect set* is a closed set with no isolated points. A set is *nowhere dense* if its closure has no interior points.

[17]A set A is of the *first category* (Baire [17]) if it is the countable union of nowhere dense sets.

[18]A set A is of the *second category* (Baire [17]) if it is not of the *first category*. Baire's theorem (see p. 32) states that *every complete metric space and every compact Hausdorff space is of the second category*.

Chapter 7
Integrals on Measure Spaces

In my opinion, a mathematician, in so far as he is a mathematician, need not preoccupy himself with philosophy – an opinion, moreover, which has been expressed by many philosophers.
–H. Lebesgue

In Chap. 5 we defined the Lebesgue integral of functions defined on $\mathbb{R}$. In this chapter we show that the theory remains valid in a much more general framework;[1] moreover, almost all proofs can be repeated word for word. The results of this chapter include integrals of several variables and integrals on probability spaces.[2]

7.1 Measures

In this section we generalize the notions of length, area and volume. We recall that by a *disjoint* set sequence we mean a sequence (A_n) of pairwise disjoint sets. To emphasize the disjointness we sometimes write $\cup^* A_n$ instead of $\cup A_n$.

We denote by 2^X the set of all subsets of a set X. The notation is motivated by the fact that if X has n elements, then 2^X has 2^n elements.

---

[1]Radon [366], Fréchet [158], Daniell [93]. In this book we consider only real-valued functions, although Bochner [46] extended the theory to Banach space-valued functions, and this has important applications among others in the theory of partial differential equations. See, e.g., Dunford–Schwartz [117], Edwards [119], Yosida [488], and Lions–Magenes [305].

[2]Kolmogorov [252].

© Springer-Verlag London 2016
V. Komornik, *Lectures on Functional Analysis and the Lebesgue Integral*,
Universitext, DOI 10.1007/978-1-4471-6811-9_7

Definition By a *semiring*[3] in a set X we mean a set system $\mathcal{P} \subset 2^X$ satisfying the following conditions:

- $\varnothing \in \mathcal{P}$;
- if $A, B \in \mathcal{P}$, then $A \cap B \in \mathcal{P}$;
- if $A, B \in \mathcal{P}$, then there exists a finite disjoint sequence $C_1, \ldots, C_n$ in $\mathcal{P}$ such that

$$A \setminus B = C_1 \cup^* \cdots \cup^* C_n.$$

Remark It follows by induction on k that $A_1 \cap \cdots \cap A_k \in \mathcal{P}$ for every finite sequence $A_1, \ldots, A_k$ in $\mathcal{P}$.

Examples

- Every σ-ring is a semiring.
- The intervals of $\mathbb{R}$ form a semiring. The bounded intervals also form a semiring.
- For any given set X and nonnegative integer k, the subsets of at most k elements of X form a semiring.
- (*Restriction*) If $\mathcal{P}$ is a semiring in X, and $Y \subset X$, then

$$\mathcal{P}_Y := \{P \in \mathcal{P} : P \subset Y\}$$

is a semiring in Y.
- (*Direct product*) If $\mathcal{P}$ is a semiring in X and $\mathcal{Q}$ is a semiring in Y, then

$$\mathcal{P} \times \mathcal{Q} := \{P \times Q : P \in \mathcal{P}, Q \in \mathcal{Q}\}$$

is a semiring in $X \times Y$.

Definitions By a *measure*[4] on X we mean a *nonnegative* set function $\mu : \mathcal{P} \to \overline{\mathbb{R}}$, defined on a *semiring* $\mathcal{P}$ in X, satisfying $\mu(\varnothing) = 0$, which is σ-*additive* in the following sense: if $(A_n) \subset \mathcal{P}$ is a disjoint set sequence and $A := \cup^* A_n \in \mathcal{P}$, then[5]

$$\mu(A) = \sum \mu(A_n). \tag{7.1}$$

In this case the triplet $(X, \mathcal{P}, \mu)$ is called a *measure space*.

---

[3] Halmos [184] introduced a slightly more restricted notion, but the present definition has become standard by now.

[4] Borel [59].

[5] Since $\mu(\varnothing) = 0$, the equality (7.1) holds for *finite* disjoint sequences as well. *Finitely additive* set functions were studied before Borel by Harnack [192], Cantor [74, pp. 229–236], Stolz [437], Peano [353, pp. 154–158] and Jordan [231, pp. 76–79].

Examples

- The length of bounded intervals is a measure on $\mathbb{R}$: if a bounded interval I is the union of a disjoint interval sequence (I_k), then $|I| = \sum |I_k|$.[6] Indeed, an elementary argument shows that $|I_1| + \cdots + |I_n| \leq |I|$ for every n; letting $n \to \infty$ this yields the inequality $\sum |I_k| \leq |I|$. The reverse inequality has been proved earlier in Proposition 4.3 (p. 155).
- (*Counting measure*) Denoting by $\mu(A)$ the number of elements of a set $A \subset X$ we get a measure on $\mathcal{P} := 2^X$.[7]
- (*Dirac measure*) For any fixed point $x \in X$ the formula

$$\delta_x(A) := \begin{cases} 1 & \text{if } x \in A, \\ 0 & \text{if } x \notin A \end{cases}$$

defines a measure on $\mathcal{P} := 2^X$.
- (*Zero measure*) The formula $\mu(A) := 0$ defines a measure on $\mathcal{P} := 2^X$.
- (*Largest measure*) The formula

$$\mu(A) := \begin{cases} 0 & \text{if } A = \varnothing, \\ \infty & \text{otherwise} \end{cases}$$

defines a measure on $\mathcal{P} := 2^X$.
- (*Zero-one measure*) Given an uncountable set X, the formula

$$\mu(A) := \begin{cases} 0 & \text{if } A \text{ is countable}, \\ 1 & \text{if } X \setminus A \text{ is countable}, \end{cases}$$

defines a measure on the σ-ring formed by the countable subsets of X and their complements.
- (*Restriction*) If μ is a measure on a semiring $\mathcal{P}$ and $Y \in \mathcal{P}$, then the restriction of μ to $\mathcal{P}_Y$ is a measure.
- (*Direct product*) If $\mu : \mathcal{P} \to \overline{\mathbb{R}}$ and $\nu : \mathcal{Q} \to \overline{\mathbb{R}}$ are two measures, then the formula

$$(\mu \times \nu)(P \times Q) := \mu(P)\nu(Q)$$

defines a measure on $\mathcal{P} \times \mathcal{Q}$.

---

[6]The statement and its proof remain valid for unbounded intervals, too.

[7]In this book we do not distinguish between different infinite cardinalities, except in an example on p. 243 and in some exercises.

- (*Finite part of a measure*) For any given measure $\varrho : \mathcal{R} \to \overline{\mathbb{R}}$,

$$\mathcal{P} := \{A \in \mathcal{R} : \varrho(A) < \infty\}$$

  is a *semiring*, and the restriction of ϱ to $\mathcal{P}$ is a measure.

Now we prove that every measure may be extended uniquely to a measure defined on a set system which is easier to manipulate. This will enable us to establish various important features of the measures.

Definition By a *ring* in a set X we mean a set system $\mathcal{R} \subset 2^X$ satisfying the following conditions:

- $\varnothing \in \mathcal{R}$;
- if $A, B \in \mathcal{R}$, then $A \setminus B \in \mathcal{R}$;
- if $A, B \in \mathcal{R}$ are disjoint sets, then $A \cup^* B \in \mathcal{R}$.

Remark If $\mathcal{R}$ is a ring, then the identity $A \cup B = (A \setminus B) \cup^* B$ shows that the disjointness is not necessary in the last condition: if $A, B \in \mathcal{R}$, then $A \cup B \in \mathcal{R}$. It follows by induction that $A := A_1 \cup \cdots \cup A_n \in \mathcal{R}$ for every finite sequence $A_1, \ldots, A_k$ in $\mathcal{R}$.

Using the identity $\cap A_n = A \setminus \cup (A \setminus A_n)$ it follows that $A_1 \cap \cdots \cap A_k \in \mathcal{R}$ for every finite sequence $A_1, \ldots, A_k$ in $\mathcal{R}$. In particular, every ring is also a semiring.

Examples

- Every σ-ring is also a ring. In particular, 2^X is a ring in X.
- The finite subsets of a set X form a ring in X.
- The finite subsets of a set X and their complements[8] form a ring in X.

Given any set system $\mathcal{A} \subset 2^X$, the intersection of all rings $\mathcal{R}$ satisfying $\mathcal{A} \subset \mathcal{R} \subset 2^X$ is a ring in X. It is called the *ring generated by* $\mathcal{A}$.

There is a simple construction of the rings generated by semirings:

Lemma 7.1 *The ring generated by a semiring $\mathcal{P}$ is formed by all finite disjoint unions of the form*

$$R = P_1 \cup^* \cdots \cup^* P_n, \quad P_1, \ldots, P_n \in \mathcal{P} \quad n = 1, 2, \ldots. \tag{7.2}$$

Proof Since every ring containing $\mathcal{P}$ contains the sets (7.2), it is sufficient to show that the system $\mathcal{R}$ of these sets is already a ring. We proceed in several steps.

(a) We have $\varnothing \in \mathcal{R}$ because $\varnothing \in \mathcal{P}$.
(b) If $R_1, \ldots, R_m \in \mathcal{R}$ are pairwise disjoint sets for some positive integer m, then $R := R_1 \cup^* \cdots \cup^* R_m \in \mathcal{R}$. Indeed, if we decompose each R_i similarly to (7.2), then we obtain a decomposition of the same form of R.

---

[8] The so-called *co-finite* sets.

(c) If $P', P \in \mathcal{P}$, then $P' \setminus P \in \mathcal{R}$ by the definition of the semiring and of $\mathcal{R}$.

(d) If $R \in \mathcal{R}$ and $P \in \mathcal{P}$, then $R \setminus P \in \mathcal{R}$. Indeed, considering a decomposition of the form (7.2) of R and using (b) and (c) we obtain that

$$R \setminus P = (P_1 \setminus P) \cup^* \cdots \cup^* (P_n \setminus P) \in \mathcal{R}.$$

(e) If $R', R \in \mathcal{R}$, then $R' \setminus R \in \mathcal{R}$. Indeed, considering a decomposition of the form (7.2) of R and applying (d) n times we obtain that

$$R' \setminus R = (\ldots (R' \setminus P_1) \setminus P_2) \ldots) \setminus P_n \in \mathcal{R}. \qquad \square$$

Proposition 7.2 *Every measure* $\mu : \mathcal{P} \to \overline{\mathbb{R}}$ *may be extended to a unique measure defined on the ring* $\mathcal{R}$ *generated by the semiring* $\mathcal{P}$.

Proof If there exists such an extension, then, still denoting it by μ, we must have

$$\mu(R) = \mu(P_1) + \cdots + \mu(P_n)$$

for every decomposition of the form (7.2). Since $P_1, \ldots, P_n \in \mathcal{P}$, this proves the uniqueness.

For the existence first we show that the above equality does indeed define an extension, i.e., if

$$R = P_1' \cup^* \cdots \cup^* P_m'$$

is another such decomposition of R, then

$$\mu(P_1) + \cdots + \mu(P_n) = \mu(P_1') + \cdots + \mu(P_k').$$

This readily follows from the additivity of $\mu : \mathcal{P} \to \overline{\mathbb{R}}$ because both sums are equal to

$$\sum_{j=1}^{n} \sum_{i=1}^{k} \mu(P_j \cap P_i').$$

The extended set function is clearly nonnegative, it remains to prove its σ-additivity. Let $R = \cup_{k=1}^{\infty} R_k$ be a disjoint union with $R, R_k \in \mathcal{R}$; we have to show that $\mu(R) = \sum \mu(R_k)$.

Replacing each R_k with a decomposition of the form (7.2) and using the definition of $\mu(R_k)$ we may assume that $R_k \in \mathcal{P}$ for every k. Now consider a decomposition of the form (7.2) of R; then we have

$$P_j = \bigcup_{k=1}^{\infty} {}^* (P_j \cap R_k)$$

for each j. Since $P_j, P_j \cap R_k \in \mathcal{P}$, and since μ is σ-additive on $\mathcal{P}$, this implies

$$\mu(P_j) = \sum_{k=1}^{\infty} \mu(P_j \cap R_k).$$

Summing these equalities we obtain the required relation:

$$\mu(R) = \sum_{j=1}^{n} \mu(P_j) = \sum_{j=1}^{n} \sum_{k=1}^{\infty} \mu(P_j \cap R_k)$$

$$= \sum_{k=1}^{\infty} \left(\sum_{j=1}^{n} \mu(P_j \cap R_k) \right)$$

$$= \sum_{k=1}^{\infty} \mu(R_k). \qquad \Box$$

Now we are ready to establish some basic properties of measures:

Proposition 7.3 *Every measure* $\mu : \mathcal{P} \to \overline{\mathbb{R}}$ *(defined on a semiring) satisfies the following conditions:*

(a) *(monotonicity) if* $A, B \in \mathcal{P}$ *and* $A \subset B$, *then* $\mu(A) \leq \mu(B)$;
(b) *(σ-subadditivity) if* $(A_n) \subset \mathcal{P}$ *is a countable cover of* $A \in \mathcal{P}$, *then* $\mu(A) \leq \sum \mu(A_n)$;
(c) *(continuity) if* $(A_n) \subset \mathcal{P}$ *is a non-decreasing set sequence and* $A := \cup A_n \in \mathcal{P}$, *then* $\mu(A_n) \to \mu(A)$;
(d) *(continuity) if* $(A_n) \subset \mathcal{P}$ *is a non-increasing set sequence with* $\mu(A_1) < \infty$ *and* $A := \cap A_n \in \mathcal{P}$, *then* $\mu(A_n) \to \mu(A)$.

Example The intervals $A_n := [n, \infty) \subset \mathbb{R}$ show that the condition $\mu(A_1) < \infty$ in (d) cannot be omitted.

Proof By the preceding proposition we may assume that $\mathcal{P}$ is a ring.

(a) Using the nonnegativity and the additivity of the measures we have

$$\mu(B) = \mu(A) + \mu(B \setminus A) \geq \mu(A).$$

(b) Setting $B_1 := A \cap A_1$ and

$$B_{n+1} := (A \cap A_{n+1}) \setminus (A_1 \cup \cdots \cup A_n), \quad n = 1, 2, \ldots$$

we have $A = \cup^* B_n$. Furthermore, $B_n \subset A_n$ and $B_n \in \mathcal{P}$ for all n (because $\mathcal{P}$ is a *ring*). We conclude by using (a):

$$\mu(A) = \sum \mu(B_n) \leq \sum \mu(A_n).$$

(c) Let $A_0 = \emptyset$, then the sets $A_k \setminus A_{k-1}$ belong to the *ring* $\mathcal{P}$. Since

$$A = \bigcup_{k=1}^{\infty}{}^* (A_k \setminus A_{k-1}) \quad \text{and} \quad A_n = \bigcup_{k=1}^{n}{}^* (A_k \setminus A_{k-1})$$

for all n, we have

$$\mu(A_n) = \sum_{k=1}^{n} \mu(A_k \setminus A_{k-1}) \to \sum_{k=1}^{\infty} \mu(A_k \setminus A_{k-1}) = \mu(A).$$

(d) Since $\mu(A_n)$ is finite for all n, changing A_n to $A_n \setminus A$ we may assume that $A = \emptyset$. The sets $A_k \setminus A_{k+1}$ belong to the *ring* $\mathcal{P}$. Since

$$A_1 = \bigcup_{k=1}^{\infty}{}^* (A_k \setminus A_{k+1}),$$

by the σ-additivity we have

$$\sum_{k=1}^{\infty} \mu(A_k \setminus A_{k+1}) = \mu(A_1).$$

Since $\mu(A_1) < \infty$ by assumption, the series is convergent, and hence

$$\sum_{k=n}^{\infty} \mu(A_k \setminus A_{k+1}) \to 0$$

as $n \to \infty$. We conclude by noticing that the last sum is equal to $\mu(A_n)$ because

$$\bigcup_{k=n}^{\infty}{}^* (A_k \setminus A_{k+1}) = A_n. \qquad \square$$

7.2   Integrals Associated with a Finite Measure

Definition A measure is *finite* if it takes only finite values.

Examples

- The finite part of a measure is a finite measure.
- Every bounded measure is finite. The length of bounded intervals shows that the converse is not always true.

For the rest of this section we fix a *semiring* $\mathcal{P}$ in a set X and a *finite* measure $\mu : \mathcal{P} \to \mathbb{R}$.

Definition By a *step function* we mean a linear combination

$$\varphi = \sum_{k=1}^{n} c_k \chi_{P_k}$$

of characteristic functions of sets in $\mathcal{P}$.

The *integral* of a step function is defined by the formula

$$\int \varphi \, d\mu := \sum_{k=1}^{n} c_k \mu(P_k).$$

Proposition 5.1 (p. 172) remains valid: by the additivity of the measure the integral does not depend on the particular representation of the step function.

Definition A set A is a *null set* if for each $\varepsilon > 0$ there exists a sequence $(P_k) \subset \mathcal{P}$ satisfying $A \subset \cup P_k$ and $\sum \mu(P_k) \leq \varepsilon$.

Equivalently, A is a null set if there exists a sequence $(P_k) \subset \mathcal{P}$ satisfying $\sum \mu(P_k) < \infty$, and covering each point $x \in A$ infinitely many times.

Proposition 4.3 (p. 155) takes the following form:

Proposition 7.4

(a) *The empty set is a null set.*
(b) *The subsets of a null set are null sets.*
(c) *The union of countably many null sets is a null set.*
(d) $P \in \mathcal{P}$ *is a null set* $\Longleftrightarrow \mu(P) = 0$.

Proof (a), (b) and (c) We may repeat the proof of Proposition 4.3.

(d) If $\mu(P) = 0$, then P is null set: we may choose $P_k = P$ for all k in the definition. Conversely, if $P \in \mathcal{P}$ is a null set, then for each $\varepsilon > 0$ there exists a sequence $(P_k) \subset \mathcal{P}$ satisfying $A \subset \cup P_k$ and $\sum \mu(P_k) \leq \varepsilon$. Using the subadditivity of the measure this implies $\mu(P) \leq \varepsilon$ for every $\varepsilon > 0$, and hence $\mu(P) = 0$.   □

Chapter 5 was written in such a way that all *theorems, propositions, corollaries and lemmas remain valid without any change.* Moreover, the proofs also remain valid with three exceptions:

- In the proof of Lemma 5.2 (p. 172) we have used the topological properties of intervals.

- In the proof of Proposition 5.16 (f) (p. 187) we have used the existence of an integrable, everywhere positive function. An example following Lemma 7.5 will show that such functions do not exist for all measures.
- In the proof of Proposition 5.19 (p. 194) we have implicitly used that the constant functions are measurable.[9] The just mentioned example will show that this is not always true either.[10]

The following alternative proofs are always valid:

Proof of Lemma 5.2 We extend μ to the generated ring $\mathcal{R}$ by Proposition 7.2.

Fix a null set $Y \subset X$ such that $\varphi_n(x) \to 0$ for every $x \in X \setminus Y$, and fix $\varepsilon > 0$ arbitrarily. Choose a set sequence $(H_i) \subset \mathcal{P}$ satisfying

$$Y \subset \cup H_i \quad \text{and} \quad \sum \mu(H_i) < \varepsilon.$$

Then the sets $S_n := H_1 \cup \cdots \cup H_n$ belong to $\mathcal{R}$,

$$S_1 \subset S_2 \subset \cdots,$$

and $\mu(S_n) < \varepsilon$ for every n.

Set

$$K_0 := \{x \in X \,:\, \varphi_1(x) > 0\}$$

and

$$K_n := \{x \in X \,:\, \varphi_n(x) > \varepsilon\}, \quad n = 1, 2, \ldots;$$

they belong to $\mathcal{R}$, and

$$K_0 \supset K_1 \supset K_2 \supset \cdots.$$

Setting $M := \max \varphi_1$ we have

$$\varphi_n \leq M \quad \text{on} \quad K_n,$$
$$\varphi_n \leq \varepsilon \quad \text{on} \quad K_0 \setminus K_n,$$
$$\varphi_n = 0 \quad \text{on} \quad X \setminus K_0.$$

---

[9]The measurability of the constant functions is equivalent to the measurability of X.
[10]In this book, following F. Riesz, we adopt a more restrictive measurability notion than usual. See Sect. 7.7 on the advantages of this choice.

Consequently,

$$0 \leq \int \varphi_n \, d\mu \leq \varepsilon\mu(K_0 \setminus K_n) + M\mu(K_n)$$

$$= \varepsilon\mu(K_0 \setminus K_n) + M\mu(K_n \cap S_n) + M\mu(K_n \setminus S_n)$$

$$\leq \varepsilon\mu(K_0) + M\varepsilon + M\mu(K_n \setminus S_n).$$

The set sequence $(K_n \setminus S_n)$ is non-increasing and

$$\mu(K_1 \setminus S_1) \leq \mu(K_0) < \infty.$$

Furthermore, its intersection is empty. Indeed, if $x \in \cap K_n$, then $\varphi_n(x) \nrightarrow 0$, so that $x \in Y$; but then $x \in S_n$ for a sufficiently large n and therefore $x \in S_n$ and $x \notin K_n \setminus S_n$.

Applying Proposition 7.3 (d) we conclude that $\mu(K_n \setminus S_n) \to 0$. Consequently, we infer from the previous estimate that

$$0 \leq \int \varphi_n \, d\mu < \big(\mu(K_0) + M + 1\big)\varepsilon$$

if n is sufficiently large.   □

Proof of Proposition 5.16 (f) If there exists a set sequence $(P_k) \subset \mathcal{P}$ such that each f_n vanishes outside $\cup P_k$ then we may repeat the proof of Chap. 5 by using the function

$$h := \sum_k \frac{\chi_{P_k}}{k^2(1 + \mu(P_k))},$$

and defining the functions g_n and g by zero outside $\cup P_k$.

The existence of such a sequence (P_k) follows from the next lemma.[11]   □

Lemma 7.5 *To each measurable function f there exists a disjoint set sequence $(P_k) \subset \mathcal{P}$ such that $f = 0$ outside $\cup^* P_k$.*[12]

Proof Choose a sequence (φ_n) of step functions converging to f a.e. By definition there exists a set sequence $(A_{0j}) \subset \mathcal{P}$ such that $\varphi_n \to f$ outside $\cup A_{0j}$.

Furthermore, by the definition of step functions there exists for each n a finite set sequence $(A_{nj}) \subset \mathcal{P}$ such that $\varphi_n = 0$ outside $\cup A_{nj}$.

We may arrange all these sets A_{0j} and A_{nj} into a set sequence (P_k). Furthermore, using the definition of a semiring we may replace each difference $P_2 \setminus P_1$, $(P_3 \setminus$

---

[11]We apply the lemma for each f_n, and we take the union of the corresponding set sequences.

[12]We sometimes express this property by saying that f *has a σ-finite support*. Using this terminology X is measurable $\Longleftrightarrow$ X is σ-finite.

$P_2) \setminus P_1, \ldots$ by a finite disjoint union of sets in $\mathcal{P}$. Then the sequence (P_k) becomes disjoint, and $f = 0$ outside $\cup^* P_k$. □

Examples

- Let μ be a finite measure on the ring of finite subsets of an uncountable set X. By Lemma 7.5 there is no measurable, strictly positive function for this measure. In particular, the non-zero constant functions are non-measurable.
- Fix a non-empty set X and consider the measure $\mu(\varnothing) := 0$ on the ring $\mathcal{P} := \{\varnothing\}$. Then only the zero function is measurable, and $\varnothing$ is the only measurable set.

Proof of Proposition 5.19 Most of the former proof remains valid. The only property to check is that if f is a measurable function and c a positive constant, then the functions $\min\{f, c\}$ and $\max\{f, -c\}$ are measurable.

For the proof we consider the sets P_k of the preceding lemma. Then $A := \cup P_k$ is measurable, hence the functions $c\chi_A$ and then the functions

$$\min\{f, c\} = \min\{f, c\chi_A\} \quad \text{and} \quad \max\{f, -c\} = \max\{f, -c\chi_A\}$$

are also measurable. □

Starting from an arbitrary finite measure defined on a semiring $\mathcal{P}$, the theory of Chap. 5 leads to a measure μ defined on the system $\mathcal{M}$ of all measurable sets. Our next result states that this is the only possible extension of the original measure to $\mathcal{M}$.

*Proposition 7.6 *Let* $v : \mathcal{N} \to \mathbb{R}$ *be another measure, defined on a semiring satisfying* $\mathcal{P} \subset \mathcal{N} \subset \mathcal{M}$. *If* $\mu = v$ *on* $\mathcal{P}$, *then* $\mu = v$ *on* $\mathcal{N}$, *too.*

Proof

(i) Every μ-null set is also a v-null set. For, if a set may be covered by a set sequence $(P_n) \subset \mathcal{P}$ of total μ-measure $< \varepsilon$, then we have

$$\sum v(P_n) = \sum \mu(P_n) < \varepsilon.$$

(ii) Now consider the two integrals associated with the measures μ and $v|_{\mathcal{P}}$. We show that every μ-integrable function f is also v-integrable, and the two integrals are equal:

$$\int f \, d\mu = \int f \, dv. \tag{7.3}$$

Since

$$\int \chi_P \, d\mu = \mu(P) = v(P) = \int \chi_P \, dv$$

for every $P \in \mathcal{P}$ by assumption, taking their linear combinations we obtain that (7.3) holds for all step functions.

The equality holds for all functions $f \in C_1(\mu)$ as well.[13] Indeed, consider a non-decreasing sequence (φ_n) of μ-step functions, converging μ-a.e. to f, and satisfying

$$\sup_n \int \varphi_n \, d\mu < \infty.$$

Then we have $\int \varphi_n \, d\mu \to \int f \, d\mu$ by definition.

Furthermore, (φ_n) converges to f also ν-a.e. by (i), and

$$\sup_n \int \varphi_n \, d\nu = \sup_n \int \varphi_n \, d\mu < \infty$$

because (7.3) has already been proved for step functions. Applying the Beppo Levi theorem we conclude that f is ν-integrable and $\int \varphi_n \, d\nu \to \int f \, d\nu$; hence (7.3) holds for f.

Finally, if f is an arbitrary μ-integrable function, then we have $f = g - h$ with suitable functions $g, h \in C_1(\mu)$. We already know that (7.3) holds for g and h; taking the difference of these equalities we see that f satisfies (7.3) as well.

(iii) It follows from (ii) that if $A \in \mathcal{N}$ and $\mu(A) < \infty$, then

$$\mu(A) = \int \chi_A \, d\mu = \int \chi_A \, d\nu = \nu(A).$$

Consider finally an arbitrary set $A \in \mathcal{N}$. Then $A \in \mathcal{M}$, hence it is μ-measurable, so that it may be covered by a disjoint sequence $(P_n) \subset \mathcal{P}$. Since $\mathcal{P} \subset \mathcal{N}$, we have

$$A \cap P_n \in \mathcal{N} \subset \mathcal{M} \quad \text{and} \quad \mu(A \cap P_n) < \infty$$

for all n. Applying the preceding equality for $A \cap P_n$ instead of A we conclude that

$$\mu(A) = \sum \mu(A \cap P_n) = \sum \nu(A \cap P_n) = \nu(A). \qquad \square$$

We end this section by characterizing the measures constructed via integrals.

Definition A measure μ, defined on a semiring $\mathcal{Q}$, is *σ-finite* if each set in $\mathcal{Q}$ has a countable cover by sets of finite measure.

---

[13]The function class C_1 was defined on p. 174.

Remark By the definition of a semiring in the σ-finite case each set in $\mathcal{Q}$ also has a countable *disjoint partition* by sets of finite measure.

Examples

- The usual Lebesgue measure in $\mathbb{R}$ is σ-finite.
- Every finite measure is σ-finite.
- Given a measure ϱ on some semiring $\mathcal{R}$, let us denote by $\mathcal{Q}$ the sets $A \in \mathcal{R}$ having a countable cover by sets $P \in \mathcal{R}$ of finite measure. Then $\mathcal{Q}$ is also a semiring. The restriction of ϱ to $\mathcal{Q}$ is called the *σ-finite* part of ϱ.
- The counting measure on an uncountable set X is *not* σ-finite. Its σ-finite part is defined on the countable subsets of X.

Consider again a finite measure defined on a semiring $\mathcal{P}$, and let μ be its extension[14] to the set system $\mathcal{M}$ of measurable sets.

*Proposition 7.7

(a) $\mathcal{M}$ is a σ-ring. The extended measure $\mu : \mathcal{M} \to \overline{\mathbb{R}}$ is σ-finite and complete.
(b) *Conversely, every σ-finite, complete measure, defined on a σ-ring may be obtained in this way.*
(c) *More generally, every σ-finite measure, defined on a semiring, is a restriction of the measure $\mu : \mathcal{M} \to \overline{\mathbb{R}}$ obtained by the extension of its finite part.*

Proof

(a) This follows from Proposition 5.18 (p. 192) and Lemma 7.5.
(c) Let $\mathcal{N}$ be a semiring and $v : \mathcal{N} \to \overline{\mathbb{R}}$ a σ-finite measure. Consider the finite part of v, i.e., the restriction of v to the semiring

$$\mathcal{P} := \{A \in \mathcal{N} : v(P) < \infty\},$$

and let $\mu : \mathcal{M} \to \overline{\mathbb{R}}$ be the extension of $v|_{\mathcal{P}}$ to the σ-ring of $v|_{\mathcal{P}}$-measurable sets. We have to show that $\mathcal{N} \subset \mathcal{M}$ and $v = \mu|_{\mathcal{N}}$.

Fix an arbitrary set $A \in \mathcal{N}$. Since v is σ-finite, there exists a disjoint set sequence $(P_n) \subset \mathcal{P}$ satisfying $A = \cup^* P_n$. Since $\mathcal{P} \subset \mathcal{M}$ and $\mathcal{M}$ is a σ-ring, $A \in \mathcal{M}$. Furthermore, since $\mu(P_n) = v(P_n)$ for every n by the definition of μ, we conclude that

$$v(A) = \sum v(P_n) = \sum \mu(P_n) = \mu(A).$$

(b) Let $\mathcal{N}$ be a σ-ring and $v : \mathcal{N} \to \overline{\mathbb{R}}$ a σ-finite, complete measure. By (c) we already know that $\mathcal{N} \subset \mathcal{M}$ and $v = \mu|_{\mathcal{N}}$. It remains to prove that $\mathcal{M} \subset \mathcal{N}$.

Fix an arbitrary $A \in \mathcal{M}$. Then χ_A is a measurable function, so that there exists a sequence (φ_n) of $\mathcal{P}$-step functions, converging to χ_A μ-a.e. In other

---

[14]We already know that this extension is unique.

words, there exists a μ-null set P_0 such that $\varphi_n \to \chi_A$ outside it. Observe that P_0 is also a ν-null set.[15]

Then

$$A_n := \left\{ x \in X \ : \ \varphi_n(x) > \frac{1}{2} \right\} \in \mathcal{N}$$

for each $n = 1, 2, \ldots$, because A_n is a union of finitely many elements of $\mathcal{P}$, $\mathcal{P} \subset \mathcal{N}$, and $\mathcal{N}$ is a ring.

Since $\mathcal{N}$ is also a σ-ring, the set

$$N := \limsup A_n := \cap_{k=1}^{\infty} \cup_{n=k}^{\infty} A_k$$

also belongs to $\mathcal{N}$.

Now observe for each $x \in X \setminus P_0$ the equivalences

$$x \in A \iff x \in A_n \text{ for infinitely many } n \iff x \in \limsup A_n.$$

It follows that

$$(A \setminus N) \cup (N \setminus A) \subset P_0,$$

i.e., A differs from $N \in \mathcal{N}$ on a ν-null set. Since ν is complete, we conclude that $A \in \mathcal{N}$.   □

7.3 Product Spaces: Theorems of Fubini and Tonelli

In classical analysis the computation of double integrals may be reduced to that of simple integrals by using the formula[16]

$$\int_{X \times Y} f(x, y) \, dx \, dy = \int_X \left(\int_Y f(x, y) \, dy \right) dx \tag{7.4}$$

$$= \int_Y \left(\int_X f(x, y) \, dx \right) dy.$$

In this section we prove that this formula remains valid for Lebesgue integrals as well.

Consider two *finite* measures $\mu : \mathcal{P} \to \mathbb{R}$ and $\nu : \mathcal{Q} \to \mathbb{R}$, where $\mathcal{P}$ is a semiring in X and $\mathcal{Q}$ is a semiring in Y. Then $\mu \times \nu : \mathcal{P} \times \mathcal{Q} \to \mathbb{R}$ is a finite measure on the

---

[15]See the beginning of the proof of Proposition 7.6: we already know that $\mathcal{P} \subset \mathcal{N} \subset \mathcal{M}$.
[16]Euler [130], Dirichlet [113], and Stolz [438, pp. 93–94].

semiring $\mathcal{P} \times \mathcal{Q}$ in $X \times Y$. In what follows we write

$$\int_{X \times Y} f(x,y) \, dx \, dy, \quad \int_X g(x) \, dx \quad \text{and} \quad \int_Y h(y) \, dy$$

instead of

$$\int f \, d(\mu \times \nu), \quad \int g \, d\mu \quad \text{and} \quad \int h \, d\nu.$$

The expressions *null set* and *a.e.* will refer to μ in X, to ν in Y, and to $\mu \times \nu$ in $X \times Y$.
The following theorem is a far-reaching generalization of the classical results:

Theorem 7.8 (Fubini)[17] *If f is integrable in $X \times Y$, then the successive integrals in (7.4) exist, and the three expressions are equal.*

Remarks

- By induction the theorem may be extended to arbitrary finite direct products of (finite) measures.
- The existence of the successive integrals does not imply their equality. Moreover, their existence and equality does not imply the integrability of f. See the examples at the end of this section.

We prepare the proof by clarifying the relationship among the null sets of the three spaces:

Lemma 7.9 *If E is a null set in $X \times Y$, then the "vertical sections"*

$$\{y \in Y \, : \, (x,y) \in E\}$$

of E are null sets in Y for almost every $x \in X$.

---

[17]Lebesgue [288] (for bounded functions), Fubini [164]. Fubini's proof was incorrect; the first correct proofs were given by Hobson [214] and de la Vallée-Poussin [464]. See Hawkins [198].

Proof Fix a sequence of "rectangles" $R_n = P_n \times Q_n$ in $\mathcal{P} \times \mathcal{Q}$, covering each point of E infinitely many times, and satisfying

$$\sum_{n=1}^{\infty} (\mu \times \nu)(R_n) < \infty.$$

By the definition of the integral of step functions we have

$$(\mu \times \nu)(R_n) = \int_{X \times Y} \chi_{R_n}(x, y) \, dx \, dy = \int_X \left(\int_Y \chi_{R_n}(x, y) \, dy \right) dx$$

(their common value is $\mu(P_n)\nu(Q_n)$), so that the series

$$\sum_{n=1}^{\infty} \int_X \left(\int_Y \chi_{R_n}(x, y) \, dy \right) dx$$

is convergent. Applying the Beppo Levi theorem we obtain that the series

$$\sum_{n=1}^{\infty} \int_Y \chi_{R_n}(x, y) \, dy$$

is convergent for a.e. $x \in X$. If x_0 is such a point, then another application of the Beppo Levi theorem implies that the series

$$\sum_{n=1}^{\infty} \chi_{R_n}(x_0, y)$$

is convergent for a.e. $y \in Y$. If y_0 is such a point, then $(x_0, y_0) \notin E$, because at the points of E we have $\sum \chi_{R_n} = \infty$. $\square$

Proof of Theorem 7.8 By symmetry we prove only the equality

$$\int_{X \times Y} f(x, y) \, dx \, dy = \int_X \left(\int_Y f(x, y) \, dy \right) dx. \tag{7.5}$$

We have to show that

- the integral $\int_Y f(x, y) \, dy$ is well defined for a.e. $x \in X$;
- the function $x \mapsto \int_Y f(x, y) \, dy$ is integrable in X;
- the two sides of (7.5) are equal.

We have seen during the proof of the preceding lemma that these properties hold true if f is the characteristic function of a "rectangle". Taking linear combinations we see that they hold for step functions as well. Since every integrable function is the difference of two step functions, it remains only to establish the three properties for functions belonging to the class C_1.

Fix $f \in C_1$ arbitrarily. Choose a non-decreasing sequence (φ_n) of step functions and a null set $E \subset X \times Y$ such that

$$\varphi_n(x, y) \nearrow f(x, y) \quad \text{for each} \quad (x, y) \in (X \times Y) \setminus E, \tag{7.6}$$

and therefore

$$\int_{X \times Y} \varphi_n(x, y) \, dx \, dy \to \int_{X \times Y} f(x, y) \, dx \, dy$$

by the definition of the integral. Since (7.5) is already known for step functions, the last relation may be rewritten in the form

$$\int_X \left(\int_Y \varphi_n(x, y) \, dy \right) dx \to \int_{X \times Y} f(x, y) \, dx \, dy. \tag{7.7}$$

Applying the Beppo Levi theorem[18] we obtain that the non-decreasing sequence of the functions

$$x \mapsto \int_Y \varphi_n(x, y) \, dy \tag{7.8}$$

converges, and hence is bounded, for a.e. $x \in X$.

Fix a point $x \in X$ where the convergence holds, and for which the section

$$\{y \in Y : (x, y) \in E\}$$

is a null set. (By the preceding lemma a.e. $x \in X$ has this property.) Then

$$\varphi_n(x, y) \nearrow f(x, y) \quad \text{for a.e.} \quad y \in Y$$

by (7.6), so that, in view of the boundedness of the functions (7.8) we may apply the Beppo Levi theorem again: the function

$$y \mapsto f(x, y)$$

is integrable, and

$$\int_Y \varphi_n(x, y) \, dy \nearrow \int_Y f(x, y) \, dy.$$

---

[18]In the proof of this theorem the application of Lemma 5.3 (p. 173) is sufficient because we consider only sequences of step functions.

We recall that this convergence holds for a.e. $x \in X$. Since the sequence of integrals

$$\int_X \left(\int_Y \varphi_n(x, y) \, dy \right) dx$$

is bounded by (7.7) and the integrability of f, a third application of the Beppo Levi theorem shows that the function

$$x \mapsto \int_Y f(x, y) \, dy$$

is integrable, and

$$\int_X \left(\int_Y \varphi_n(x, y) \, dy \right) dx \rightarrow \int_X \left(\int_Y f \, dy \right) dx. \tag{7.9}$$

The equality (7.5) follows from (7.7) and (7.9).   □

Fubini's theorem remains valid for generalized (infinite-valued) integrals:

Theorem 7.10 (Tonelli)[19] *If the integral of a function f exists in $X \times Y$, then the successive integrals in (7.4) also exist, and the three quantities are equal.*

Remarks

- Like that of Fubini, Tonelli's theorem holds for arbitrary finite direct products of measures as well.
- We recall that every nonnegative, measurable function has an integral.

Proof Considering the positive and negative parts of f, at least one of them is integrable, hence satisfies the assumptions of Fubini's theorem. Therefore it is sufficient to investigate the case of a nonnegative, measurable function f.

Applying Lemma 7.5 we fix a non-decreasing sequence (A_n) of sets of finite measure such that $f = 0$ outside $\cup A_n$. Then the functions

$$\varphi_n := \chi_{A_n} \min\{f, n\}$$

are integrable in $X \times Y$ by Proposition 5.16 (e) (p. 187), and $\varphi_n \nearrow f$ a.e. by construction. We may therefore choose a null set E in $X \times Y$ such that

$$\varphi_n(x, y) \nearrow f(x, y) \quad \text{for each} \quad (x, y) \in (X \times Y) \setminus E.$$

---

[19]Tonelli [457].

Let us observe that formally this relation is identical with (7.6). We may therefore repeat the preceding proof with two small changes:

- instead of the Beppo Levi theorem (or Lemma 5.3) we apply the generalized Beppo Levi theorem, i.e., Proposition 5.17 (e) (p. 190);
- the validity of the equality for φ_n (instead of f) now follows from Fubini's theorem.   □

Examples The following examples show the optimality of the assumptions of Theorems 7.8 and 7.10.[20]

- The formula

$$f(x,y) := \begin{cases} 1 & \text{if } x < y < x+1, \\ -1 & \text{if } x-1 < y < x, \\ 0 & \text{otherwise} \end{cases}$$

defines a measurable function $f : \mathbb{R}^2 \to \mathbb{R}$ whose integral is not defined, although the successive integrals in (7.4) exist, and are equal (to zero).[21]
- Let μ be the counting measure on the finite subsets of $\mathbb{R}$. Furthermore, let $\nu(A) = 0$ and $\nu(\mathbb{R} \setminus A) = 1$ for every finite subset of $\mathbb{R}$. For the characteristic function f of the set

$$D := \{(x,x) \,:\, x \in \mathbb{R}\}$$

the two successive integrals in (7.4) exist, and they are equal to 0 and 1, respectively.

Observe that f is non-measurable by Lemma 7.5, hence its integral is undefined.

7.4 Signed Measures: Hahn and Jordan Decompositions

Consider the integral associated with a finite measure defined on a semiring. Let us denote by $\mathcal{M}$ the σ-ring of measurable sets, and by $\mu : \mathcal{M} \to \overline{\mathbb{R}}$ the corresponding extended measure.

Equivalently, in view of Proposition 7.7 (p. 223), let $\mu : \mathcal{M} \to \overline{\mathbb{R}}$ be a *σ-finite, complete measure on a σ-ring* $\mathcal{M}$.

---

[20]The former counterexamples of Cauchy [81, p. 394], Thomae [452] and du Bois-Reymond [53] were based on the smallness of the class of Riemann integrable functions.

[21]Further counterexamples are given in Exercise 7.8 below, p. 253.

It is natural to define the integrals on a set $A \in \mathcal{M}$ by the formula

$$\int_A f \, d\mu := \int f \chi_A \, d\mu$$

when the right-hand side integral is defined.

Let us generalize the *indefinite integrals*:

Proposition 7.11 *If a measurable function f has an integral, then the formula*

$$\nu(A) := \int_A f \, d\mu$$

defines a σ-additive set function $\nu : \mathcal{M} \to \overline{\mathbb{R}}$ with $\nu(\varnothing) = 0$.

Proof Taking the positive and negative parts of f we may assume that f is nonnegative. Then the result follows from Proposition 5.17 (f) (p. 190). $\qquad\square$

The proposition motivates the following definitions:

Definitions

- By a *signed measure* we mean a σ-additive set function ν, satisfying $\nu(\varnothing) = 0$.
- The signed measure in the above proposition is called the *indefinite integral* of f with respect to μ.

Examples

- Every measure is a signed measure.
- The difference of two measures, at least one of which is finite, is a signed measure.
- (Smolyanov)[22] Consider the following ring on an infinite set X:

$$\mathcal{R} := \{ A \subset X \ : \ A \quad \text{or} \quad X \setminus A \quad \text{is finite} \} \, .$$

The formulas

$$\mu(A) := |A| \, , \quad \mu(X \setminus A) := - |A| \, ,$$

where $|A|$ denotes the number of elements of a finite set A, define a signed measure on $\mathcal{R}$.

If a signed measure ν is defined by an indefinite integral as in Proposition 7.11, then the indefinite integrals ν_+, ν_- associated with f_+, f_- are measures, and $\nu = \nu_+ - \nu_-$. Furthermore, ν_+ and ν_- are concentrated on the disjoint sets $\{ f > 0 \}$ and $\{ f < 0 \}$, and at least one of the two measures is bounded.

---

[22]See Gurevich–Silov [175, p. 180].

Fig. 7.1 Hahn
decomposition

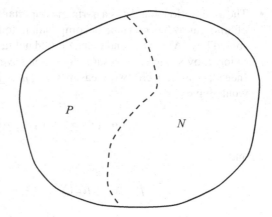

These properties remain valid for *all* signed measures, defined on σ-*rings*. Thanks to the following theorem many questions about signed measures may be reduced to the study of measures.

Theorem 7.12 *Let μ be a signed measure on a σ-ring $\mathcal{M}$.*

(a) *(Hahn decomposition)*[23] *There exists a decomposition $X = P \cup^* N$ such that $A \cap P, A \cap N \in \mathcal{M}$,*

$$\mu(A \cap P) \geq 0 \quad and \quad \mu(A \cap N) \leq 0$$

for every $A \in \mathcal{M}$. (See Fig. 7.1.)

(b) *(Jordan decomposition)*[24] *There exist two measures μ_+, μ_- on $\mathcal{M}$, satisfying the equality $\mu = \mu_+ - \mu_-$, concentrated on disjoint sets, and such that at least one of them is bounded.*

Remarks

- If $\mu = \mu_+ - \mu_-$ is a Jordan decomposition, then at least one of the measures μ_+ and μ_- is bounded. For otherwise there would exist two disjoint sets A, B with $\mu_+(A) = \mu_-(B) = \infty$ and $\mu_-(A) = \mu_+(B) = 0$, and then $\mu_+(A \cup B) - \mu_-(A \cup B)$ would not be defined.

---

[23] Hahn [180, p. 404].
[24] Jordan [229]. The decomposition is clearly unique.

- The assumption that $\mathcal{M}$ is a σ-ring is important: for example, the signed measure of Smolyanov has no Hahn decomposition. Indeed, for such a decomposition we should have $N = \varnothing$,[25] and then μ could not take negative values.
- Smolyanov's signed measure does not have a Jordan decomposition either. Indeed, if there were two measures μ_+, μ_- such that $\mu = \mu_+ - \mu_-$, then we would have

$$\mu_+(X) \geq \mu_+(A) \geq \mu(A) = |A|$$

and

$$\mu_-(X) \geq \mu_-(X \setminus A) \geq -\mu(X \setminus A) = |A|$$

for each finite set A. This would imply $\mu_+(X) = \mu_-(X) = \infty$ and then $\mu_+(X) - \mu_-(X)$ would not be defined.
- The preceding remarks imply that Smolyanov's finite signed measure cannot be extended to a signed measure defined on a σ-ring.[26] This contrasts with the case of finite *measures*.

The following lemma prepares the proof of the theorem.

Definition Let μ be a signed measure on $\mathcal{M}$. A set $A \in \mathcal{M}$ is called *negative* if $\mu(B) \leq 0$ for every subset of A, belonging to $\mathcal{M}$.

Lemma 7.13 *Let $\mu : \mathcal{M} \to \overline{\mathbb{R}}$ be a signed measure on a σ-ring $\mathcal{M}$.*

(a) *If $A, B \subset \mathcal{M}$ and $B \subset A$, then*

$$\mu(A) < \infty \Longrightarrow \mu(B) < \infty \quad and \quad \mu(A) > -\infty \Longrightarrow \mu(B) > -\infty.$$

(b) *If μ is finite, then it is bounded.*
(c) *μ is bounded from below or from above.*
(d) *If $A \in \mathcal{M}$ and $\mu(A) < 0$, then there exists a negative subset A' of A such that $\mu(A') \leq \mu(A)$.*

We will often use property (b) in the sequel.

Proof

(a) This follows from the equality $\mu(A) = \mu(B) + \mu(A \setminus B)$ because the sum is defined by definition.

---

[25]For otherwise we would have for every one-point set $A \subset N$ the impossible inequalities $1 = \mu(A) = \mu(A \cap N) \leq 0$.
[26]This also follows from Lemma 7.13 (c) below.

(b) If, for example, $\sup \mu = \infty$, then we may define recursively a set sequence (A_n) satisfying

$$\mu(A_n) > 1 + \sum_{k<n} \mu(A_k), \quad n = 1, 2, \ldots .$$

Then the sets $B_n := A_n \setminus \cup_{k<n} A_k$ are disjoint and $\mu(B_n) > 1$ for every n, so that $\mu(\cup B_n) = \infty$. Hence μ is not finite.

(c) For otherwise, by the proof of (b) there would be two sets satisfying $\mu(B) = \infty$ and $\mu(C) = -\infty$. Then $\mu(B \cup C)$ would not be defined: we cannot have $\mu(B \cup C) < \infty$ because $\mu(B) = \infty$, and we cannot have $\mu(B \cup C) > -\infty$ either because $\mu(C) = -\infty$.

(d) If A is a negative set, then we may take $A' := A$. Otherwise let k_1 be the smallest positive integer for which A has a subset A_1 satisfying $\mu(A_1) \geq 1/k_1$. We have

$$\mu(A) = \mu(A_1) + \mu(A \setminus A_1),$$

whence $\mu(A \setminus A_1) \leq \mu(A)$.[27]

If $A \setminus A_1$ is a negative set, then we may take $A' := A \setminus A_1$. Otherwise let k_2 be the smallest positive integer for which $A \setminus A_1$ has a subset A_2 satisfying $\mu(A_2) \geq 1/k_2$.

Continuing we obtain either a suitable negative set of the form

$$A' := A \setminus (A_1 \cup \cdots \cup A_n)$$

after a finite number of steps, or an infinite disjoint sequence $(A_n) \subset \mathcal{M}$, satisfying $\mu(A_n) \geq 1/k_n$ for all n with suitable positive integers k_n.

In the latter case we have

$$\sum \frac{1}{k_n} \leq \sum \mu(A_n) = \mu(\cup^* A_n) < \infty;$$

the last inequality follows by applying (a) with $B := \cup^* A_n \subset A$. It follows that $k_n \to \infty$.

Set $A' := A \setminus \cup^* A_n$, then $A' \in \mathcal{M}$ and

$$\mu(A) = \mu(A') + \mu(\cup^* A_n).$$

Consequently, $\mu(A') \leq \mu(A)$.

It remains to show that $B \in \mathcal{M}$ and $B \subset A'$ imply $\mu(B) \leq 0$. Since $k_n \to \infty$, we have $k_n \geq 2$ and (by construction) $\mu(B) < 1/(k_n - 1)$ for all sufficiently large n. Letting $n \to \infty$ we conclude that $\mu(B) \leq 0$.   □

---

[27] We may have equality if $\mu(A) = -\infty$.

Proof of Theorem 7.12

(a) By Lemma 7.13 (c) we may assume for example that μ does not take the value $-\infty$. Set

$$a = \inf \mu(A),$$

where A runs over the negative sets in $\mathcal{M}$; since $\varnothing$ is a negative set, $a \leq 0$.

Let (A_n) be a sequence of negative sets satisfying $\mu(A_n) \to a$. Then $N := \cup A_n \in \mathcal{M}$ is also a negative set, and $\mu(N) = a$. Since μ does not take the value $-\infty$, this implies that $a > -\infty$, i.e., a is finite.

Let $P = X \setminus N$, then $X = P \cup^* N$. Let $A \in \mathcal{M}$. Since $N \in \mathcal{M}$, we have

$$A \cap N \in \mathcal{M} \quad \text{and} \quad A \cap P = A \setminus (A \cap N) \in \mathcal{M},$$

and $\mu(A \cap N) \leq 0$ because N is negative. It remains to prove that $\mu(A \cap P) \geq 0$.

Assume on the contrary that $\mu(A \cap P) < 0$. Applying the preceding lemma, $A \cap P$ has a *negative* subset A' satisfying $\mu(A') \leq \mu(A \cap P)$. But then $N \cup^* A'$ is also negative, and the inequality

$$\mu(N \cup^* A') = \mu(N) + \mu(A') = a + \mu(A') < a$$

contradicts the definition of a.

(b) Assume again that μ does not take the value $-\infty$, and consider the Hahn decomposition $X = P \cup^* N$ with $N \in \mathcal{M}$, obtained in (a).

The formulas

$$\mu_+(A) := \mu(A \cap P) \quad \text{and} \quad \mu_-(A) := -\mu(A \cap N) \qquad (7.10)$$

define two measures satisfying $\mu = \mu_+ - \mu_-$, and concentrated on the disjoint sets P and N.

The measure μ_- is bounded because

$$\mu_-(A) = -\mu(A \cap N) \leq -\mu(N) = -a < \infty$$

for all $A \in \mathcal{M}$. $\square$

Remarks

• We stress that at least one of the two sets of the Hahn decomposition $X = P \cup^* N$ belongs to $\mathcal{M}$.

• It follows from the formulas (7.10) that

$$\mu_+(A) := \max \{\mu(B) \; : \; B \in \mathcal{M}, B \subset A\}$$

and

$$\mu_-(A) := -\min\{\mu(B) \,:\, B \in \mathcal{M}, B \subset A\}.$$

This alternative definition of the Jordan decomposition does not use the Hahn decomposition.

Definition The measures μ_+, μ_- are called the *positive and negative parts* of μ. The measure $|\mu| := \mu_+ + \mu_-$ is called the *total variation measure* of μ.

7.5 Lebesgue Decomposition

We have seen at the end of Sect. 6.2 that every function of bounded variation is the sum of an absolutely continuous and a singular function. We generalize this result for measures.

Similarly to the Hahn and Jordan decompositions, *in this section we consider only measures defined on σ-rings.* Hence the finite and bounded measures are the same.

Definitions Let μ, ν and σ be three measures on a σ-ring $\mathcal{N}$ in X.

- We say that ν is *absolutely continuous* with respect to μ, and we write $\nu \ll \mu$, if

$$\mu(A) = 0 \Longrightarrow \nu(A) = 0.$$

- We say that μ and σ are *singular*, and we write $\sigma \perp \mu$, if there is a partition $X = M \cup^* S$ of X such that

$$A \in \mathcal{N} \quad \text{and} \quad A \subset S \quad \Longrightarrow \quad \mu(A) = 0,$$
$$A \in \mathcal{N} \quad \text{and} \quad A \subset M \quad \Longrightarrow \quad \sigma(A) = 0.$$

Thus μ and σ are concentrated on the disjoint sets M and S.

In some cases an equivalent ε–δ definition holds:

*Lemma 7.14 *Let ν be absolutely continuous with respect to μ.*[28] *If ν is finite, then for every $\varepsilon > 0$ there exists a $\delta > 0$, that*

$$\mu(A) < \delta \Longrightarrow \nu(A) < \varepsilon.$$

---

[28] We recall that they are defined on a σ-ring.

Example The indefinite integral[29] ν of the function $x \mapsto 1/x$ with respect to the usual Lebesgue measure μ in $(0, 1)$ shows that the boundedness assumption cannot be omitted in the lemma.

Proof Assume on the contrary that there exist $\varepsilon > 0$ and a sequence (A_n) satisfying $\mu(A_n) < 2^{-n}$ and $\nu(A_n) \geq \varepsilon$ for every n. Then

$$A := \limsup A_n := \cap_{m=1}^{\infty} \cup_{n=m}^{\infty} A_n$$

satisfies $\mu(A) = 0$ and $\nu(A) \geq \varepsilon$, contradicting the relation $\nu \ll \mu$.

Indeed, the sets $B_m := A_m \cup A_{m+1} \cup \ldots$ form a *non-increasing* sequence such that

$$\mu(B_m) < \sum_{n=m}^{\infty} 2^{-n} = 2^{1-m} \quad \text{and} \quad \nu(B_m) \geq \nu(A_m) \geq \varepsilon$$

for all m. Since $\nu(B_1) < \infty$, letting $m \to \infty$ we get[30]

$$\mu(\cap B_m) = 0 \quad \text{and} \quad \nu(\cap B_m) \geq \varepsilon. \qquad \square$$

In order to state the main result of this section, we strengthen the σ-finiteness property:

Definition A measure $\varphi : \mathcal{N} \to \overline{\mathbb{R}}$ is *strongly σ-finite* if there exists a countable set sequence $(P_n) \subset \mathcal{N}$ such that $\varphi(P_n)$ is finite for all n, and $\varphi(A) = 0$ for all $A \in \mathcal{N}$, disjoint from $\cup P_n$.

If this is the case, we may assume that the sequence (P_n) is disjoint.

Examples

- Every *finite* measure φ is strongly σ-finite. Indeed, it suffices to choose a sequence $(P_n) \subset \mathcal{N}$ satisfying $\varphi(P_n) \to \sup \varphi$. If $A \in \mathcal{N}$ is disjoint from $\cup P_n$, then

$$\varphi(P_n) + \varphi(A) = \varphi(P_n \cup^* A) \leq \sup \varphi$$

  and hence $\varphi(A) \leq \sup \varphi - \varphi(P_n)$ for all n. Since $\sup \varphi < \infty$ (because every finite measure on a σ-ring is bounded), letting $n \to \infty$ we conclude that $\varphi(A) = 0$.
- If X is φ-measurable, then φ is strongly σ-finite by Lemma 7.5 (p. 220).
- The counting measure (p. 213) on the σ-ring of the countable subsets of an uncountable set X is σ-finite, but not strongly σ-finite.

---

[29]See Proposition 7.11, p. 230.
[30]See Proposition 7.3, p. 216.

Theorem 7.15 (Lebesgue Decomposition)[31] *Let μ and φ be two measures on a σ-ring $\mathcal{N}$. If φ is strongly σ-finite, then it has a unique decomposition $\varphi = \nu + \sigma$ with two measures ν and σ satisfying*

$$\nu \ll \mu \quad and \quad \sigma \perp \mu.$$

Proof Proof of existence for bounded measures φ. Set

$$\alpha := \sup\{\varphi(S) \; : \; S \in \mathcal{N} \quad and \quad \mu(S) = 0\} < \infty.$$

The upper bound is attained. Indeed, consider a maximizing sequence (S_n): $\mu(S_n) = 0$ for all n, and $\varphi(S_n) \to \alpha$. Then $S := \cup S_n$ belongs to $\mathcal{N}$, $\mu(S) = 0$ and $\varphi(S) = \alpha$.

The formulas

$$\sigma(A) := \varphi(A \cap S) \quad and \quad \nu(A) := \varphi(A \setminus S)$$

define two measures σ and ν on $\mathcal{N}$ such that $\varphi = \nu + \sigma$. Furthermore, if $A \in \mathcal{N}$, then

$$\sigma(A \setminus S) = \varphi((A \setminus S) \cap S) = \varphi(\varnothing) = 0,$$

so that $\sigma \perp \mu$ with $M := X \setminus S$.

If $\mu(A) = 0$, then $\mu(A \cup S) = 0$, and hence $\varphi(A \cup S) \le \alpha = \varphi(S)$ by the definition of S. Consequently,

$$0 \le \nu(A) = \varphi(A \setminus S) = \varphi(A \cup S) - \varphi(S) \le 0,$$

whence $\nu(A) = 0$. This proves the relation $\nu \ll \mu$.

Proof of Existence in the General Case. Fix a disjoint sequence $(P_n) \subset \mathcal{N}$ such that $\varphi(P_n) < \infty$ for all n, and $\varphi = 0$ outside $P := \cup^* P_n$. Applying the preceding step for each P_n, we obtain a sequence of sets $S_n \subset P_n$ satisfying $\mu(S_n) = 0$ and the implications

$$A \subset P_n \quad and \quad \mu(A) = 0 \quad \Longrightarrow \quad \varphi(A \setminus S_n) = 0.$$

Set $S = \cup^* S_n$, and define

$$\sigma(A) := \varphi(A \cap S), \quad \nu(A) := \varphi(A \setminus S)$$

---

[31]Lebesgue [295, pp. 232–249].

for all $A \in \mathcal{N}$. We have $\varphi = \nu + \sigma$. Furthermore, we have $\sigma \perp \mu$ because

$$\mu(S) = \sum \mu(S_n) = 0, \quad \text{and} \quad \sigma(A \setminus S) = \varphi(\varnothing) = 0$$

for all $A \in \mathcal{N}$.

Finally, we have $\nu \ll \mu$ because if $A \in \mathcal{N}$ and $\mu(A) = 0$, then

$$\nu(A) = \varphi(A \setminus S)$$

$$= \sum \varphi((A \cap P_n) \setminus S)$$

$$= \sum \varphi((A \cap P_n) \setminus S_n)$$

$$= 0.$$

In the last step we have $\varphi((A \cap P_n) \setminus S_n) = 0$ by the choice of S_n.

Uniqueness. We may assume by decomposition that φ is bounded. We have to show that if two measures ν' and σ' on $\mathcal{N}$ satisfy

$$\nu' \ll \mu, \quad \sigma' \perp \mu \quad \text{and} \quad \varphi = \nu' + \sigma',$$

then the signed measure $\varrho := \nu' - \nu = \sigma - \sigma'$ vanishes identically.

Consider the corresponding partitions $X = M \cup^* S = M' \cup^* S'$. For each $A \in \mathcal{N}$ we have

$$\mu(A \cap (S \cup S')) \leq \mu(S) + \mu(S') = 0;$$

by the absolute continuity of ν' and ν this yields $\varrho(A \cap (S \cup S')) = 0$.

On the other hand, the inclusion $A \setminus (S \cup S') \subset M \cap M'$ implies that

$$\sigma(A \setminus (S \cup S')) = \sigma'(A \setminus (S \cup S')) = 0,$$

and therefore $\varrho(A \setminus (S \cup S')) = 0$. Consequently,

$$\varrho(A) = \varrho(A \setminus (S \cup S')) + \varrho(A \cap (S \cup S')) = 0. \qquad \qquad \square$$

Remark The proof shows that $S \in \mathcal{N}$.

Example The strong σ-finiteness condition cannot be omitted: for example, the counting measure φ has no Lebesgue decomposition with respect to the zero-one measure μ.[32]

Indeed, assume on the contrary that there is such a decomposition $\varphi = \nu + \sigma$, where μ and σ are concentrated on the disjoint sets $M, S \subset X$. If A is a countable set,

---

[32] See p. 213.

then the relation $\nu \ll \mu$ implies that $\nu(A) = 0$, and therefore $\sigma(A) = \varphi(A) = |A|$. Hence $S = X$, and thus $\mu(X) = 0$ by the definition of singularity, contradicting the definition of μ.

Remark The above definitions of absolute continuity, singularity and strong σ-finiteness remain meaningful for signed measures.[33] Theorem 7.15 may be generalized to the case where μ is still a measure but φ is a strongly σ-finite *signed measure*: there exists a unique decomposition $\varphi = \nu + \sigma$ with *signed measures* ν and σ satisfying $\nu \ll \mu$ and $\sigma \perp \mu$.

Indeed, applying the theorem to the positive and negative parts of φ we obtain four measures $\nu_\pm, \sigma_\pm$ satisfying the relations

$$\varphi_+ = \nu_+ + \sigma_+, \quad \varphi_- = \nu_- + \sigma_-,$$

$$\nu_+ \ll \mu, \quad \nu_- \ll \mu,$$

and two partitions $X = M_+ \cup *S_+ = M_- \cup *S_-$ with $S_\pm \in \mathcal{N}$ such that σ_+, σ_- and μ are concentrated on S_+, S_- and $M := M_+ \cap M_-$, respectively.

It follows that

- $\nu := \nu_+ - \nu_- \ll \mu$;
- μ and $\sigma := \sigma_+ - \sigma_-$ are concentrated on M and $S := S_+ \cup S_- = X \setminus M$, respectively;
- $\varphi = \nu + \sigma$.

The proof of the uniqueness of the decomposition, given above, remains valid for signed measures.

7.6 The Radon–Nikodým Theorem

As usual, we consider the integral associated with a finite measure defined on a semiring in X. We denote by $\mathcal{M}$ the σ-ring of measurable sets, and by $\mu : \mathcal{M} \to \overline{\mathbb{R}}$ the extended measure.

Equivalently, in view of Proposition 7.7 (p. 223), let $\mu : \mathcal{M} \to \overline{\mathbb{R}}$ be a σ-finite, complete measure on a σ-ring $\mathcal{M}$.

In this section the expressions "integrable", "absolutely continuous", "a.e." will be meant with respect to μ.

If f is a nonnegative, integrable function, then its indefinite integral

$$\nu(A) := \int_A f \, d\mu, \quad A \in \mathcal{M} \tag{7.11}$$

---

[33] However, the present definition of absolute continuity is interesting only if μ is a measure.

is a bounded measure by Proposition 7.11 (p. 230). Moreover, ν is absolutely continuous because

$$\mu(A) = 0 \Longrightarrow f\chi_A = 0 \text{ a.e.} \Longrightarrow \nu(A) = \int f\chi_A \, d\mu = 0.$$

The converse often holds true:

Theorem 7.16 (Radon–Nikodým)[34] *If μ is strongly σ-finite, then the formula (7.11) establishes a one-to-one correspondence between*

- *nonnegative, integrable functions*

and

- *absolutely continuous, bounded measures.*

Definition The function f of the theorem is called the *Radon–Nikodým derivative* of ν with respect to μ, and it is denoted by $d\nu/d\mu$.

Example Let us explain the terminology. If $F : [a, b] \to \mathbb{R}$ is an absolutely continuous function on a compact interval $[a, b]$ as discussed in Chap. 6, then the formula

$$\nu(I) := F(d) - F(c), \quad I = [c, d)$$

defines a signed measure on the semiring of half-open intervals $[a, b)$, and

$$\nu(I) = \int_I F' \, dx$$

for all these intervals by Theorem 6.5 (p. 204). Hence Theorem 7.16 is a far-reaching generalization of the Lebesgue–Vitali theorem, itself a generalization of the Newton–Leibniz formula.

Proof of Theorem 7.16[35]

It remains to show that every absolutely continuous, bounded measure ν is the indefinite integral of a unique nonnegative, integrable function f.

It is sufficient to consider the case where $\mu(X) < \infty$. Indeed, in the general case there exists a disjoint set sequence $(X_n) \subset \mathcal{M}$ such that $\mu(X_n) < \infty$ for all n,

---

[34]Radon [366, pp. 1342–1351] and Nikodým [342, pp. 167–179]. We recall from the preceding section that the strong σ-finiteness condition is satisfied if μ is a finite measure or if X is measurable.

[35]See also an alternative proof of von Neumann [339, pp. 124–130], based on the orthogonal projection in Hilbert spaces.

and $\mu = 0$ outside $\cup^* X_n$. Applying the result for each X_n, we obtain nonnegative, integrable functions f_n on X_n, satisfying

$$\nu(A) := \int_A f_n \, d\mu \qquad (7.12)$$

for all measurable sets $A \subset X_n$. Defining $f := f_n$ on X_n for all n, and $f := 0$ outside $\cup^* X_n$, we obtain a nonnegative, measurable function satisfying (7.11). Moreover, f is integrable because

$$\int |f| \, d\mu = \sum_n \int_{X_n} f_n \, d\mu = \sum_n \nu(X_n) = \nu(X) < \infty.$$

The uniqueness of f follows from the uniqueness of each f_n because (7.11) implies the relations (7.12) with $f_n = f|_{X_n}$, and from the fact that every measurable set outside $\cup^* X_n$ is a null set.

In view of this remark we assume henceforth that $\mu(X) < \infty$.

Proof of the Uniqueness. Two different integrable functions f and g have different indefinite integrals. Indeed, at least one of the sets $A := \{f > g\}$ and $B := \{f < g\}$ has a positive measure. If for example $\mu(A) > 0$, then

$$\int_A f \, d\mu - \int_A g \, d\mu = \int_A (f - g) \, d\mu > 0$$

and therefore

$$\int_A f \, d\mu \neq \int_A g \, d\mu.$$

We prove a technical lemma: If $\nu \neq 0$, then there exist $A \in \mathcal{M}$ and $\varepsilon > 0$ such that $\mu(A) > 0$, and

$$\varepsilon\mu(B) \leq \nu(B) \quad \text{for all measurable subsets} \quad B \subset A.$$

For the proof we consider for each $n = 1, 2, \ldots$ the Hahn decomposition of the signed measure $\nu - n^{-1}\mu$, and we set

$$P = \cup P_n, \quad N = \cap N_n.$$

Since $\nu - n^{-1}\mu$ is bounded from above, we have $P_n \in \mathcal{M}$ for every n.[36] It remains to find some n with $\mu(P_n) > 0$ because then we may choose $A := P_n$ and $\varepsilon := 1/n$.

We have $\nu(B) = 0$ for every measurable set $B \subset N$ because $\mu(B) < \infty$, and

$$0 \leq \nu(B) \leq \frac{1}{n}\mu(B)$$

---

[36]See the remark on p. 238

for all n because $N \subset N_n$. Since $\nu \neq 0$, $\nu(P) > 0$, and then $\mu(P) > 0$ by the absolute continuity of ν.

Finally, since

$$0 < \mu(P) \leq \sum \mu(P_n)$$

by σ-subadditivity, we have $\mu(P_n) > 0$ for at least one n.

Proof of the Existence. Let us denote by $\mathcal{F}$ the family of nonnegative, integrable functions f satisfying

$$\int_A f \, d\mu \leq \nu(A)$$

for all $A \in \mathcal{M}$. Since ν is bounded and $0 \in \mathcal{F}$, the formula

$$\alpha := \sup_{f \in \mathcal{F}} \int f \, d\mu$$

defines a finite, nonnegative number.

The upper bound is attained. For the proof we choose a maximizing sequence $(f_n) \in \mathcal{F}$ satisfying

$$\int f_n \, d\mu \to \alpha.$$

Then $g_n := \max\{f_1, \dots, f_n\} \in \mathcal{F}$ for each n. Indeed, every set $A \in \mathcal{M}$ has a partition $A_1 \cup^* \cdots \cup^* A_n$ such that $g_n = f_j$ on each A_j, and then

$$\int_A g_n \, d\mu = \sum_{j=1}^n \int_{A_j} f_j \, d\mu \leq \sum_{j=1}^n \nu(A_j) = \nu(A).$$

Applying the Beppo Levi theorem, the functions g_n converge a.e. to a nonnegative, integrable function f. Applying the Fatou lemma (or again the Beppo Levi theorem) for the sequences $(\chi_A g_n)$, we infer from the inequalities $\int_A g_n \, d\mu \leq \nu(A)$ that $f \in \mathcal{F}$. Finally, the relations $f_n \leq g_n \leq f$ and $\int f_n \, d\mu \to \alpha$ imply the equality $\int f \, d\mu = \alpha$.

To end the proof we show that the measure

$$\nu_0(A) := \nu(A) - \int_A f \, d\mu, \quad A \in \mathcal{M}$$

vanishes identically. Assume on the contrary that $\nu_0 \neq 0$. Then by the above lemma there exist $A \in \mathcal{M}$ and $\varepsilon > 0$ satisfying $\mu(A) > 0$, and

$$\varepsilon \mu(A \cap B) \leq \nu(A \cap B) - \int_{A \cap B} f \, d\mu$$

for all $B \in \mathcal{M}$. Since $f \in \mathcal{F}$ implies

$$0 \leq \nu(B \setminus A) - \int_{B \setminus A} f \, d\mu,$$

adding the two equalities we get

$$\varepsilon \mu(A \cap B) \leq \nu(B) - \int_B f \, d\mu,$$

i.e.,

$$\int_B f + \varepsilon \chi_A \, d\mu \leq \nu(B)$$

for all $B \in \mathcal{M}$. Hence $f + \varepsilon \chi_A \in \mathcal{F}$. This, however, is impossible because

$$\int f + \varepsilon \chi_A \, d\mu = \int f \, d\mu + \varepsilon \mu(A) = \alpha + \varepsilon \mu(A) > \alpha. \qquad \square$$

Example We show[37] that the strong σ-finiteness assumption cannot be omitted in Theorem 7.16 (p. 240).

Let $Z = X \times Y$ with two uncountable sets X, Y satisfying $\operatorname{card} X > \operatorname{card} Y$. A set $L \subset Z$ is called a *vertical line* if there exists an $x \in X$ such that both sets $L \setminus (\{x\} \times Y)$ and $(\{x\} \times Y) \setminus L$ are countable.

Similarly, a set $L \subset Z$ is called a *horizontal line* if there exists a $y \in Y$ such that both sets $L \setminus (X \times \{y\})$ and $(X \times \{y\}) \setminus L$ are countable.

The countable unions of vertical lines, horizontal lines and points form a σ-ring $\mathcal{M}$. Denoting by $\mu(A)$ the number of lines contained in A, we obtain a complete, σ-finite[38] (but not strongly σ-finite) measure $\mu : \mathcal{M} \to \overline{\mathbb{R}}$, for which the null sets are the countable sets.

Denoting by $\nu(A)$ the number of vertical lines contained in A, we obtain another measure $\nu : \mathcal{M} \to \overline{\mathbb{R}}$, satisfying $\nu \leq \mu$ and hence $\nu \ll \mu$. We claim that the Radon–Nikodým derivative $\partial \nu / \partial \mu$ does not exist.

Assume on the contrary that there exists a measurable function $f : Z \to \overline{\mathbb{R}}$ satisfying

$$\nu(L) = \int_L f \, d\mu \quad \text{for every line} \quad L. \tag{7.13}$$

---

[37] See Halmos [184, pp. 131–132]. In this example we use the notion of cardinality of infinite sets, but we need only the simplest results: see, e.g., Halmos [186] or Vilenkin [467, 468].

[38] Because every measurable set is covered by countably many lines.

By the measurability condition f is constant a.e. on each line L, and $\int_L f\, d\mu$ is equal to this constant. Therefore we infer from (7.13) that $f = 1$ a.e. on every vertical line, and $f = 0$ a.e. on every horizontal line. This implies the inequalities

$$\operatorname{card} X \le \operatorname{card} \{x \in Z \,:\, f(x) = 1\} \le \operatorname{card} Y,$$

contradicting the choice of X and Y.

We may further generalize the preceding theorem for unbounded and even signed measures ν:

---

*Theorem 7.17 *If μ is strongly σ-finite, then the formula (7.11) establishes a one-to-one correspondence between*

- *the functions f having an integral*

and

- *the absolutely continuous signed measures ν.*

Remark It is easy to see that

$$\nu \quad \text{is a measure} \quad \Longleftrightarrow \quad f \quad \text{is nonnegative.}$$

Indeed, if $f \ge 0$, then ν is a measure because $f\chi_A \ge 0$ for every $A \in \mathcal{M}$, and therefore $\nu(A) = \int f\chi_A\, d\mu \ge 0$. Conversely, if $f < 0$ on some set A of positive measure, then $\nu(A) = \int f\chi_A\, d\mu < 0$, and therefore ν is not a measure.

Proof of Theorem 7.17 It follows again from Proposition 7.11 that if f has an integral, then the indefinite integral is an absolutely continuous signed measure.

It remains to prove that each absolutely continuous signed measure ν is the indefinite integral of a unique measurable function f. Similarly to the preceding proof we may assume that $\mu(X) < \infty$.

Proof of the Uniqueness of f. Let f and g be two different functions whose integrals are defined. We have to find a set A such that $\mu(A) > 0$, and either $f > g$ on A or $f < g$ on A.

Assume by symmetry that $B := \{f > g\}$ is not a null set. Since $f > -\infty$ and $g < \infty$ on B, setting

$$A_k := \{x \in B \,:\, f(x) > -k \quad \text{and} \quad g(x) < k\}$$

we have

$$\cup_k A_k = B.$$

Since $0 < \mu(B) \le \mu(X) < \infty$, there exists a k such that $0 < \mu(A_k) < \infty$. Then

$$\int_{A_k} f \, d\mu \ge -k\mu(A_k) > -\infty \quad \text{and} \quad \int_{A_k} g \, d\mu \le k\mu(A_k) < \infty.$$

Consequently, the integral $\int_{A_k} (f - g) \, d\mu$ exists,[39] and hence

$$\int_{A_k} f \, d\mu - \int_{A_k} g \, d\mu = \int_{A_k} (f - g) \, d\mu > 0.$$

A technical lemma:[40] *if ν is an absolutely continuous measure, then there exists a disjoint sequence (F_n) of sets of finite ν-measure such that for each measurable set A, disjoint from $F := \cup F_n$, we have either $\mu(A) = 0$ or $\nu(A) = \infty$ (or both).*

For the proof we denote by $\mathcal{A}$ the σ-ring of measurable sets having a countable cover by sets of finite ν-measure. The upper bound

$$\alpha := \sup\{\mu(B) \,:\, B \in \mathcal{A}\} \le \mu(X) < \infty$$

is attained on some set $F \in \mathcal{A}$ because if $(B_n) \subset \mathcal{A}$ and $\mu(B_n) \to \alpha$, then $F := \cup B_n \in \mathcal{A}$ and $\mu(B_n) \le \mu(F)$ for all n, i.e., $\mu(F) = \alpha$.

Consider a measurable set A, disjoint from F and satisfying $\nu(A) < \infty$. Since $F \cup^* A \in \mathcal{A}$, we have

$$\alpha \ge \mu(F \cup^* A) = \mu(F) + \mu(A) = \alpha + \mu(A);$$

since α is finite, we conclude that $\mu(A) = 0$.

Proof of the Existence When ν is a Measure. Consider the disjoint set sequence (F_n) of the previous step, and set $E := X \setminus \cup F_n$. Apply the already proved result for each F_n, and denote by f_n the corresponding Radon–Nikodým derivatives.

Setting $f := f_n$ on each F_n and $f := \infty$ on E we get a nonnegative, measurable function. Each $A \in \mathcal{M}$ is the disjoint union of the sets $A \cap F_n$ and $A \cap E$, and

$$\nu(A \cap F_n) = \int_{A \cap F_n} f_n \, d\mu$$

for every n by the choice of f_n. It remains to show that

$$\nu(A \cap E) = \int_{A \cap E} \infty \, d\mu.$$

Indeed, then adding all these equalities we obtain (7.11).

---

[39] See Proposition 5.17 (d), p. 190.
[40] See Hewitt–Stromberg [207, p. 317].

If $\nu(A \cap E) = \infty$, then $\mu(A \cap E) > 0$ by the absolute continuity of ν, and hence $\int_{A \cap E} \infty \, d\mu = \infty$. Otherwise we have $\mu(A \cap E) = 0$ by the definition of E; hence clearly $\int_{A \cap E} \infty \, d\mu = 0$, while $\nu(A \cap E) = 0$ by the absolute continuity.

Proof of Existence when ν is a Signed Measure. Applying the preceding result to the measures ν_+, ν_- of the Jordan decomposition $\nu = \nu_+ - \nu_-$, we obtain two nonnegative, measurable functions f_+, f_- satisfying (7.11) with $f_\pm$ and $\nu_\pm$ instead of f and ν. Since at least one of the measures ν_+ and ν_- is bounded,[41] at least one of the functions f_+, f_- is integrable, so that the function $f := f_+ - f_-$ and the integral $\int f \, d\mu$ are defined. Taking the difference of the equalities for ν_+ and ν_- we obtain (7.11) for f and ν. $\qquad\square$

Using the Radon–Nikodým theorem we may greatly generalize the change of variable formula of integration[42]:

Proposition 7.18 *Assume that μ is strongly σ-finite, and let $\nu \ll \mu$ be an absolutely continuous measure. Then*

$$\int g \frac{d\nu}{d\mu} \, d\mu = \int g \, d\nu \qquad (7.14)$$

whenever the right-hand integral exists.

Proof We may assume as usual that $\mu(X) < \infty$.

We write $f := d\nu/d\mu$ for brevity.

(i) The set $X_0 := \{x \in X \ : \ f(x) = 0\}$ satisfies the equality

$$\nu(X_0) = \int_{X_0} f \, d\mu = \int_{X_0} 0 \, d\mu = 0$$

and hence

$$\int_{X_0} gf \, d\mu = \int_{X_0} 0 \, d\mu = 0 = \int_{X_0} g \, d\nu$$

for all ν-measurable functions g.[43] Therefore, changing X to $X \setminus X_0$ we may assume that $f > 0$. Then the μ-null sets and ν-null sets are the same by (7.11), so that we may use the expression a.e. without mentioning the corresponding measure μ or ν.

(ii) Since μ is bounded, every ν-step function is also a μ-step function, and hence every ν-measurable function is also μ-measurable.

---

[41]See Lemma 7.13 (b), p. 232.

[42]The proposition extends classical results of Euler [130, p. 303], Lagrange [280, p. 624] and Jacobi [224, p. 436].

[43]They are also ν-measurable because $\mu(X) < \infty$.

If g is the characteristic function of a set $A \in \mathcal{M}$, then (7.14) reduces to the equality (7.11). Taking linear combinations it follows that (7.14) holds for all ν-step functions.

If (g_n) is a sequence of ν-step functions satisfying $g_n \nearrow g$ a.e., then $(g_n f)$ is a sequence of μ-measurable functions satisfying $g_n f \nearrow g f$ a.e. Applying the generalized Beppo Levi theorem[44] to the sequences $(g_n - g_1)$ and $(g_n f - g_1 f)$ we get the equality (7.14).

In the general case the equality (7.14) holds for g_+ and g_- instead of g. Taking the difference of these equalities we get (7.14) for g. $\qquad\qquad$ □

7.7 * Local Measurability

As usual, we consider an integral associated with a finite measure defined on a semiring $\mathcal{P}$. We denote by $\mathcal{M}$ the σ-ring of measurable sets and by $\mu : \mathcal{M} \to \overline{\mathbb{R}}$ the extended measure.

In the terminology of this chapter the constant functions are not necessarily measurable. In such cases the non-zero constant functions have no integral, and the measure of X is not defined either. We are going to extend the notions of the integral and the measure so as to deal with these cases in particular.

Definition A function f is *locally measurable* if $f \chi_P$ is measurable for every $P \in \mathcal{P}$.

Remarks

- Measurability implies local measurability.
- The constant functions are locally measurable. If they are also measurable, then the notions of measurability and local measurability coincide. This is the case for $X = \mathbb{R}$, studied in Chap. 5, more generally for $X = \mathbb{R}^N$, and for the probability measures.
- If f is locally measurable, then the product fg is measurable for every measurable function g. For step functions g this follows at once from the definition. In the general case we choose a sequence (φ_n) of step functions converging to g a.e. Then the functions $f \varphi_n$ are measurable, and they converge to fg a.e., so that fg is measurable as well.

An easy adaptation of the proof of Proposition 5.16 (p. 187) leads to the following

Proposition 7.19

(a) *The constant functions are locally measurable.*
(b) *If f is locally measurable, and $f = g$ a.e., then g is locally measurable.*

---

[44]Proposition 5.17 (e), p. 190.

(c) *If* $F : \mathbb{R}^N \to \mathbb{R}$ *is continuous, and* $f_1, \dots, f_N$ *are finite-valued, locally measurable functions, then the composite function* $h := F(f_1, \dots, f_N)$ *is locally measurable.*

 In particular, if f, g *are finite-valued, locally measurable functions, then* $|f|$, $f + g, f - g, fg, \max\{f, g\}$ *and* $\min\{f, g\}$ *are locally measurable as well.*

(d) *If* f *is locally measurable and* $f \neq 0$ *a.e., then* $1/f$ *is locally measurable.*

(e) *If* f *is locally measurable, g is integrable, and* $|f| \leq g$ *a.e., then* f *is integrable.*

(f) *If a sequence of locally measurable functions converges to* f *a.e., then* f *is also locally measurable.*

Next we generalize the integral:

Definition Let f be a locally measurable function.

- If f is nonnegative and non-integrable, then we define $\int f \, dx := \infty$.
- If at least one of f_+ and f_- is integrable, then we define

$$\int f \, dx := \int f_+ \, dx - \int f_- \, dx.$$

Remarks

- If f is measurable, then the new definition reduces to the earlier one.
- If neither f_+ nor f_- is integrable, then the right-hand sum is undefined.
- We still keep the adjective "integrable" for the case where the integral is finite.

Proposition 5.17 (p. 190) on the integration rules remains valid; we only have to use the *local* measurability of h in the proof of (d) instead of its measurability.

After the integral we generalize the measure:

Definition A set A is *locally measurable* if its characteristic function is locally measurable, i.e., if $A \cap P \in \mathcal{M}$ for every $P \in \mathcal{P}$.

Remark The fundamental set X is always locally measurable.[45]

The following notion will be useful in the sequel:

Definition A *σ-algebra* in X is a σ-ring containing X. Explicitly, a set system $\mathcal{M}$ in X is a σ-algebra if the following conditions are satisfied:

- $\emptyset \in \mathcal{M}$;
- if $A \in \mathcal{M}$, then $X \setminus A \in \mathcal{M}$;
- if (A_n) is a disjoint sequence in $\mathcal{M}$, then $\cup^* A_n \in \mathcal{M}$.

Examples

- $\{\emptyset, X\}$ and 2^X are σ-algebras in X.

---

[45] We recall from Lemma 7.5 (p. 220) that X is measurable $\Longleftrightarrow$ it has a countable cover by sets of $\mathcal{P}$ (and hence of finite measure).

- The usual Lebesgue measurable sets of $\mathbb{R}$ form a σ-algebra.
- The countable subsets of an uncountable set X form a σ-ring, but not a σ-algebra.

An easy adaptation of the proof of Proposition 5.19 (p. 194) leads to

Proposition 7.20

(a) *The locally measurable sets form a σ-algebra.*
(b) *f is locally measurable $\Longleftrightarrow$ the sets*

$$\{f > c\}, \quad \{f < c\}, \quad \{f \geq c\}, \quad \{f \leq c\}$$

are locally measurable for all $c \in \overline{\mathbb{R}}$.

Remark The local measurability of $\{f > c\}$ for all $c \in \mathbb{R}$ already implies the local measurability of f. This follows from the relations

$$\{f > -\infty\} = \cup_{n=1}^{\infty}\{f > -n\},$$
$$\{f > \infty\} = \emptyset \in \overline{\mathcal{M}},$$
$$\{f \geq c\} = \cap_{n=1}^{\infty}\{f > c - 1/n\},$$
$$\{f < c\} = X \setminus \{f \geq c\},$$
$$\{f \leq c\} = X \setminus \{f > c\}.$$

Three similar statements are obtained by changing $\{f > c\}$ to $\{f < c\}$, $\{f \geq c\}$ or $\{f \leq c\}$.

We extend the measure μ to the *σ-algebra* $\overline{\mathcal{M}}$ of locally measurable sets by setting

$$\overline{\mu}(A) := \int \chi_A \, d\mu.$$

Observe that $\overline{\mu}(A) = \infty$ for every $A \in \overline{\mathcal{M}} \setminus \mathcal{M}$.

Now we clarify the relationship between integrals and arbitrary measures. The following result complements Proposition 7.7 (p. 223):

*Proposition 7.21

(a) *$\overline{\mathcal{M}}$ is a σ-algebra, and $\overline{\mu} : \overline{\mathcal{M}} \to \overline{\mathbb{R}}$ is complete.*
(b) *Every measure, defined on a semiring, is the restriction of the measure $\overline{\mu} :$ $\overline{\mathcal{M}} \to \overline{\mathbb{R}}$ associated with its finite part.*

Proof

(a) We already know that $\overline{\mathcal{M}}$ is a σ-algebra. The completeness of $\overline{\mu} : \overline{\mathcal{M}} \to \overline{\mathbb{R}}$ follows from that of $\mu : \mathcal{M} \to \overline{\mathbb{R}}$ because $\overline{\mu}(A) = \infty$ and thus $\overline{\mu}(A) \neq 0$ for all $A \in \overline{\mathcal{M}} \setminus \mathcal{M}$.

(b) Let $\nu : \mathcal{N} \to \overline{\mathbb{R}}$ be a measure on a semiring,

$$\mathcal{P} := \{A \in \mathcal{N} \: : \: \nu(P) < \infty\},$$

and $\overline{\mu} : \overline{\mathcal{M}} \to \overline{\mathbb{R}}$ the measure obtained by the usual extension of $\mu := \nu|_{\mathcal{P}}$. We have to show that $\mathcal{N} \subset \overline{\mathcal{M}}$ and $\nu(A) = \overline{\mu}(A)$ for every $A \in \mathcal{N}$.

First we observe the implication

$$A \in \mathcal{N} \quad \text{and} \quad P \in \mathcal{P} \Longrightarrow A \cap P \in \mathcal{P}. \tag{7.15}$$

Indeed, since $\mathcal{P} \subset \mathcal{N}$ and $\mathcal{N}$ is a semiring, we have $A \cap P \in \mathcal{N}$. Furthermore, $\nu(A \cap P) \le \nu(P) < \infty$ and therefore $A \cap P \in \mathcal{P}$.

Since $\mathcal{P} \subset \mathcal{M}$, (7.15) implies that every $A \in \mathcal{N}$ is locally measurable, i.e., $\mathcal{N} \subset \overline{\mathcal{M}}$.

It remains to show that $\nu(A) = \overline{\mu}(A)$ for every $A \in \mathcal{N}$. We distinguish the cases $A \in \mathcal{M}$ and $A \in \overline{\mathcal{M}} \setminus \mathcal{M}$.

If $A \in \mathcal{N} \cap \mathcal{M}$, then A has a disjoint cover by sets $P_n \in \mathcal{P}$. Changing each P_n to $A \cap P_n$ by (7.15), we may also assume that $A = \cup^* P_n$. Since $\overline{\mu}(P_n) = \nu(P_n)$ for every n by the definition of $\overline{\mu}$, it follows that

$$\overline{\mu}(A) = \sum \overline{\mu}(P_n) = \sum \nu(P_n) = \nu(A).$$

If $A \in \mathcal{N}$ and $A \in \overline{\mathcal{M}} \setminus \mathcal{M}$, then $\overline{\mu}(A) = \infty$ by the definition of $\overline{\mu}$. Furthermore, $A \notin \mathcal{P}$ because $\mathcal{P} \subset \mathcal{M}$, and therefore $\nu(A) = \infty$ by the definition of $\mathcal{P}$. Hence $\overline{\mu}(A) = \nu(A)$ again.   □

Remark In view of part (b) of the proposition we may speak about the integral associated with an arbitrary measure, meaning the integral associated with its finite part.

By the results of this section it is tempting to use local measurability and the measure $\overline{\mu} : \overline{\mathcal{M}} \to \overline{\mathbb{R}}$ instead of measurability and the measure $\mu : \mathcal{M} \to \overline{\mathbb{R}}$.[46] The following observations, however, convinced the author to return to the original definitions of Fréchet and Riesz[47]:

• Tonelli's theorem on successive integration (p. 228) does not hold for *locally* measurable functions having an integral: the function f in the last example of Sect. 7.4 is locally measurable.

---

[46]Indeed, this choice is taken by most contemporary textbooks by defining measurability using inverse images. While Hausdorff's elegant characterization of continuous functions by inverse images of open or closed sets is extremely useful in topology, the analogous definition of measurability leads to several annoying counterexamples.

[47]Fréchet [158] and Riesz–Sz.-Nagy [394].

- Proposition 7.6 (p. 221) on the unique extension of measures does not remain valid for the σ-algebra $\overline{\mathcal{M}}$. To see this we consider the zero measure μ on the semiring $\mathcal{P}$ of finite subsets of an uncountable set X. Then $\overline{\mathcal{M}} = 2^X$, and

$$\overline{\mu}(A) = \begin{cases} 0 & \text{if } A \text{ is countable,} \\ \infty & \text{if } A \text{ is uncountable.} \end{cases}$$

But the zero measure on 2^X is also an extension of μ!

Moreover, the two measures already differ on the smallest σ-algebra $\mathcal{N}$ containing $\mathcal{M}$, i.e., on the family of countable subsets and their complements. In fact,[48] there are infinitely many other extensions of μ to $\mathcal{N}$: the formula

$$\mu_\alpha(A) = \begin{cases} 0 & \text{if } A \text{ is countable,} \\ \alpha & \text{if } X \setminus A \text{ is countable} \end{cases}$$

defines an extension of μ for each $0 \le \alpha \le \infty$.

- The first part of the Radon–Nikodým theorem remains valid for locally measurable functions: if a locally measurable function has an integral, then the formula

$$\nu(A) := \int_A f \, d\mu$$

defines an absolutely continuous signed measure $\nu : \mathcal{M} \to \mathbb{R}$, and even $\nu : \overline{\mathcal{M}} \to \overline{\mathbb{R}}$.

However, in the counterexample on p. 243 the Radon–Nikodým derivative $f = d\nu/d\mu$ does not exist, even if we allow f to be only locally measurable.

7.8 Exercises

Exercise 7.1 For each measure μ introduced in the examples on p. 213, determine its finite part, the σ-ring $\mathcal{M}$ of measurable sets, and the σ-algebra $\overline{\mathcal{M}}$ of locally measurable sets.

Exercise 7.2 Construct a nonnegative and additive, but not σ-additive function on the σ-algebra of all subsets of a countably infinite set X.

Exercise 7.3 Construct a measurable set in $\mathbb{R}^2$ whose projections onto the coordinate axes are non-measurable.

---

[48] L. Czách, private communication, 2005.

Exercise 7.4 (Outer Measure)[49] Given a finite measure μ on a semiring $\mathcal{P}$ in X, we set

$$\mu^*(A) := \inf \sum_{k=1}^{\infty} \mu(P_k)$$

for each $A \subset X$ where the infimum is taken over all sequences $(P_k) \subset \mathcal{P}$ such that $A \subset \cup_k P_k$.

(i) Show that μ^* is an *outer measure:* a nonnegative, σ-subadditive function on 2^X, i.e,

$$\mu^*(A) \le \sum_{n=1}^{\infty} \mu^*(A_n) \quad \text{whenever} \quad A \subset \bigcup_{n=1}^{\infty} A_n.$$

(ii) Prove that

$$\mu^*(A \cup B) + \mu^*(A \cap B) \le \mu^*(A) + \mu^*(B)$$

for all $A, B \subset X$.

(iii) Prove that $A \subset X$ is measurable $\iff$

$$\mu^*(B) = \mu^*(B \cap A) + \mu^*(B \setminus A)$$

for all $B \subset X$.

Exercise 7.5 (Riemann–Stieltjes Integral)[50] Let us be given two functions $f, g : [a, b] \to \mathbb{R}$ on a compact interval. For each finite subdivision $I = \{x_0, \xi_1, x_1, \ldots, x_{n-1}, \xi_n, x_n\}$ of the segment $[a, b]$, where

$$a = x_0 < \xi_1 < x_1 < \cdots < x_{n-1} < \xi_n < x_n = b,$$

we set

$$\delta(I) := \min_k (x_k - x_{k-1}),$$

and we define the corresponding Riemann–Stieltjes sum by the formula

$$S(I) := \sum_{k=1}^{n} f(\xi_k) \left(g(x_k) - g(x_{k-1}) \right).$$

---

[49] Carathéodory [77]. See also Burkill [68], Halmos [184], and Natanson [332].
[50] Stieltjes [435].

If $S(I)$ converges to a finite limit L as $\delta(I) \to 0$, then we say that f is integrable with respect to g, and we write

$$f \in R(g), \quad \int f\, dg = L.$$

Prove the following properties:

 (i) If f is continuous and g has bounded variation, then $f \in R(g)$.
 (ii) If $f \in R(g)$, then $g \in R(f)$, and

$$\int f\, dg + \int g\, df = [fg]_a^b.$$

Exercise 7.6 For which values of α does the limit

$$\lim_{h \searrow 0} \int_h^1 x^\alpha\, d \sin \frac{1}{x}$$

exist?

Exercise 7.7 Give an example of a strongly σ-finite measure that is not finite, and for which X is not measurable.

Exercise 7.8 Construct measurable functions $f_i : \mathbb{R}^2 \to \mathbb{R}$ with the following properties:

 (i) The successive integrals of f_1 in (7.4) exist, and are equal to zero.
 (ii) The successive integrals of f_2 are equal to 0 and ∞, respectively.
 (iii) The successive integrals of f_3 are equal to 0 and 1, respectively.
 (iv) One of the successive integrals of f_4 is equal to 0, and the other is undefined.

Taking linear combinations of the functions $f_i(x, y)$ and $f_i(y, x)$ show that no conclusion can be made of the successive integrals if $f : \mathbb{R}^2 \to \mathbb{R}$ is a measurable function whose integral is not defined.

Exercise 7.9 (Hausdorff Dimension)[51] Given a set $A \subset \mathbb{R}$ and positive real numbers s, δ, let

$$H_\delta^s(A) := \inf \left\{ \sum_{i=1}^{\infty} |I_i|^s \right\},$$

---

[51] Hausdorff [196]. See, e.g., Falconer [134]. Some number-theoretical applications are given in de Vries–Komornik [101] and Komornik–Kong–Li [259].

where the infimum is taken over the countable covers of A by intervals of length $|I_i| \leq \delta$, and let

$$H^s(A) := \sup_{\delta > 0} H^s_\delta(A).^{52}$$

Prove the following results:

(i) $H^s_\delta(A) \nearrow H^s(A)$ as $\delta \searrow 0$.
(ii) H^s is an outer measure on $\mathbb{R}$.[53]
(iii) There exists $d \in [0, \infty]$ such that $H^s(A) = \infty$ if $s < d$, and $H^s(A) = 0$ if $s > d$. It is called the *Hausdorff dimension* of A.
(iv) Let $S_i : \mathbb{R} \to \mathbb{R}$ be a similarity with a scaling constant $c_i \in (0, 1)$, for $i = 1, \ldots, m$. If a non-empty compact set K is the disjoint union of $S_1(K), \ldots, S_m(K)$, then the Hausdorff dimension d of K is the solution of the equation $c_1^d + \cdots + c_m^d = 1$.
(v) The Hausdorff dimension of Cantor's ternary set is equal to $\ln 2 / \ln 3 \approx 0.63$.

---

[52]More generally, we may consider countable covers by sets of diameter $\operatorname{diam} I_i \leq \delta$ in a metric space.
[53]Carathéodory's construction (Exercise 7.4) yields the s-dimensional *Hausdorff measure*.

Part III
Function Spaces

We may resist everything, except temptation.
–O. Wilde

Functional analysis started by studying (in today's terminology) the space $C(I)$ of continuous functions defined on a compact interval. The idea of *function spaces* had already appeared in the doctoral dissertation of Riemann [370]. Dini [109] proved that for *monotone* sequences of continuous functions pointwise convergence is necessarily uniform. Ascoli [12] gave a sufficient condition for the compactness of a set in $C(I)$. This forms the basis for Peano's theorem (1886) on the solvability of differential equations of the form $x' = f(t,x)$ where f is merely continuous. (The Lipschitz condition serves only for the uniqueness of the solution.) Arzelà [8] proved that Ascoli's condition is also necessary.

Weierstrass [483] proved the density of polynomials in $C(I)$. Le Roux [299] and Volterra [472–475] obtained theorems of existence and uniqueness for a wide class of integral equations. Fredholm [150] discovered that the general theory of integral equations is much simpler than previously believed. Riesz [379] gave an elegant description of the dual space of $C(I)$ by using Stieltjes integrals.

Cantor influenced Borel [58], Baire [17] and Lebesgue [287, 288] to widen the classes of sets and functions to be investigated. In his Ph.D. under the supervision of Hadamard, Fréchet [154] introduced the metric spaces and the notions of compactness, completeness and separability. Riesz [373, 374, 376] and Fischer [146] proved the completeness of the spaces of Lebesgue integrable functions, Riesz [375, 379] and Fréchet [155] characterized the duals of these spaces, and the discipline started to grow exponentially.

The following works contain more complete studies of the historical development: [37, 45, 61, 106, 117, 203, 327, 365, 394, 421].

This last part of our book also serves as a synthesis: while Parts I and II are largely independent, here we build upon both.

We did not resist the *temptation* to give multiple proofs of some theorems: either we could not choose among them or because they enlighten the problem from different angles, and thus contribute to the deeper understanding of the interconnections between different branches of analysis.

Chapter 8
Spaces of Continuous Functions

From the point of view of Mathematics the XIXth century could be called the century of the Theory of functions.... (V. Volterra, 1900)

In this chapter the letter K always denotes a *compact Hausdorff space*. We recall from topology that the continuous functions $f : K \to \mathbb{R}$ form a Banach space $C(K)$ with respect to the norm

$$\|f\|_\infty := \max_{t \in K} |f(t)|,$$

and that norm convergence is uniform convergence on K. We will only present some basic results.[1]

Except for some uninteresting degenerate cases, the spaces $C(K)$ are *not* reflexive:

Examples

- Set $I := [0, 1]$, and consider in $X := C(I)$ the closed affine subspace

$$M := \left\{ f \in C(I) \ : \ f(0) = 0 \quad \text{and} \quad \int_0^1 f(t)\, dt = 1 \right\}.$$

We claim that M has no element of minimal norm, so that the distance $\mathrm{dist}(0, M)$ is not attained.

---

[1]Gillman–Jerison [169] and Semadeni [421] treat many further topics.

© Springer-Verlag London 2016
V. Komornik, *Lectures on Functional Analysis and the Lebesgue Integral*,
Universitext, DOI 10.1007/978-1-4471-6811-9_8

Fig. 8.1 Graph of f_n

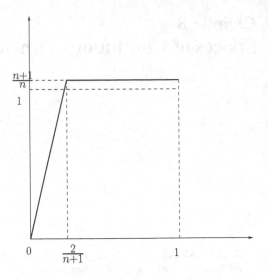

To prove this, first we observe that $\text{dist}(0, M) \geq 1$ because

$$1 = \int_0^1 f(t)\, dt \leq \int_0^1 \|f\|_\infty \, dt = \|f\|_\infty \qquad (8.1)$$

for all $f \in M$. Furthermore, the formula (see Fig. 8.1)

$$f_n(t) := \frac{n+1}{n} \min\left\{\frac{(n+1)t}{2}, 1\right\}, \quad n = 1, 2, \ldots$$

defines a sequence $(f_n) \subset M$ satisfying $\|f_n\|_\infty = (n+1)/n \to 1$, so that in fact $\text{dist}(0, M) = 1$.

But this distance is not attained because the inequality in (8.1) is strict for every $f \in M$ because of the continuity of f and the condition $f(0) = 0$. Applying Proposition 2.1 (p. 55) we conclude that $C(I)$ is not reflexive.

- Set $I = [-1, 1]$, and consider on $X := C(I)$ the linear functional

$$\varphi(f) := \int_{-1}^1 (\text{sign } t) f(t)\, dt.$$

The obvious estimate

$$|\varphi(f)| \leq \int_{-1}^1 |f(t)|\, dt \leq 2\,\|f\|_\infty \qquad (8.2)$$

shows that φ is continuous, and $\|\varphi\| \leq 2$.

Fig. 8.2 Graph of g_n

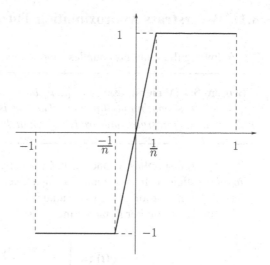

Furthermore, the formula[2] (see Fig. 8.2)

$$g_n(t) := \text{med}\,\{-1, nt, 1\}$$

defines a sequence $(g_n) \subset X$ satisfying $\|g_n\|_\infty = 1$ for all n, and $\varphi(g_n) \to 2$; this implies that in fact $\|\varphi\| = 2$.

But the norm $\|\varphi\|$ is not attained, because $|\varphi(f)| < 2\,\|f\|_\infty$ for all non-zero functions $f \in X$. Indeed, we could have equality in (8.2) only if $(\text{sign}\,t)f(t)$ were constant in $[-1, 1]$, but this condition excludes all non-zero continuous functions.

Applying Proposition 2.1 again, we conclude that $C(I)$ is not reflexive.

• The spaces $C(I)$ are not only non-reflexive: they are not even dual spaces.[3] Indeed, it follows from the Banach–Alaoglu and Krein–Milman theorems that the closed unit ball C of every dual Banach space is spanned by its extremal points.

This is not satisfied for the closed unit ball C of $C(I)$: its only extremal points are the constant functions 1 and -1, and their closed convex hull contains only constant functions, while C contains non-constant functions as well.

Later (on p. 298) we will also give a direct proof of the non-reflexivity.

Despite their non-reflexivity, these spaces occur in many applications. This justifies their study in this chapter.

---

[2] We recall that med $\{x, y, z\}$ denotes the middle number among x, y and z.

[3] See Gelbaum–Olmsted [168]. The situation is similar to that of c_0; see p. 140.

8.1   Weierstrass Approximation Theorems

The following theorem has countless applications:

Theorem 8.1 (Weierstrass)[4] *Let* $[a, b]$ *be a bounded, closed interval, and* $f :$ $[a, b] \rightarrow \mathbb{R}$ *a continuous function. There exists a sequence* (p_n) *of algebraic polynomials, converging uniformly to* f *on* $[a, b]$.

The theorem implies at once that $C([a, b])$ is separable: the polynomials with *rational* coefficients form a countable, dense set.

The following proof is due to Landau.[5]

Fix a positive number R and define $q : \mathbb{R} \rightarrow \mathbb{R}$ by the formula (see Fig. 8.3)

$$q(t) := \begin{cases} R^2 - t^2 & \text{if } |t| \leq R, \\ 0 & \text{if } |t| \geq R. \end{cases}$$

Lemma 8.2 *For each fixed* $\delta > 0$ *we have*

$$\frac{\int_{|t|>\delta} q(t)^n \, dt}{\int_{-\infty}^{\infty} q(t)^n \, dt} \rightarrow 0 \quad as \quad n \rightarrow \infty.$$

Proof The case $\delta \geq R$ is obvious. Assuming henceforth that $\delta < R$, we observe that q is a continuous even function, positive and decreasing in $(0, R)$, and vanishing

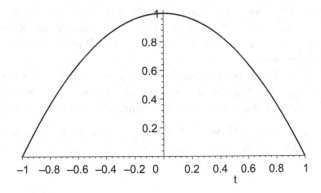

Fig. 8.3 Graph of q for $R = 1$

---

[4]Weierstrass [483], p. 5.
[5]Landau [283]. See Proposition 8.16 and Exercise 8.3 below (pp. 282, 300) for other proofs.

outside $(-R, R)$. Therefore

$$\int_{|t|>\delta} q(t)^n \, dt < (2R - 2\delta)q(\delta)^n < 2Rq(\delta)^n$$

and

$$\int_{-\infty}^{\infty} q(t)^n \, dt > \int_{|t|\leq\delta/2} q(t)^n \, dt > \delta q(\delta/2)^n,$$

so that

$$0 \leq \frac{\int_{|t|>\delta} q(t)^n \, dt}{\int_{-\infty}^{\infty} q(t)^n \, dt} \leq \frac{2R}{\delta}\left(\frac{q(\delta)}{q(\delta/2)}\right)^n.$$

Since $0 < q(\delta) < q(\delta/2)$, the last expression tends to zero as $n \to \infty$. □

Proof of Theorem 8.1 By adding an affine polynomial if necessary, we may assume that $f(a) = f(b) = 0$. Then we may extend f by zero to a continuous function defined on $\mathbb{R}$. The extended function is uniformly continuous, so that

$$\omega(f, \delta) := \sup \{|f(x) - f(t)| : |x - t| \leq \delta\} \to 0$$

as $\delta \searrow 0$.[6]

Let us consider the function q of the preceding lemma with R to be chosen later, and set

$$c_n = \int_{-\infty}^{\infty} q(t)^n \, dt \quad \text{and} \quad Q_n(t) = c_n^{-1}q(t)^n$$

for all $n = 1, 2, \ldots$ and $t \in \mathbb{R}$. Then we have

$$Q_n \geq 0 \quad \text{in} \quad \mathbb{R}, \tag{8.3}$$

$$Q_n(t) = 0 \quad \text{if} \quad |t| \geq R, \tag{8.4}$$

$$\int_{-\infty}^{\infty} Q_n(t) \, dt = 1, \tag{8.5}$$

$$\int_{|t|>\delta} Q_n(t) \, dt \to 0 \quad \text{as} \quad n \to \infty, \quad \text{for each} \quad \delta > 0; \tag{8.6}$$

see Fig. 8.4.

---

[6]$\omega(f, \delta)$ is called the *uniform continuity modulus* of f.

Fig. 8.4 Graphs of Q_1, Q_2
and Q_3 for $R = 1$

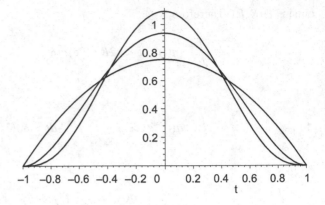

Fig. 8.4 Graphs of Q_1, Q_2 and Q_3 for $R = 1$

We claim that the functions

$$p_n(x) := \int_{-\infty}^{\infty} f(t) Q_n(x - t)\, dt$$

converge to f uniformly in $\mathbb{R}$.

Indeed, applying (8.3) and (8.5) we have

$$|f(x) - p_n(x)| = \left| \int_{-\infty}^{\infty} (f(x) - f(t)) Q_n(x - t)\, dt \right| \tag{8.7}$$

$$\leq \int_{|x-t| \leq \delta} |f(x) - f(t)| Q_n(x - t)\, dt$$

$$+ \int_{|x-t| > \delta} |f(x) - f(t)| Q_n(x - t)\, dt$$

$$\leq \omega(f, \delta) + 2 \|f\|_\infty \int_{|s| > \delta} Q_n(s)\, ds$$

for each x.

For any fixed $\varepsilon > 0$ choose $\delta > 0$ such that $\omega(f, \delta) < \varepsilon/2$, and then using (8.6) choose N such that

$$2 \|f\|_\infty \int_{|s| > \delta} Q_n(s)\, ds < \varepsilon/2 \quad \text{for all} \quad n \geq N.$$

Then we conclude from (8.7) that $|f(x) - p_n(x)| < \varepsilon$ for all $x \in \mathbb{R}$ and $n \geq N$.

We complete the proof by showing that the *restriction* of p_n to $[a, b]$ is a polynomial if we choose $R \geq b - a$ at the beginning of the proof. Applying (8.4), using the fact that f vanishes outside $[a, b]$, and taking into account that $[a, b] \subset [x - R, x + R]$ for every $a \leq x \leq b$, we obtain the following equality for each $a \leq x \leq b$:

$$p_n(x) = \int_{-\infty}^{\infty} f(t) Q_n(x - t)\, dt$$

$$= \int_{x-R}^{x+R} f(t) c_n^{-1} (R^2 - (x - t)^2)^n\, dt$$

$$= \int_a^b f(t) c_n^{-1} (R^2 - (x - t)^2)^n\, dt.$$

Since

$$c_n^{-1} (R^2 - (x - t)^2)^n = \sum_{j=0}^{2n} a_j(t) x^j$$

with suitable polynomials $a_j(t)$, it follows that

$$p_n(x) = \sum_{j=0}^{2n} b_j x^j \quad \text{with} \quad b_j = \int_a^b f(t) a_j(t)\, dt. \qquad \square$$

Remark The above proof was perhaps the first example of *regularization by convolution*, a technique widely used today to establish density theorems in various functions spaces.[7]

Weierstrass also proved a similar result for periodic functions. The 2π-periodic continuous functions form a closed subspace $C_{2\pi}$ in the Banach space $\mathcal{B}(\mathbb{R})$, hence $C_{2\pi}$ is also a Banach space with respect to the norm $\|\cdot\|_\infty$.[8]

Definition A *trigonometric polynomial* is a finite linear combination of the functions

$$1, \cos t, \sin t, \cos 2t, \sin 2t, \cos 3t, \sin 3t, \dots.$$

[7] See the references in the footnote of Sect. 9.3 below, p. 320.
[8] We recall that in this book by a *subspace* without adjective we always mean a *linear* subspace.

Fig. 8.5 Graph of q for
$R = 1$

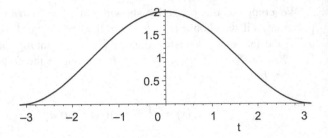

Remark Using the three identities[9]

$$2 \cos kt \, \cos mt = \cos(k - m)t + \cos(k + m)t,$$

$$2 \sin kt \, \sin mt = \cos(k - m)t - \cos(k + m)t,$$

$$2 \sin kt \, \cos mt = \sin(k - m)t + \sin(k + m)t$$

it is easy to show that the trigonometric polynomials form not only a vector space, but also an algebra: the product of two trigonometric polynomials is again a trigonometric polynomial.

Theorem 8.3 (Weierstrass)[10] *For each $f \in C_{2\pi}$ there exists a sequence (p_n) of trigonometric polynomials converging uniformly to f on $\mathbb{R}$.*

The following proof is due to de la Vallée-Poussin.[11]

Proof Introducing the function

$$q(t) := \begin{cases} 1 + \cos t & \text{if } |t| \leq \pi, \\ 0 & \text{if } |t| \geq \pi \end{cases}$$

(see Fig. 8.5), and repeating the preceding proof with $R = \pi$ we obtain that $p_n \to f$ uniformly in $\mathbb{R}$.

---

[9]Several proofs of this chapter could be simplified by adopting the complex framework, and using Euler's formula $e^{ix} = \cos x + i \sin x$. For example, the trigonometric polynomials would be simply the algebraic polynomials of e^{it}, and the single identity $e^{u+v} = e^u e^v$ would suffice instead of these three real identities.

[10]Weierstrass [483]. See Theorem 8.11 and a remark following Proposition 8.21 below (pp. 276, 288) for other proofs.

[11]de la Vallée-Poussin [463]. His work was motivated by that of Landau.

It remains to show that p_n is a trigonometric polynomial. This follows from the following computation:

$$p_n(x) = \int_{-\infty}^{\infty} f(t) Q_n(x - t) \, dt$$

$$= c_n^{-1} \int_{x-\pi}^{x+\pi} f(t)(1 + \cos(x - t))^n \, dt$$

$$= c_n^{-1} \int_{-\pi}^{\pi} f(t)(1 + \cos(x - t))^n \, dt$$

$$= c_n^{-1} \int_{-\pi}^{\pi} f(t)(1 + \cos x \cos t + \sin x \sin t)^n \, dt$$

$$= a_0 + \sum_{k=1}^{n} a_k \cos kx + b_k \sin kx,$$

where a_k and b_k are suitable real numbers. The third equality follows from the 2π-periodicity of the function under the integral sign, while the last one from the repeated application of the three trigonometric identities of the preceding remark.

□

Remark Jackson [221], [222] investigated the error of the approximation as a function of the regularity of the approximated function. Müntz [329], Szász [445], Clarkson and Erdős [90] proved important generalizations of Theorem 8.1. See also Achieser [1], Cheney [85], Jackson [223], Natanson [333], Rudin [405].

8.2 * The Stone–Weierstrass Theorem

Stone proved a far-reaching generalization of the Weierstrass approximation theorems.

Definition A subspace M of $C(K)$ is a *subalgebra* if $f, g \in M$ imply $fg \in M$.

Theorem 8.4 (Stone–Weierstrass)[12] *Let K be a compact topological space and M a subalgebra of $C(K)$. Assume that M contains the constant functions, and separates the points of K: for any two distinct points $x, y \in K$ there exists an $h \in M$ such that $h(x) \neq h(y)$. Then M is dense $C(K)$.*

---

[12]Stone [440], [441].

Examples

- Let K be a compact interval in $\mathbb{R}$. The restrictions of the algebraic polynomials to K form a subalgebra M satisfying the conditions of Theorem 8.4. Hence Theorem 8.1 is a special case of Theorem 8.4.
- More generally, if K is a compact set in $\mathbb{R}^N$, then the algebraic polynomials of N variables form a subalgebra M satisfying the conditions of Theorem 8.4.
- Let K be the unit circle in $\mathbb{R}^2$. Setting $T(s) := (\cos s, \sin s)$, the function $f \mapsto f \circ T$ establishes an isometric isomorphism between the Banach spaces $C(K)$ and $C_{2\pi}$. Furthermore, the algebraic polynomials of two variables correspond to the trigonometric polynomials. Thus Theorem 8.3 also follows from Theorem 8.4.

In the proof we use the notion of vector lattices (see p. 172).

Proof of Theorem 8.4 First step.   If $f_n \to f$ and $g_n \to g$ $C(K)$, then $f_n g_n \to fg$ because

$$\|fg - f_n g_n\|_\infty \le \|f - f_n\|_\infty \|g\|_\infty + \|f_n\|_\infty \|g - g_n\|_\infty \to 0.$$

Hence the closure $\overline{M}$ of the subalgebra M is still a subalgebra of $C(K)$.

Second step.   We show that the closed subalgebra $\overline{M}$ is a *vector lattice*. Fix $h \in \overline{M}$ arbitrarily and fix a number $T > \|h\|_\infty$. By Theorem 8.1 there exist polynomials p_n satisfying $p_n(x) \to |x|$ uniformly in $[-T, T]$. Then $p_n \circ h \in \overline{M}$, and $p_n \circ h \to |h|$ uniformly in K, so that $|h| \in \overline{M}$.

The following proposition completes the proof of the theorem.   □

Proposition 8.5 (Kakutani–Krein)[13] *Let K be a compact topological space and $M \subset C(K)$ a vector lattice. Assume that $1 \in M$, and that M separates the points of K. Then M is dense in $C(K)$.*

Proof Fixing $f \in C(K)$ and $\varepsilon > 0$ arbitrarily, we have to find $g \in M$ satisfying $\|f - g\|_\infty < \varepsilon$.

First step.   For each fixed $x \in K$ there exists a function $f_x \in M$ satisfying

$$f_x > f - \varepsilon \quad \text{on} \quad K, \quad \text{and} \quad f_x(x) = f(x).$$

Indeed, by our assumption for each $y \in K$ there exists a function $f_{xy} \in M$ equal to f at x and y. Then the open sets

$$U_y := \{z \in K : f_{xy}(z) > f(z) - \varepsilon\}, \quad y \in K$$

---

[13]Kakutani [240, pp. 1004–1005], Krein–Krein [268].

cover the compact set K, because $y \in U_y$ for every y. If

$$K = U_{y_1} \cup \cdots \cup U_{y_n}$$

is a finite subcover, then the function

$$f_x := \max \{f_{xy_1}, \ldots, f_{xy_n}\}$$

has the required properties.

Second step.   There exists a function $g \in M$ satisfying

$$f - \varepsilon < g < f + \varepsilon \quad \text{on} \quad K,$$

and hence the inequality $\|f - g\|_\infty < \varepsilon$.

For the proof we consider the functions $f_x \in M$ obtained in the first step. The open sets

$$V_x := \{z \in K : f_x(z) < f(z) + \varepsilon\}, \quad x \in K$$

cover the compact set K, because $x \in V_x$ for every x. If

$$K = V_{x_1} \cup \cdots \cup V_{x_m}$$

is a finite subcover, then the function

$$g := \min \{f_{x_1}, \ldots, f_{x_m}\}$$

has the required properties.   □

The following interesting application will be useful later[14]:

Proposition 8.6 (Stone)[15] *Let K be a compact set in a topological space X, and assume that the points of K may be separated by the continuous functions $h : X \to \mathbb{R}$. Then every continuous function $f : K \to \mathbb{R}$ may be extended to a continuous function $F : X \to \mathbb{R}$.*

Proof The restrictions of the continuous functions $F : X \to \mathbb{R}$ to K form a vector lattice M in $C(K)$, containing the constant functions. By our assumption M satisfies the conditions of the Kakutani–Krein theorem, and hence it is dense in $C(K)$. It remains to prove that M is closed.

Let $(f_n) \subset M$ converge uniformly on K to some function f. We have to find a continuous function $F : X \to \mathbb{R}$ such that $F = f$ on K.

[14]See the proof of Lemma 8.27, p. 297.
[15]Stone [441]. This is a version of similar theorems of Urysohn [461] and Tietze [453].

Taking a subsequence if necessary, we may assume that

$$|f_{n+1} - f_n| \leq 2^{-n} \quad \text{on} \quad K$$

for every n.[16]

By the definition of M the functions f_1 and $f_{n+1} - f_n$ have continuous extensions F_1 and G_n to X. Furthermore, we may assume that

$$|G_n| \leq 2^{-n} \quad \text{on} \quad K$$

for every n: change G_n to

$$\text{med}\,\{-2^{-n}, G_n, 2^{-n}\}$$

if necessary. Then the function series

$$F_1 + \sum_{n=1}^{\infty} G_n$$

converges uniformly to some function $F : X \to \mathbb{R}$. We conclude that F is continuous, and $F = f$ on K.   □

8.3   Compact Sets. The Arzelà–Ascoli Theorem

In this section we characterize the compact sets of $C(K)$. Since in complete metric spaces the compact sets coincide with the totally bounded[17] closed sets, it is sufficient to characterize the totally bounded sets.

Definitions Consider a family of functions $\mathcal{F} \subset C(K)$.

- $\mathcal{F}$ is *pointwise bounded* if $\{f(t) : f \in \mathcal{F}\}$ is bounded in $\mathbb{R}$ for each $t \in K$.
- $\mathcal{F}$ is *equicontinuous* if for each $\varepsilon > 0$ and $t \in K$ there is a neighborhood V of t such that $|f(s) - f(t)| < \varepsilon$ for all $s \in V$ and $f \in \mathcal{F}$.

Proposition 8.7 (Arzelà–Ascoli)[18] *A family of functions $\mathcal{F} \subset C(K)$ is totally bounded $\iff$ it is pointwise bounded and equicontinuous.*

---

[16]We have already used this technique when proving the Riesz Lemma 5.13, p. 184.

[17]We recall that a set A is *totally bounded* or *precompact* if for each $r > 0$ it has a finite cover by balls of radius r.

[18]Ascoli [12] (pp. 545–549, sufficiency for $K = [0, 1]$), Arzelà [8] (necessity), [9] (simplified treatment), [10], Fréchet [154] (general case).

Proof First let $\mathcal{F}$ be totally bounded. Then it is also bounded in norm, i.e., uniformly bounded on K, and hence pointwise bounded as well.

To show the equicontinuity, it suffices to find for any fixed $t \in K$ and $r > 0$ a neighborhood V of t such that

$$|f(t) - f(s)| < 3r \quad \text{for all} \quad f \in \mathcal{F} \quad \text{and} \quad s \in V. \tag{8.8}$$

Let us cover $\mathcal{F}$ with finitely many balls of radius r:

$$\mathcal{F} \subset B_r(f_1) \cup \cdots \cup B_r(f_m)$$

with $f_1, \ldots, f_m \in \mathcal{F}$.

Since each f_i is continuous at t, we may choose a neighborhood V_i of t such that

$$|f_i(t) - f_i(s)| < r \quad \text{for all} \quad s \in V_i.$$

Then (8.8) is satisfied with $V := V_1 \cap \cdots \cap V_m$.

Indeed, for any given $f \in \mathcal{F}$ and $s \in V$, choosing i such that $\|f - f_i\| < r$, we have

$$|f(t) - f(s)| \leq |f(t) - f_i(t)| + |f_i(t) - f_i(s)| + |f_i(s) - f(s)| < r + r + r.$$

Conversely, if $\mathcal{F}$ is equicontinuous, then by the compactness of K we may find for each fixed $r > 0$ finitely many points $t_1, \ldots, t_m \in K$ and their neighborhoods $V_1, \ldots, V_m$ such that $K = V_1 \cup \cdots \cup V_m$, and

$$|f(t) - f(t_i)| < r \quad \text{whenever} \quad f \in \mathcal{F} \quad \text{and} \quad t \in V_i.$$

If, moreover, $\mathcal{F}$ is pointwise bounded, then the set

$$\{(f(t_1), \ldots, f(t_m)) : f \in \mathcal{F}\}$$

is bounded $\mathbb{R}^m$, and also totally bounded there.[19] There exist therefore finitely many functions $f_1, \ldots, f_n \in \mathcal{F}$ such that[20]

$$\{(f(t_1), \ldots, f(t_m)) : f \in \mathcal{F}\} \subset \bigcup_{j=1}^{n} B_r(f_j(t_1), \ldots, f_j(t_m)).$$

---

[19] We recall that the bounded and totally bounded sets are the same in all finite-dimensional normed spaces.

[20] In this formula the balls are taken in $\mathbb{R}^m$.

We complete the proof by showing that[21]

$$\mathcal{F} \subset B_{3r}(f_1) \cup \cdots \cup B_{3r}(f_n).$$

For any given $f \in \mathcal{F}$ first we choose f_j satisfying

$$(f(t_1), \ldots, f(t_m)) \in B_r(f_j(t_1), \ldots, f_j(t_m)).$$

Next, for any given $t \in K$ we choose i such that $t \in V_i$. Then we have

$$\left| f(t) - f_j(t) \right| \le |f(t) - f(t_i)| + \left| f(t_i) - f_j(t_i) \right| + \left| f_j(t_i) - f_j(t) \right| < r + r + r,$$

whence $f \in B_{3r}(f_j)$. $\square$

8.4 Divergence of Fourier Series

By the *Fourier series* of a function $f \in C_{2\pi}$ we mean the function series[22]

$$\frac{a_0}{2} + \sum_{k=1}^{\infty} a_k \cos kx + b_k \sin kx,$$

with the *Fourier coefficients* a_k, b_k defined by the formulas

$$a_k := \frac{1}{\pi} \int_{-\pi}^{\pi} f(t) \cos kt \, dt \quad \text{and} \quad b_k := \frac{1}{\pi} \int_{-\pi}^{\pi} f(t) \sin kt \, dt.$$

Remark $C_{2\pi}$ is a Euclidean space with respect to the scalar product $(f, g) := \int_{-\pi}^{\pi} fg \, dt$. A simple computation shows that the mth partial sum of the Fourier series is the orthogonal projection of f onto the subspace $\mathcal{T}_m$ of the *trigonometric polynomials of order* $\le m$, spanned by the functions

$$1, \cos t, \sin t, \cos 2t, \sin 2t, \cos 3t, \sin 3t, \ldots, \cos mt, \sin mt.$$

See Sect. 1.4, p. 24.

---

[21] We recall that $r > 0$ was chosen arbitrarily at the beginning.

[22] Daniel Bernoulli [38], Fourier [148]. Using complex numbers the Fourier series would take the simpler form $\sum_{k=-\infty}^{\infty} c_k e^{ikx}$.

Following Fourier's revolutionary treatise, many works were devoted to the convergence of Fourier series[23]:

- Dirichlet and Jordan[24] proved (among others) that if $f \in C_{2\pi}$ has bounded variation, then its Fourier series converges to f uniformly.
- Lipschitz and Dini[25] proved (among others) that if $f \in C_{2\pi}$, then its Fourier series converges to $f(a)$ at each point a where f is differentiable.

It remained an open question for fifty years whether mere continuity already ensures the convergence of the Fourier series. Finally, a counterexample was found:

Proposition 8.8 (du Bois-Reymond)[26] *There exists an $f \in C_{2\pi}$ whose Fourier series does not converge pointwise to f.*

Remarks

- However, Carleson proved that the Fourier series of each $f \in C_{2\pi}$ converges to f a.e. everywhere.[27]
- On the other hand, Kahane and Katznelson[28] proved that for each null set E there exists a function $f \in C_{2\pi}$ that diverges at the points of E.

First we establish two lemmas.

Lemma 8.9 (Dirichlet)[29] *The partial sums*

$$(S_m f)(x) := \frac{a_0}{2} + \sum_{k=1}^{m} a_k \cos kx + b_k \sin kx$$

of the Fourier series of a function $f \in C_{2\pi}$ may be written in the closed form

$$(S_m f)(x) = \frac{1}{2\pi} \int_{-\pi}^{\pi} D_m(x-t) f(t)\, dt,$$

with the Dirichlet *kernel $D_m \in C_{2\pi}$ defined by the formula*[30]

$$D_m(2s) := \frac{\sin(2m+1)s}{\sin s}.$$

---

[23] A fascinating historical account is given by Kahane [237].

[24] Dirichlet [112], Jordan [229].

[25] Lipschitz [308] and Dini [107], [110]. See a short proof in Exercise 8.5, p. 301.

[26] du Bois-Reymond [49], [51]. A simpler explicit counterexample was given later by Fejér [139], [140]. We prove here the mere existence of such functions.

[27] Carleson [78]. This was a long-standing open problem of Lusin [313]. See also the remark following Corollary 9.6 below (p. 314) concerning L^p convergence.

[28] Kahane and Katznelson [238]. See also Edwards [120], Katznelson [245] and Zygmund [493] for many further results.

[29] Dirichlet [112].

[30] For $\sin s = 0$ we replace the right-hand side by its limit $(2m+1)$.

Fig. 8.6 Graph of D_0

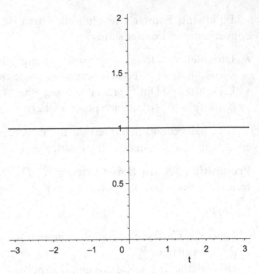

Fig. 8.7 Graph of D_1

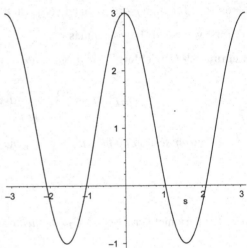

See Figs. 8.6, 8.7, 8.8, and 8.9.

Proof Since

$$(S_m f)(x) = \frac{a_0}{2} + \sum_{k=1}^{m} a_k \cos kx + b_k \sin kx$$

$$= \frac{1}{2\pi} \int_{-\pi}^{\pi} \left(1 + 2 \sum_{k=1}^{m} \cos kx \cos kt + \sin kx \sin kt\right) f(t) \, dt$$

$$= \frac{1}{2\pi} \int_{-\pi}^{\pi} \left(1 + 2 \sum_{k=1}^{m} \cos k(x-t)\right) f(t) \, dt,$$

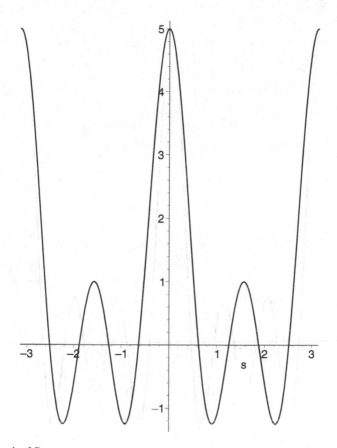

Fig. 8.8 Graph of D_2

it is sufficient to prove the identity

$$1 + 2\sum_{k=1}^{m}\cos 2ks = \frac{\sin(2m+1)s}{\sin s}.$$

The case $m = 0$ is obvious. The general case follows by induction, using the trigonometric identities

$$2\sin s\cos 2(m+1)s = \sin(2m+3)s - \sin(2m+1)s, \quad m = 0, 1, \dots. \qquad \square$$

Now we introduce the linear functionals

$$\varphi_m(f) := (S_m f)(0)$$

on the Banach space $C_{2\pi}$.

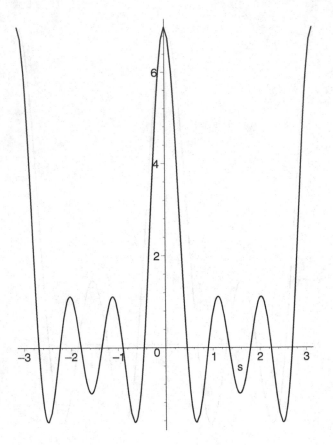

Fig. 8.9 Graph of D_3

Lemma 8.10 *The linear functionals φ_m are continuous, and $\|\varphi_m\| \to \infty$ as $m \to$* ∞.

Proof Since

$$|a_k|, \ |b_k| \leq 2 \, \|f\|_\infty \, ,$$

we deduce from the definition of S_m that

$$\|S_m f\|_\infty \leq \left(2m + \frac{1}{2}\right) \cdot 2 \, \|f\|_\infty = (4m+1) \, \|f\|_\infty \, ;$$

hence $\|\varphi_m\| \leq 4m + 1 < \infty$.

On the other hand, the formula

$$f(2s) := (\text{sign} \sin s) \sin(2m+1)s$$

defines a function $f \in C_{2\pi}$ satisfying $\|f\|_\infty = 1$ and

$$
\begin{aligned}
\varphi_m(f) &= \frac{1}{2\pi} \int_{-\pi}^{\pi} D_m(-t)f(t)\, dt & &= \frac{1}{\pi} \int_{-\pi/2}^{\pi/2} D_m(-2s)f(2s)\, ds \\
&= \frac{1}{\pi} \int_{-\pi/2}^{\pi/2} \frac{\sin^2(2m+1)s}{|\sin s|}\, ds & &= \frac{2}{\pi} \int_0^{\pi/2} \frac{\sin^2(2m+1)s}{\sin s}\, ds \\
&> \frac{2}{\pi} \int_0^{\pi/2} \frac{\sin^2(2m+1)s}{s}\, ds & &= \frac{2}{\pi} \int_0^{(2m+1)\pi/2} \frac{\sin^2 s}{s}\, ds \\
&> \frac{2}{\pi} \sum_{j=1}^{m} \int_{(j-1)\pi}^{j\pi} \frac{\sin^2 s}{s}\, ds & &> \frac{2}{\pi} \int_0^{\pi} \sum_{j=1}^{m} \frac{\sin^2 s}{j\pi}\, ds \\
&= \frac{1}{\pi} \sum_{j=1}^{m} \frac{1}{j}.
\end{aligned}
$$

Hence,

$$
\|\varphi_m\| \ge \varphi_m(f) > \frac{1}{\pi} \sum_{j=1}^{m} \frac{1}{j} \to \infty. \qquad\qquad \square
$$

Remarks

- We note for later reference that the test functions used in the proof are even.
- Fejér[31] has established the more precise asymptotic formulas

$$
\|\varphi_m\| = \frac{4}{\pi^2} \log m + O(1), \quad m \to \infty.
$$

Proof of Proposition 8.8 Assume on the contrary that $\varphi_m(f) \to f(0)$ for each $f \in C_{2\pi}$. Then applying the Banach–Steinhaus theorem (p. 81) with $X = C_{2\pi}$ and $Y = \mathbb{R}$ we obtain $\sup \|\varphi_m\| < \infty$, contradicting the preceding lemma. $\qquad\qquad \square$

8.5 Summability of Fourier Series. Fejér's Theorem

> Thought is only a flash in the middle of a long night, but this flash is everything.
> (H. Poincaré)

---

[31]Fejér [141]. See also Edwards [120] or Zygmund [493].

The counterexample of du Bois-Reymond made obvious the difficulties of representing continuous functions by Fourier series. Minkowski even asked whether the Fourier series of a continuous function may converge pointwise to another function.[32] The long period of stagnation ended when Fejér discovered the following remarkable

---

Theorem 8.11 (Fejér)[33] *Given any* $f \in C_{2\pi}$, *the mean values*

$$\sigma_n f := \frac{1}{n+1} \sum_{m=0}^{n} S_m f, \quad n = 0, 1, \dots$$

converge to f uniformly on $\mathbb{R}$.

Remarks The theorem has important consequences:

- It provides a new proof of the second approximation theorem of Weierstrass.
- It implies that the Fourier series of $f \in C_{2\pi}$ cannot converge at any point x to a value different from $f(x)$.[34] Indeed, this follows from a classical result of Cauchy[35]: if $a_n \to a$ for a numerical sequence, then we also have $(a_1 + \cdots + a_n)/n \to a$.

First we prove a lemma:

Lemma 8.12 *We have*

$$(\sigma_n f)(x) = \frac{1}{2\pi} \int_{-\pi}^{\pi} F_n(x - t) f(t) \, dt$$

with the Fejér kernel $F_n \in C_{2\pi}$ *defined by the formula*[36]

$$F_n(2s) := \frac{1}{n+1} \frac{\sin^2 (n+1)s}{\sin^2 s}.$$

Let us compare Figs. 8.10, 8.11, 8.12, and 8.13 and Figs. 8.6, 8.7, 8.8, and 8.9 on p. 274: the *positivity* of the Fejér kernel has a great importance.

---

[32]See Hawkins [198]. An analogous phenomenon for Taylor series has been known since Cauchy [80, p. 230].

[33]Fejér [137, 138]. He also investigated pointwise convergence for discontinuous functions f. Lebesgue [292] extended his results to Lebesgue integrable functions.

[34]Thereby he has answered Minkowski's question. Banach [20] has shown that Minkowski's phenomenon occurs for a slight modification of the trigonometric system.

[35]Cauchy [79].

[36]For $\sin s = 0$ the right-hand side is replaced by its limit $(n + 1)$.

8.5 Summability of Fourier Series. Fejér's Theorem

277

Fig. 8.10 Graph of F_0

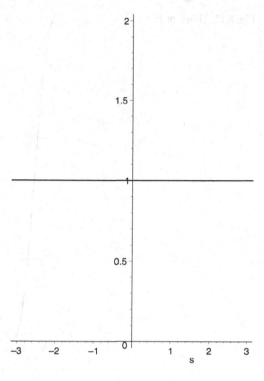

Fig. 8.11 Graph of F_1

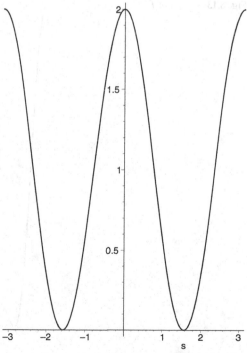

Fig. 8.12 Graph of F_2

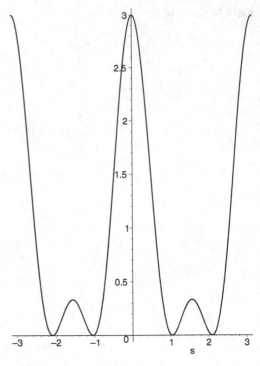

Fig. 8.13 Graph of F_3

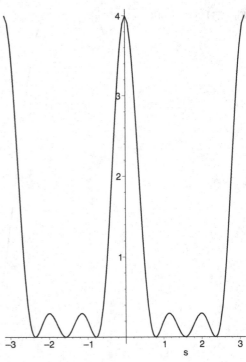

Proof By the definition of the operators σ_n it suffices to prove the equalities

$$F_n = \frac{D_0 + \cdots + D_n}{n+1},$$

or equivalently that

$$\frac{\sin^2(n+1)s}{\sin^2 s} = \sum_{m=0}^{n} \frac{\sin(2m+1)s}{\sin s}.$$

They follow by a direct computation:

$$\sum_{m=0}^{n}(\sin s)\sin(2m+1)s = \frac{1}{2}\sum_{m=0}^{n}(\cos 2ms - \cos(2m+2)s)$$

$$= \frac{1 - \cos(2n+2)s}{2}$$

$$= \sin^2(n+1)s. \qquad \square$$

Proof of Theorem 8.11 We obtain the relations

$$\sigma_n 1 = 1, \quad \sigma_n \cos = \frac{n}{n+1}\cos \quad \text{and} \quad \sigma_n \sin = \frac{n}{n+1}\sin$$

directly from the definitions. Hence $\|f - \sigma_n f\|_\infty \to 0$ for the three functions $f = 1$, cos and sin.

If $f \geq 0$, then $\sigma_n f \geq 0$ by the positivity of the Fejér kernels. Therefore we may conclude by applying Proposition 8.13 below. $\qquad \square$

Definition A linear map $L : C_{2\pi} \to C_{2\pi}$ is *positive* if $f \geq 0 \implies Lf \geq 0$.

Proposition 8.13 (Korovkin)[37] *Consider a sequence of positive linear maps $L_n :$ $C_{2\pi} \to C_{2\pi}$. If $\|f - L_n f\|_\infty \to 0$ for the three functions $f = 1, \cos, \sin$, then the relation $\|f - L_n f\|_\infty \to 0$ holds in fact for all $f \in C_{2\pi}$.*

We prove a more general theorem in the next section.

8.6 * Korovkin's Theorems. Bernstein Polynomials

Let us investigate the positive linear maps $L : C(K) \to C(K)$ for an arbitrary compact topological space.

---

[37] Korovkin [263]. Many applications are given in Korovkin [264].

Definition L is *positive* if $f \geq 0 \Longrightarrow Lf \geq 0$.

Remarks If L is a positive linear map, then

- L is *monotone*: $Lf \leq Lg$ whenever $f \leq g$: this follows at once from the linearity of L;
- L is *continuous* with $\|L\| = \|L1\|_\infty$. Indeed, using the monotonicity we infer from the inequalities $-\|f\|_\infty \leq f \leq \|f\|_\infty$ that

$$-\|f\|_\infty (L1) \leq Lf \leq \|f\|_\infty (L1),$$

and hence $\|Lf\|_\infty \leq \|L1\|_\infty \|f\|_\infty$ for all f. Since equality holds for $f = 1$, we conclude that $\|L\| = \|L1\|_\infty$.

Let K be a compact topological space and $h_1, \ldots, h_m \in C(K)$. Assume that the functions h_j separate the points of K: for any two distinct points $x, y \in K$ there exists a j such that $h_j(x) \neq h_j(y)$.

Consider a sequence of positive linear maps $L_n : C(K) \to C(K)$.

Proposition 8.14 (Freud)[38]

If $\|f - L_n f\|_\infty \to 0$ *for the functions*

$$f = 1, h_1, \ldots, h_m \quad and \quad f = h_1^2 + \cdots + h_m^2, \tag{8.9}$$

then $\|f - L_n f\|_\infty \to 0$ *for all* $f \in C(K)$.

Example If K is a compact set in $\mathbb{R}^m$, then we may apply the proposition to the projections $h_j(x) := x_j, j = 1, \ldots, m$.

Proof Fix $f \in C(K)$ and $\varepsilon > 0$ arbitrarily.

First step.   For each $N = 1, 2, \ldots$, let us denote by U_N the set of pairs $(x, y) \in K \times K$ satisfying the inequality

$$|f(x) - f(y)| < \varepsilon + N \sum_{j=1}^{m} |h_j(x) - h_j(y)|^2. \tag{8.10}$$

These sets are open by the continuity of the functions f and h_j, and they form an increasing set sequence. Furthermore, since

$$\sum_{j=1}^{m} |h_j(x) - h_j(y)|^2 > 0$$

[38]Freud [153]. See Altomare and Campiti [5] for a very complete review of the subject.

whenever $x \neq y$ (by the separation condition), they cover $K \times K$. The latter space being compact, there exists a positive integer N such that (8.10) is satisfied for *all $x, y \in K$.*

Second step. For any fixed $x \in K$, (8.10) implies the inequality

$$|f(x)(L_n 1)(y) - (L_n f)(y)| \leq \varepsilon (L_n 1)(y)$$

$$+ N \sum_{j=1}^{m} h_j^2(x)(L_n 1)(y) - 2N \sum_{j=1}^{m} h_j(x)(L_n h_j)(y)$$

$$+ NL_n \left(\sum_{j=1}^{m} h_j^2 \right)(y)$$

for all $y \in K$.

Choosing $y = x$ and applying the triangle inequality this yields the following estimate:

$$|f - L_n f| \leq |f| \cdot |1 - L_n 1| + \varepsilon (L_n 1)$$

$$+ N \sum_{j=1}^{m} h_j^2 (L_n 1) - 2N \sum_{j=1}^{m} h_j (L_n h_j) + NL_n \left(\sum_{j=1}^{m} h_j^2 \right).$$

Letting $n \to \infty$, the right-hand side tends to ε uniformly by our assumption, and hence

$$\| f - L_n f \|_\infty < 2\varepsilon$$

for all sufficiently large n. □

Corollary 8.15 (Bohman–Korovkin)[39] *Let I be a compact interval, and consider a sequence of positive linear maps $L_n : C(I) \to C(I)$.*

If the relation $\| f - L_n f \|_\infty \to 0$ holds for the three functions $f(x) = 1, x, x^2$, then it holds in fact for all $f \in C(I)$.

Proof We apply the preceding example with $K = I$ and $m = 1$. □

Now we return to the last statement of the preceding section.

---

[39]Bohman [47], Korovkin [263].

Fig. 8.14 $x_1^2 + x_2^2 = 1$

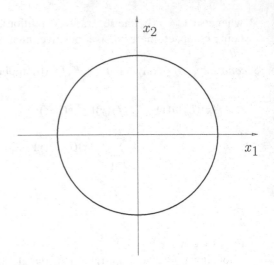

Proof of Proposition 8.13 We apply the preceding example to the unit circle K of $\mathbb{R}^2$. (See Fig. 8.14.) Since $x_1^2 + x_2^2 = 1$ on K, we have only three test functions instead of four. Hence, if a sequence of positive linear maps $L_n : C(K) \to C(K)$ satisfies $\|f - L_n f\|_\infty \to 0$ for the three functions $f(x) := 1, x_1, x_2$, then the relation $\|f - L_n f\|_\infty \to 0$ holds in fact for all $f \in C(I)$.

Now we recall (p. 266) that the map $f \mapsto f \circ T$, where $T(s) := (\cos s, \sin s)$, is an isometric isomorphism between the Banach spaces $C(K)$ and $C_{2\pi}$. Furthermore, $f \geq 0 \iff f \circ T \geq 0$, and the map transforms the functions $f(x) = 1, x_1, x_2$ into $f(T(s)) = 1, \cos s, \sin s$. Hence the result obtained for K is equivalent to Proposition 8.13.   ☐

As another application of Korovkin's theorems, we give a new proof of the first approximation theorem of Weierstrass.[40] Let $I = [0, 1]$ for simplicity, and introduce for each $f \in C(I)$ the *Bernstein polynomials*[41]

$$(B_n f)(x) := \sum_{k=0}^{n} \binom{n}{k} f\left(\frac{k}{n}\right) x^k (1 - x)^{n-k}, \quad x \in I, \quad n = 1, 2, \ldots.$$

Proposition 8.16 (Bernstein)[42] *The Bernstein polynomials $B_n f$ converge uniformly to f on I for each $f \in C(I)$.*

---

[40]Theorem 8.1, p. 260.

[41]Bernstein's proof is probabilistic, based on the law of large numbers.

[42]Bernstein [39]. His result answered a question of Borel [60, pp. 79–82].

Proof The operators B_n are clearly positive linear on $C(I)$. Let us also observe that[43] $B_n 1 = 1$ and B_n id $=$ id for every n via the binomial theorem:

$$(B_n 1)(x) = \sum_{k=0}^{n} \binom{n}{k} x^k (1-x)^{n-k}$$

$$= (x + 1 - x)^n$$

$$= 1$$

and

$$(B_n \text{ id})(x) = \sum_{k=0}^{n} \binom{n}{k} \frac{k}{n} x^k (1-x)^{n-k}$$

$$= \sum_{k=1}^{n} \binom{n-1}{k-1} x^k (1-x)^{n-k}$$

$$= x(x + 1 - x)^{n-1}$$

$$= x.$$

In view of the Bohman–Korovkin theorem (p. 281) it suffices to show that $B_n(\text{id}^2)$ converges uniformly to id^2 on $[0, 1]$. For this we first note that

$$B_n\left(\text{id}^2 - \frac{\text{id}}{n}\right)(x) = \sum_{k=0}^{n} \binom{n}{k} \frac{k(k-1)}{n^2} x^k (1-x)^{n-k}$$

$$= \frac{n-1}{n} \sum_{k=2}^{n} \binom{n-2}{k-2} x^k (1-x)^{n-k}$$

$$= \frac{n-1}{n} x^2.$$

Hence

$$B_n(\text{id}^2) = \frac{n-1}{n} \text{id}^2 + \frac{1}{n} \text{id}$$

and therefore

$$\|\text{id}^2 - B_n(\text{id}^2)\|_\infty = \frac{1}{n} \|\text{id}^2 - \text{id}\|_\infty \to 0. \qquad \square$$

---

[43] We denote by id the identity map of I.

8.7 * Theorems of Haršiladze–Lozinski, Nikolaev and Faber

The main theorem of this section reveals a deep common reason for many divergence theorems. As in Sect. 8.4, we denote by $\mathcal{T}_m$ the vector space of trigonometric polynomials of order $\leq m$, and we denote by $S_m f$ the mth partial sum of the Fourier series of f.

Theorem 8.17 (Haršiladze–Lozinski)[44] *Consider a sequence of continuous linear maps $L_m : C_{2\pi} \to C_{2\pi}$. If L_m is a projection onto $\mathcal{T}_m$ for each m, then there exists a function $f \in C_{2\pi}$ such that $\|f - L_m f\|_\infty \not\to 0$.*

The main ingredient of the proof is an optimality property of Fourier series:

Proposition 8.18 (Lozinski)[45] *If a continuous linear map $L_m : C_{2\pi} \to C_{2\pi}$ is a projection onto $\mathcal{T}_m$, then $\|L_m\| \geq \|S_m\|$.*

Indeed, in view of the Banach–Steinhaus theorem (p. 81), Theorem 8.17 follows from this proposition and from the fact that $\|S_m\| \to \infty$, proved in Lemma 8.10 (p. 273).

Proof of Proposition 8.18 For each real number s the formula

$$(T_s f)(x) := f(x + s)$$

defines in $C_{2\pi}$ a continuous linear operator of norm one. It suffices to establish the following identity[46]:

$$(S_m f)(x) = \frac{1}{2\pi} \int_{-\pi}^{\pi} (T_{-s} L_m T_s f)(x) \, ds, \quad x \in \mathbb{R}, \quad f \in C_{2\pi}. \tag{8.11}$$

Indeed, since

$$|(T_{-s} L_m T_s f)(x)| \leq \|T_{-s} L_m T_s f\|_\infty$$
$$\leq \|T_{-s}\| \cdot \|L_m\| \cdot \|T_{-s}\| \cdot \|f\|_\infty$$
$$= \|L_m\| \cdot \|f\|_\infty$$

for all f, s and x, (8.11) implies $\|S_m f\|_\infty \leq \|L_m\| \cdot \|f\|_\infty$ for all f, and hence $\|S_m\| \leq \|L_m\|$.

---

[44]Lozinski [311].
[45]Lozinski [311].
[46]Marcinkiewicz [314], Lozinski [310].

It is sufficient to prove (8.11) for the functions[47]

$$f_k(x) = \cos kx \quad (k = 0, 1, \ldots) \quad \text{and} \quad g_k(x) = \sin kx \quad (k = 1, 2, \ldots).$$

Indeed, then the identity will hold for all trigonometric polynomials by linearity, and then for all $f \in C_{2\pi}$ by the Weierstrass approximation theorem because all operators occurring in (8.11) are continuous.

If $f \in \mathcal{T}_m$, then $T_s f \in \mathcal{T}_m$. Hence $L_m T_s f = T_s f$ and therefore

$$\frac{1}{2\pi} \int_{-\pi}^{\pi} (T_{-s} L_m T_s f)(x) \, ds = \frac{1}{2\pi} \int_{-\pi}^{\pi} f(x) \, ds = f(x) = (S_m f)(x).$$

It remains to prove that

$$\int_{-\pi}^{\pi} (T_{-s} L_m T_s f_k)(x) \, ds = \int_{-\pi}^{\pi} (T_{-s} L_m T_s g_k)(x) \, ds = 0$$

for all $k > m$ and $x \in \mathbb{R}$. We deduce from the identities

$$\cos k(x + s) = \cos ks \cos kx - \sin ks \sin kx$$

and

$$\sin k(x + s) = \sin ks \cos kx + \cos ks \sin kx$$

that

$$T_s f_k = (\cos ks) f_k - (\sin ks) g_k \quad \text{and} \quad T_s g_k = (\sin ks) f_k + (\cos ks) g_k.$$

Consequently,

$$\int_{-\pi}^{\pi} (T_{-s} L_m T_s f_k)(x) \, ds$$

$$= \int_{-\pi}^{\pi} (\cos ks)(L_m f_k)(x - s) - (\sin ks)(L_m g_k)(x - s) \, ds$$

and

$$\int_{-\pi}^{\pi} (T_{-s} L_m T_s g_k)(x) \, ds$$

$$= \int_{-\pi}^{\pi} (\sin ks)(L_m f_k)(x - s) + (\cos ks)(L_m g_k)(x - s) \, ds.$$

---

[47]The proof may be simplified by using complex numbers. See Exercise 8.10, p. 303.

For any fixed x, $(L_m f_k)(x - s)$ and $(L_m g_k)(x - s)$ are trigonometric polynomials of order $\leq m$ in s. Since $k > m$, they are therefore orthogonal to the functions $\cos ks$ and $\sin ks$, so that the right-hand side of both identities vanishes. $\square$

Next we establish an algebraic variant of Theorem 8.17. For this we need a variant of Proposition 8.18, where we replace $C_{2\pi}$ and $\mathcal{T}_m$ by the subspaces $\tilde{C}_{2\pi}$ and $\tilde{\mathcal{T}}_m$ formed by the *even* functions. Let us denote the restriction of S_m to $\tilde{C}_{2\pi}$ by $\tilde{S}_m$, and observe that $\tilde{S}_m : \tilde{C}_{2\pi} \to \tilde{C}_{2\pi}$.

Proposition 8.19 *If a continuous linear map* $L_m : \tilde{C}_{2\pi} \to \tilde{C}_{2\pi}$ *is a projection onto* $\tilde{\mathcal{T}}_m$, *then* $\|L_m\| \geq \|\tilde{S}_m\| / 2$.

Proof Using the notations of the preceding proof it suffices to prove the following identity:

$$(\tilde{S}_m f)(x) = \frac{1}{2\pi} \int_{-\pi}^{\pi} (T_{-s} L_m (T_{-s} + T_s) f)(x) \, ds$$

for all $f \in \tilde{C}_{2\pi}$ and $x \in \mathbb{R}$. Indeed, this will imply

$$\|\tilde{S}_m f\| \leq 2 \|L_m\| \cdot \|f\|$$

for all $f \in \tilde{C}_{2\pi}$.

Since the functions f_k span $\tilde{C}_{2\pi}$, it suffices to prove the identity for these functions. We infer from the trigonometric identity

$$\cos k(x - s) + \cos k(x + s) = 2 \cos ks \cos kx$$

that

$$(T_{-s} + T_s) f_k = (2 \cos ks) f_k,$$

and hence

$$\tilde{R}_m f(x) := \frac{1}{2\pi} \int_{-\pi}^{\pi} (T_{-s} L_m (T_{-s} + T_s) f)(x) \, ds = \frac{1}{2\pi} \int_{-\pi}^{\pi} (2 \cos ks)(L_m f_k)(x - s) \, ds.$$

If $k > m$, then for each fixed x, $(L_m f_k)(x - s)$ is a trigonometric polynomial of order $< k$ in s, and thus orthogonal to $\cos ks$. Therefore $\tilde{R}_m f_k = 0 = \tilde{S}_m f_k$.

If $k \leq m$, then $L_m f_k = f_k$, so that

$$(\tilde{R}_m f_k)(x) = \frac{1}{2\pi} \int_{-\pi}^{\pi} 2 \cos ks \cos k(x - s) \, ds$$

$$= \frac{1}{2\pi} \int_{-\pi}^{\pi} \cos kx + \cos k(x - 2s) \, ds$$

$$= \cos kx$$

$$= f_k(x)$$

$$= \tilde{S}_m f_k(x)$$

again. □

Let us denote by $\mathcal{P}_m$ the vector space of algebraic polynomials of degree $\leq m$.

Theorem 8.20 (Haršiladze–Lozinski)[48] *Consider a sequence of continuous linear maps* $L_m : C_I \to C_I$, *where* I *is a compact interval. If* L_m *is a projection onto* $\mathcal{P}_m$ *for each* m, *then there exists an* $f \in C_I$ *such that* $\|f - L_m f\|_\infty \not\to 0$.

Proof Let $I = [-1, 1]$ for simplicity of notation, and consider the isometric isomorphism $T : f \mapsto f \circ \cos$ between the Banach spaces $C(I)$ and $\tilde{C}_{2\pi}$. Since

$$f \in \mathcal{P}_m \iff Tf \in \tilde{\mathcal{T}}_m,$$

we deduce from the preceding proposition that

$$\|L_m\| = \|TL_m T^{-1}\| \geq \|\tilde{S}_m\| / 2.$$

Let us observe that $\|\tilde{S}_m\| \to \infty$ by the *proof* of Lemma 8.10 (p. 273), because in the proof only even test functions were used. Therefore we may conclude by applying the Banach–Steinhaus theorem (p. 81). □

We end this section with two further famous results. Given a compact interval $I = [a, b]$, we may ask the following natural questions:

- Does there exist a *weight function*[49] on some compact interval $J \supset I$ such that, considering the corresponding orthonormal sequence of polynomials p_n, the Fourier series $\sum (f, p_n) p_n$ converges uniformly to f on I for every $f \in C(J)$?
- Given a system of points $x_{m,0} < \cdots < x_{m,m}$ in I for $m = 0, 1, \ldots$, we may define for each $f \in C(I)$ a sequence of Lagrange interpolation polynomials $L_m f$ such that $L_m = f$ in the points $x_{m,0}, \ldots, x_{m,m}$. Is there a choice of points $x_{m,k}$ such that $L_m f$ converges uniformly to f for every $f \in C(I)$?

---

[48]Lozinski [311].

[49]By a *weight function* we mean a positive, integrable function. If w is a weight function on a compact interval J, then we may define a scalar product on the vector space $\mathcal{P}$ of algebraic polynomials by the formula $(p, q) := \int_J pqw \, dt$, and we may apply the Gram–Schmidt orthogonalization (Proposition 1.15, p. 28) for the sequence of functions $1, \text{id}, \text{id}^2, \ldots$ to obtain a sequence of *orthogonal polynomials* satisfying $\deg p_k = k$ for every $k = 0, 1, \ldots$.

In case of a positive answer we would obtain a natural proof of the Weierstrass approximation theorem. But the answer is negative:

Proposition 8.21

(a) *(Nikolaev)*[50] *For any given weight function there exists an* $f \in C(J)$ *such that* $\sum (f, p_n)p_n$ does not converge uniformly *to f on I.*
(b) *(Faber)*[51] *For any given point system* $(x_{m,k})$ *there exists an* $f \in C(I)$ *such that* $L_m f$ does not converge uniformly *to f on I.*

Proof (a) The continuous linear projections

$$L_m f := \sum_{n=0}^{m} (f, P_n)P_n$$

satisfy the conditions of Theorem 8.20.
(b) These operators L_m also satisfy the conditions of Theorem 8.20.   □

Remarks Historically, the theorems of du Bois Reymond and Faber paved the way to the discovery of the Banach–Steinhaus theorem. Let us mention three further results related to Faber's theorem.

• (Fejér)[52] Let us choose for $x_{m,0}, \ldots, x_{m,m} \in [-1, 1] =: I$ the zeros of the corresponding Chebyshev polynomial, and for $f \in C(I)$ let $H_m f$ denote the Hermite interpolation polynomial of degree $\leq 2m + 1$, satisfying the equalities $(H_m f)(x_{m,k}) = f(x_{m,k})$ and $(H_m f)'(x_{m,k}) = 0$. Then $H_m f$ converges uniformly to f.
• (Erdős–Turán)[53] If w is a weight function on I and $x_{m,0}, \ldots, x_{m,m}$ are the zeros of the corresponding mth orthogonal polynomial, then $L_m f$ converges to f in the weaker norm associated with the scalar product $(p, q) := \int_I pqw \, dt$.
• (Erdős–Vértesi)[54] For any given system of points $x_{m,k}$ there exists a function $f \in C(I)$ such that $\limsup |L_n f(x)| = \infty$ for almost every $x \in I$. Not only do we not have uniform convergence, but we even have *divergence almost everywhere*!

[50]Nikolaev [346]. However, we will see later (Corollary 9.6, p. 314) that the answer is *affirmative* for the *weaker* norm associated with the scalar product.
[51]Faber [133].
[52]Fejér [142]; see also Cheney [85]. In this way, Hermite interpolation can be used to prove the Weierstrass approximation theorem.
[53]Erdős–Turán [124].
[54]Erdős–Vértesi [125].

8.8   * Dual Space. Riesz Representation Theorem

Let K be a compact Hausdorff space. Using measure theory we may characterize the dual of $C(K)$.

Definition Let us denote by $\mathcal{B}$ the smallest σ-ring containing all sets of the form $\{f = 0\}$, where f runs over $C(K)$. The elements of $\mathcal{B}$ are called *Baire sets*.[55]

Remarks

- $\mathcal{B}$ is even a σ-algebra. Moreover, if $g \in C(K)$ and $c \in \mathbb{R}$, then the *level sets*

$$\{g = c\}, \quad \{g \le c\}, \quad \{g \ge c\}$$

and their complements

$$\{g \ne c\}, \quad \{g > c\}, \quad \{g < c\}$$

are also Baire sets, because

$$\{g = c\} = \{g - c = 0\},$$
$$\{g \le c\} = \{(g - c)^+ = 0\}$$

and

$$\{g \ge c\} = \{(g - c)^- = 0\}.$$

- In fact, $\mathcal{B}$ contains all open, closed or compact sets of K. This follows from the Tietze–Urysohn theorem of topology because every compact Hausdorff space is normal. See, e.g., Kelley [247].

Definition By a *(signed) Baire measure* we mean a *finite* (signed) measure defined on $\mathcal{B}$.

Examples For any fixed $a \in K$ the Dirac measure at a is a Baire measure.

The Baire measures have an important *regularity property*: they may be well approximated by both open and closed sets:

Proposition 8.22 *Let μ be a Baire measure, $A \in \mathcal{B}$ and $\varepsilon > 0$. There exist a closed set F and an open set G in $\mathcal{B}$ such that*

$$F \subset A \subset G \quad and \quad \mu(G \setminus F) < \varepsilon. \tag{8.12}$$

---

[55]Baire [17].

Proof Let us denote temporarily by $\tilde{B}$ the family of Baire sets having the property (8.12). We have to show that $\tilde{B}$ is a σ-algebra containing all sets $\{f = 0\}$ with $f \in C(K)$.

If $A = \{f = 0\}$ for some $f \in C(K)$, then the formulas

$$F := A, \quad G_n := \{|f| < 1/n\}, \quad n = 1, 2, \ldots$$

define a closed set $F \in B$ and open sets $G_n \in B$ satisfying $F \subset A \subset G_n$ for all n.

Since the set sequence (G_n) is non-increasing and

$$\bigcap_{n=1}^{\infty}(G_n \setminus F) = \bigcap_{n=1}^{\infty}\{0 < |f| < 1/n\} = \varnothing,$$

Proposition 7.3 (p. 216) implies that $\mu(G_n \setminus F) < \varepsilon$ if n is sufficiently large.

It remains to prove the σ-algebra property. Choosing the constant functions $f = 0$ and $f = 1$ we see that K and $\varnothing$ belong to B. Moreover, since they are both open and closed, they belong to $\tilde{B}$ as well: we may choose $F = G = \varnothing$ and $F = G = K$.

If $A \in \tilde{B}$, then $K \setminus A \in \tilde{B}$. Indeed, if F and G satisfy (8.12), then $K \setminus G$ is closed, $K \setminus F$ is open, both belong to B,

$$K \setminus G \subset K \setminus A \subset K \setminus F \quad \text{and} \quad \mu\big((K \setminus F) \setminus (K \setminus G)\big) = \mu(G \setminus F) < \varepsilon.$$

Finally, if (A_n) is a disjoint sequence in $\tilde{B}$, then $A := \cup^* A_n \in \tilde{B}$. For the proof, for any fixed $\varepsilon > 0$ we choose closed sets $F_n \in B$ and open sets $G_n \in B$ such that

$$F_n \subset A_n \subset G_n \quad \text{and} \quad \mu(G_n \setminus F_n) < 2^{-n-1}\varepsilon$$

for all n. Then $G := \cup_{n=1}^{\infty} G_n$ is open, the sets $F^N := \cup_{n=1}^{N} F_n$ are closed for all $N = 1, 2, \ldots$, all belong to B, $F^N \subset A \subset G$, and

$$\mu(G \setminus F^N) \le \left(\sum_{n=1}^{N}\mu(G_n \setminus F_n)\right) + \sum_{n>N}\mu(G_n) < \frac{\varepsilon}{2} + \sum_{n>N}\mu(G_n).$$

Since

$$\sum_{n=1}^{\infty}\mu(G_n) < \sum_{n=1}^{\infty}(\mu(A_n) + 2^{-n}\varepsilon) = \mu(A) + \varepsilon < \infty,$$

it follows that $\mu(G \setminus F^N) < \varepsilon$ if N is sufficiently large. □

Setting $\|\mu\| := |\mu|\,(K)$ the signed Baire measures form a normed space $M(K)$,[56] and the formula

$$(j\mu)(f) := \int f\,d\mu$$

defines a continuous linear map $j : M(K) \to C(K)'$ of norm ≤ 1.

The only non-trivial property is the triangle inequality. For the proof we consider two measures μ, μ' and the corresponding Hahn decompositions $K = P \cup^* N$ and $K = P' \cup^* N'$. Setting

$$A := (P \cap P') \cup^* (N \cap N') \quad \text{and} \quad B := (P \cap N') \cup^* (N \cap P')$$

we have the following relations:

$$
\begin{aligned}
\|\mu + \mu'\| &= |\mu + \mu'|\,(K) \\
&= |\mu + \mu'|\,(A) + |\mu + \mu'|\,(B) \\
&= (|\mu| + |\mu'|)\,(A) + |\mu + \mu'|\,(B) \\
&\leq (|\mu| + |\mu'|)\,(A) + (|\mu| + |\mu'|)\,(B) \\
&= (|\mu| + |\mu'|)\,(K) \\
&= \|\mu\| + \|\mu'\|\,.
\end{aligned}
$$

The main result of this section states that *every* linear functional on $C(K)$ may be obtained in this way, and that $M(K)$ is *complete*.

Theorem 8.23 (Riesz)[57] *If K is a compact topological space, then j is an isometric isomorphism between $M(K)$ and $C(K)'$.*

Remark It is not necessary to assume the Hausdorff property of K: identifying two points x, y if $h(x) = h(y)$ for every $h \in C(K)$, we may reduce the theorem to the case where any two distinct points may be separated by a continuous function. Henceforth we assume this property.[58]

We proceed in several steps.

---

[56]Here $|\mu|$ denotes the total variation of μ; see p. 231.

[57]Riesz [377] $(K = [0, 1])$, Radon [366] $(K \subset \mathbb{R}^N$, p. 1333), Banach [25] and Saks [410] (compact metric spaces), Markov [315] $(C_b(K)$ certain non-compact spaces), Kakutani [240] (compact topological spaces). See also the beautiful simple proof of Riesz for $K = [0, 1]$: Riesz and Sz.-Nagy [394, Sect. 50].

[58]We will need it only during the proof of Lemma 8.27 below, in order to apply Proposition 8.6.

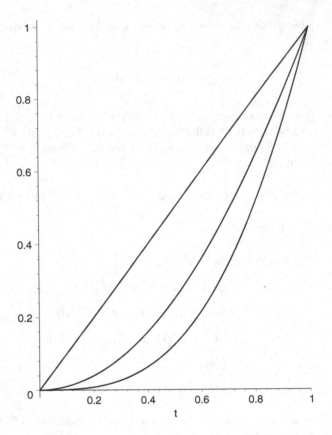

Fig. 8.15 Theorem of Dini

Proposition 8.24 (Dini)[59] *If a non-increasing sequence* $(f_n) \subset C(K)$ *tends to zero pointwise, then the convergence is uniform.*

Proof For any fixed $\varepsilon > 0$ we have to find a positive integer N such that $\|f_n\|_\infty < \varepsilon$ for all $n \geq N$.

For each $t \in K$ there exists an index n_t such that $f_{n_t}(t) < \varepsilon$; by continuity the inequality $f_{n_t} < \varepsilon$ remains valid in some open neighborhood V_t of t. Since K is compact, a finite number of such neighborhoods, say $V_{t_1}, \ldots, V_{t_m}$, already cover K.

Choose $N := \max\{n_{t_1}, \ldots, n_{t_m}\}$, let $n \geq N$, and consider a point $s \in K$. Then s belongs to some neighborhood V_{t_i}, and therefore

$$0 \leq f_n(s) \leq f_{n_{t_i}}(s) < \varepsilon$$

by the non-increasingness of the sequence (f_n).   □

---

[59]Dini [109, Sect. 99]. See the graphs of the functions $f_n(t) := t^n$ for $n = 1, 2, 3$ in Fig. 8.15, and let $K = [0, a], 0 < a < 1$.

Fig. 8.16 An "interval"
$[f, g)$

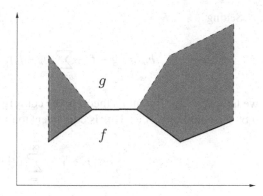

Lemma 8.25 *For each* positive *linear functional* $\varphi : C(K) \to \mathbb{R}$ *there exists a* Baire measure $\mu \in M(K)$ *such that* $\varphi = j\mu$.

Proof Following Kindler[60] we introduce the "intervals"

$$[f, g) := \{(x, t) \in K \times \mathbb{R} : f(x) \leq t < g(x)\}$$

for all functions $f, g \in C(K)$ satisfying $f \leq g$.[61] They form a semiring $\mathcal{P}$ in $K \times \mathbb{R}$,[62] and the formula

$$\nu([f, g)) := \varphi(g - f)$$

defines a finite, additive set function on $\mathcal{P}$, satisfying $\nu(\varnothing) = 0$.

  This set function is also σ-additive, and hence a measure. For the proof we consider an arbitrary countable decomposition $[f, g) = \cup^*[f_n, g_n)$. We have

$$[f(x), g(x)) = \cup^*[f_n(x), g_n(x))$$

for each $x \in K$, and therefore

$$g(x) - f(x) = \sum_{n=1}^{\infty} g_n(x) - f_n(x),$$

because the length of ordinary intervals is a measure.

---

[60] Kindler [248].

[61] See Fig. 8.16.

[62] The proof is similar to that of ordinary intervals.

Setting

$$h_m := g - f - \sum_{n=1}^{m} (g_n - f_n), \quad m = 1, 2, \ldots$$

we have $h_m \searrow 0$. By Dini's theorem the convergence is uniform, and then $\varphi(h_m) \to 0$ by the continuity of φ. This is equivalent to the σ-additivity relation

$$\nu([f, g)) = \sum_{n=1}^{\infty} \nu([f_n, g_n)).$$

Applying Proposition 5.18 (p. 192) we extend ν to a measure defined on the σ-ring $\mathcal{M}$ of measurable sets, still denoted by ν.[63]

If $f \in C(K)$ and c is a positive real number, then the set

$$\{f = 0\} \times [0, c) = \bigcap_{n=1}^{\infty} [\min\{n|f|, c\}, c)$$

belongs to $\mathcal{M}$. Since $\mathcal{B}$ is the *smallest* σ-algebra containing the sets $\{f = 0\}$, this implies that

$$A \in \mathcal{B} \Longrightarrow A \times [0, 1) \in \mathcal{M}.$$

Consequently, the formula

$$\mu(A) := \nu(A \times [0, 1))$$

defines a Baire measure $\mu \in M(K)$.[64] It remains to prove that $\varphi(f) = \int f \, d\mu$ for all $f \in C(K)$.

Given $f \in C(K)$, the continuous functions

$$f_n(x) := \text{med}\{0, n(f(x) - 1), 1\}, \quad x \in K, \quad n = 1, 2, \ldots$$

form a non-decreasing sequence converging to the characteristic function $\chi_{\{f > 1\}}$. Hence

$$\{f > 1\} \times [0, c) = \bigcup_{n=1}^{\infty} [0, cf_n)$$

---

[63] $\mathcal{M}$ is even a σ-algebra.
[64] The finiteness follows from the relation $\mu(K) = \varphi(1) < \infty$.

for each positive number c, and therefore

$$\nu(\{f > 1\} \times [0, c)) = \lim_{n \to \infty} \nu([0, cf_n)) = \lim_{n \to \infty} \varphi(cf_n)$$

$$= c \lim_{n \to \infty} \varphi(f_n) = c \lim_{n \to \infty} \nu([0, f_n))$$

$$= c\nu(\{f > 1\} \times [0, 1)) = c\mu(\{f > 1\}).$$

By the additivity of the measures ν and μ this implies the more general relations

$$\nu(\{a < f \le b\} \times [0, c)) = c\mu(\{a < f \le b\}) \qquad (8.13)$$

for all numbers $0 < a < b$.[65]

Now we use (8.13) to prove the equalities $\varphi(f) = \int f \, d\mu$. Separating the positive and negative parts of f we may assume that $f \ge 0$. Then the "interval" $[0, f)$ is the union of the non-decreasing sequence of sets

$$B_n := \sum_{i=1}^{n2^n} \left\{ \frac{i}{2^n} < f \le \frac{i+1}{2^n} \right\} \times \left[0, \frac{i}{2^n} \right),$$

and therefore

$$\varphi(f) = \nu([0, f)) = \lim_{n \to \infty} \nu(B_n)$$

$$= \lim_{n \to \infty} \frac{i}{2^n} \sum_{i=1}^{n2^n} \mu \left(\left\{ \frac{i}{2^n} < f \le \frac{i+1}{2^n} \right\} \right) = \int f \, d\mu. \qquad \square$$

Lemma 8.26 *Every continuous linear functional $\varphi \in C(K)'$ is the difference of two positive linear functionals.*

Proof We denote by $C_+(K)$ the set of nonnegative functions in $C(K)$, and for $f \in C_+(K)$ we define

$$\psi(f) := \sup \{\varphi(f') : f' \in C_+(K) \text{ and } f' \le f \text{ on } K\}.$$

Then all $f, g \in C_+(K)$ and $c \ge 0$ satisfy the following conditions:

$$\varphi(f) \le \psi(f);$$

$$0 \le \psi(f) \le \|\varphi\| \cdot \|f\| < \infty;$$

---

[65] We apply the preceding identity to f/a and f/b, and we take the differences of the resulting equalities.

$$\psi(cf) = c\psi(f) \quad \text{for all} \quad c \ge 0;$$
$$\psi(f + g) = \psi(f) + \psi(g).$$

Only the last relation is not obvious: for the proof it suffices to establish for each fixed $\varepsilon > 0$ the inequalities

$$\psi(f + g) \ge \psi(f) + \psi(g) - 2\varepsilon \quad \text{and} \quad \psi(f + g) \le \psi(f) + \psi(g) + \varepsilon.$$

To prove the first one we choose two functions $0 \le f' \le f$ and $0 \le g' \le g$ satisfying

$$\varphi(f') > \psi(f) - \varepsilon \quad \text{and} \quad \varphi(g') > \psi(g) - \varepsilon.$$

Then we have

$$\psi(f + g) \ge \varphi(f' + g') = \varphi(f') + \varphi(g') > \psi(f) + \psi(g) - 2\varepsilon.$$

To prove the second one we choose a function $0 \le h' \le f + g$ satisfying

$$\varphi(h') > \psi(f + g) - \varepsilon.$$

Setting

$$f' := \min\{f, h'\} \quad \text{and} \quad g' := h' - f'$$

we have[66]

$$0 \le f' \le f \quad \text{and} \quad 0 \le g' \le g,$$

and therefore

$$\psi(f + g) < \varphi(h') + \varepsilon = \varphi(f') + \varphi(g') + \varepsilon \le \psi(f) + \psi(g) + \varepsilon.$$

Now we extend ψ to a positive linear map on $C(K)$ by setting[67]

$$\Psi(f) := \psi(f^+) - \psi(f^-).$$

Only the additivity is not obvious. This follows from the additivity of ψ, by using the nonnegative function

$$h := f^+ + g^+ - (f + g)^+ = f^- + g^- - (f + g)^-$$

---

[66] We have $g'(x) = 0$ if $f(x) \ge h'(x)$, and $0 \le g'(x) = h'(x) - f(x) \le g(x)$ otherwise.
[67] As usual, f^+ and f^- denote the positive and negative parts of f.

as follows:

$$\Psi(f + g) = \psi((f + g)^+) - \psi((f + g)^-)$$
$$= \psi((f + g)^+) + \psi(h) - \psi((f + g)^-) - \psi(h)$$
$$= \psi(f^+ + g^+) - \psi(f^- + g^-)$$
$$= \psi(f^+) + \psi(g^+) - \psi(f^-) - \psi(g^-)$$
$$= \Psi(f) + \Psi(g).$$

We complete the proof of the lemma by observing that, as a result of the inequality $\varphi \leq \psi$, $\Psi - \varphi$ is also a *positive* linear functional on $C(K)$. $\square$

It follows from the preceding two lemmas that the linear map $j : M(K) \to C(K)'$ is surjective. The next lemma completes the proof of Theorem 8.23:

Lemma 8.27 *The linear map* $j : M(K) \to C(K)'$ *is an isometry.*

Proof We already know that j is continuous, and $\|j\| \leq 1$. It remains to prove the inequality $\|j\mu\| \geq \|\mu\|$ for each μ.

Fix $\mu \in M(K)$ and $\varepsilon > 0$ arbitrarily, and consider the Hahn decomposition $K = P \cup^* N$ of μ. By Proposition 8.22 (p. 289) there exist two disjoint closed sets $P' \subset P$ and $N' \subset N$ satisfying

$$\left|\mu(P \setminus P')\right| < \varepsilon \quad \text{and} \quad \left|\mu(N \setminus N')\right| < \varepsilon.$$

The function

$$g(t) := \begin{cases} 1 & \text{if } t \in P', \\ -1 & \text{if } t \in N' \end{cases}$$

is clearly continuous on $P' \cup^* N'$. Applying Proposition 8.6 (p. 267), g may be extended to a function $f \in C(K)$. Changing f to med $\{-1, f, 1\}$ if necessary, we may also assume that $|f| \leq 1$ on K.[68] Then $\|f\| \leq 1$, and

$$\|j\mu\| \geq (j\mu)(f)$$

$$= \int_{P'} f \, d\mu + \int_{N'} f \, d\mu + \int_{P \setminus P'} f \, d\mu + \int_{N \setminus N'} f \, d\mu$$

---

[68]If K is metrizable, then we may define f explicitly by the formula

$$f(t) := \frac{\text{dist}(t, N') - \text{dist}(t, P')}{\text{dist}(t, N') + \text{dist}(t, P')}.$$

$$\geq \mu(P') - \mu(N') - 2\varepsilon$$
$$\geq \mu(P) - \mu(N) - 4\varepsilon$$
$$= \|\mu\| - 4\varepsilon.$$

Letting $\varepsilon \to 0$ we conclude that $\|j\mu\| \geq \|\mu\|$.   □

Example Using Theorem 8.23 we may prove directly the non-reflexivity of $C([0,1])$.[69] Given any $\mu \in M(K)$ with $K := [0,1]$, the formulas

$$m(t) := \mu([0,t]), \quad t \in [0,1]$$

and

$$\Phi(\mu) := \sum_{0 < t < 1} m(t+) - m(t-)$$

define a continuous linear functional Φ on $M(K)$.[70]

We claim that Φ is not represented by any function $f \in C(K)$. Assume on the contrary that there exists an $f \in C(K)$ satisfying

$$\Phi(\mu) = \int_0^1 f \, d\mu$$

for all $\mu \in M(K)$. Applying this to the Dirac measures $\mu := \delta_t$, we obtain $m = \chi_{[t,1]}$, and hence $f(t) = 1$ for each $0 < t < 1$. But then $\int_0^1 f \, d\mu = 1$ for the usual Lebesgue measure, while $\Phi(\mu) = 0$ because now $m(t) \equiv t$ is continuous.

Remark Using the Dirac measures we may also show that the dual of $C([0,1])$ is *non-separable*. For the proof first we observe that if $0 \leq a < b \leq 1$, then $\|\delta_a - \delta_b\| = 2$. Indeed, we have

$$|(\delta_a - \delta_b)(f)| = |f(a) - f(b)| \leq 2 \|f\|_\infty$$

for all $f \in C([0,1])$, so that $\|\delta_a - \delta_b\| \leq 2$. On the other hand, choosing

$$f(t) := \mathrm{med} \left\{ -1, \frac{2t - a - b}{b - a}, 1 \right\}$$

---

[69]We follow Riesz–Sz.-Nagy [394].

[70]The second formula is meaningful because m has bounded variation and hence at most countably many discontinuities.

(make a figure) we have $\|f\|_\infty = 1$, so that[71]

$$\|\delta_a - \delta_b\| \geq |(\delta_a - \delta_b)(f)| = |f(a) - f(b)| = 2.$$

It follows that $C([0, 1])'$ contains uncountably many pairwise disjoint open balls:

$$B_1(\delta_a), \quad a \in [0, 1],$$

and no countable set may meet each of them.[72]

8.9   Weak Convergence

We recall that the strong convergence in $C(K)$ is uniform convergence on K. Now we characterize the weak convergence[73]:

Proposition 8.28 *If* $f_n, f \in C(K)$, *then the following conditions are equivalent:*

(a) $f_n \rightharpoonup f$;
(b) *the sequence* (f_n) *is uniformly bounded, and converges* pointwise *to* f.

Proof If f_n converges weakly to f in $C(K)$, then (f_n) is bounded in norm by Proposition 2.24 (p. 82), i.e., it is uniformly bounded. Furthermore, using the Dirac measures $\delta_t \in C(K)'$ we see that $\delta_t(f_n) \to \delta_t(f)$, i.e., $f_n(t) \to f(t)$ for each $t \in K$.

Conversely, if (f_n) is uniformly bounded, and converges pointwise to f, then

$$\int f_n \, d\mu \to \int f \, d\mu$$

for every $\mu \in M(K)$ by Lebesgue's dominated convergence theorem (p. 181). In view of Theorem 8.23 (p. 291) this means that f_n converges weakly to f.

Example Using the proposition we may give yet another proof of the non-reflexivity of $C([0, 1])$. The formula $f_n(t) := t^n$ defines a uniformly bounded sequence (f_n) in $C([0, 1])$, converging pointwise to the *non-continuous* function

$$f(t) := \begin{cases} 0 & \text{if } 0 \leq t < 1, \\ 1 & \text{if } t = 1. \end{cases}$$

(See Fig. 8.15, p. 292.)

---

[71] Komornik–Yamamoto [261, 262] apply such estimates to inverse problems.
[72] Compare this with the proof of the non-separability of ℓ^∞, p. 74.
[73] For the characterization of the weakly compact sets of $C(K)$ see, e.g., Dunford–Schwartz [117].

Hence no subsequence of (f_n) can converge pointwise to any continuous function, i.e, (f_n) has no weakly convergent subsequence. In view of Theorem 2.30 (p. 90) this implies that $C([0, 1])$ is not reflexive.

8.10 Exercises

Exercise 8.1 Consider[74] the polynomials $q_0(x) = 1$ and

$$q_n(x) := \frac{1}{2}\left(q_{n-1}(x)^2 + 1 - x^2\right), \quad n = 1, 2, \ldots.$$

(i) Prove by induction that

$$q_n \geq 0 \quad \text{and} \quad q_n \geq q_{n+1} \quad \text{in} \quad [-1, 1] \quad \text{for all} \quad n.$$

(ii) Prove that $q_n(x) \to 1 - |x|$ uniformly in $[-1, 1]$.
(iii) Deduce from the preceding result that $|x|$ is the uniform limit of a suitable sequence of polynomials in each compact interval $[a, b]$.

Exercise 8.2 Prove that for any given finite subdivision $a = x_1 < \cdots < x_n = b$ of $I := [a, b]$, the functions $x \mapsto |x - x_i|$, $i = 1, \ldots, n$, form a basis of the vector space L of continuous functions $f : I \to \mathbb{R}$ which are linear in each subinterval (x_i, x_{i+1}).

Exercise 8.3 Prove the Weierstrass approximation theorem in the following way[75]:

(i) Each $f \in C(I)$ may be approximated uniformly by continuous and piecewise linear functions.
(ii) Prove the theorem for piecewise linear functions by applying the preceding two exercises.

Exercise 8.4 Let $I := [a, b]$ be a compact interval, $f \in C(I)$, and denote by $\mathcal{P}_n$ the subspace of $C(I)$ formed by the polynomials of degree $\leq n$, $n = 0, 1, \ldots$. Prove the following[76]:

(i) $\mathcal{P}_n$ has a closest element p to f. Set $d := \|f - p\|_\infty$.
(ii) There exist at least $n + 2$ consecutive values where $f(x) - p(x) = \pm d$, with alternating signs.
(iii) The closest polynomial p is unique.

[74]Visser [469], see Sz.-Nagy [448, p. 77.].
[75]Lebesgue [286, 296].
[76]Chebyshev [83], Borel [60].

Exercise 8.5 (Convergence of Fourier Series) Given $f \in C_{2\pi}$, set[77]

$$\hat{f}(n) = \frac{1}{2\pi} \int_{-\pi}^{\pi} f(x)e^{-inx}\, dx$$

and

$$S_{m,n}(x) = \sum_{k=-m}^{n} \hat{f}(k)e^{ikx}.$$

We are going to show that *if* $f \in C_{2\pi}$ *is differentiable at* x_0, *then* $S_{m,n}(x_0) \to f(x_0)$ *as* $m, n \to \infty$.[78] Prove the following:

(i) If $g \in C_{2\pi}$, then $\hat{g}(n) \to 0$ as $n \to \pm\infty$.
(ii) If $x_0 = 0, f(0) = 0$ and $f'(0)$ exists, then

$$f(x) = (e^{ix} - 1)g(x)$$

with some $g \in C_{2\pi}$.
(iii) Deduce from the last equality that

$$S_{m,n}(0) = \sum_{k=-m}^{n} \hat{f}(k) = \hat{g}(-m-1) - \hat{g}(n) \to 0$$

as $m, n \to \infty$.
(iv) Prove the general case by a translation argument.

Exercise 8.6 Prove Ascoli's theorem (p. 268) for compact *metric* spaces K as follows. Let $(f_n) \subset C(K)$ be a pointwise bounded and equicontinuous sequence of functions.

(i) Choose a countable dense set $\{x_j\} \subset K$ and prove the existence of a subsequence $(f_{n_k}) \subset (f_n)$ converging at each x_j.
(ii) Prove that (f_{n_k}) converges at each point of K.
(iii) Prove that the convergence is uniform, and hence the limit function is continuous.

---

[77] For brevity we use the complex notation.
[78] Chernoff [86]. The method is quite general and leads to an improvement of the classical theorems of Lipschitz and Dini. It was motivated by an earlier simple proof of the Fourier inversion theorem by Richards [369].

Exercise 8.7 (A Nowhere Differentiable Continuous Function)[79] Set[80]

$$a_0(x) := \text{dist}(x, \mathbb{Z}), \quad a_k(x) := 2^{-k}a_0(2^k x) \quad \text{and} \quad f(x) := \sum_{k=0}^{\infty} a_k(x)$$

for $x \in \mathbb{R}$. Prove the following:

(i) $f : \mathbb{R} \to \mathbb{R}$ is a continuous, one-periodic function.
(ii) For any fixed $x \in \mathbb{R}$ choose a sequence (m_n) of integers such that

$$y_n := m_n 2^{-n-1} \leq x \leq (m_n + 1)2^{-n-1} =: z_n, \quad n = 1, 2, \ldots.$$

Show that if f is differentiable in x, then

$$\lim_{n\to\infty} \frac{f(z_n) - f(y_n)}{z_n - y_n} = f'(x).$$

(iii) Show that

$$\frac{a_k(z_n) - a_k(y_n)}{z_n - y_n} = \pm 1 \quad \text{if} \quad k \leq n, \quad \text{and} \quad = 0 \quad \text{otherwise.}$$

(iv) Conclude that the fractions $\frac{f(z_n)-f(y_n)}{z_n-y_n}$ are alternatively odd and even integers, and hence their sequence is divergent.

Exercise 8.8 (Peano Curve)[81] We prove that *there exists a continuous map of the unit interval* $[0, 1]$ *onto the unit square* $[0, 1] \times [0, 1]$.

We recall that Cantor's ternary set C consists of those points $t \in [0, 1]$ which can be written in the form

$$t = 2\left(\frac{t_1}{3} + \frac{t_2}{3^2} + \cdots + \frac{t_n}{3^n} + \cdots\right)$$

with suitable integers $t_n \in \{0, 1\}$. Set

$$f_1(t) := \frac{t_1}{2} + \frac{t_3}{2^2} + \cdots + \frac{t_{2n-1}}{2^n} + \cdots$$

---

[79]The first examples were due to Bolzano [55] around 1832 (published only in 1930) and Weierstrass [480, 481]. See also Bolzano [57], Russ [407], Jarník [227, p. 37], du Bois-Reymond [50], Dini [108], Hawkins [198].
[80]Takagi [449]. His example was rediscovered by van der Waerden [477]. See also Billingsley [44], Shidfar–Sabetfakhiri [422], McCarthy [319].
[81]Peano [354]. The following proof is due to Lebesgue [297, pp. 44–45]. An interesting variant of this proof is due to Schoenberg [417]. See also Aleksandrov [4].

and

$$f_2(t) := \frac{t_2}{2} + \frac{t_4}{2^2} + \cdots + \frac{t_{2n}}{2^n} + \cdots .$$

Prove the following:

(i) $f := (f_1, f_2)$ maps C *onto* $[0, 1] \times [0, 1]$.
(ii) f is uniformly continuous.
(iii) f_1, f_2 may be extended to continuous functions of $[0, 1]$ into $[0, 1]$.
(iv) f is Hölder continuous (this last step is not necessary for the proof of the theorem).

Exercise 8.9 Prove Lemmas 8.9 and 8.12 (pp. 271, 276) on the Dirichlet and Fejér kernels by using complex exponentials.

Exercise 8.10 Simplify the proof of Lozinski's Proposition 8.18 (p. 284) by using complex exponentials.

Exercise 8.11 (Schauder Basis)[82] A *Schauder basis* of a normed space X is a sequence $(f_n) \subset X$ such that each $f \in X$ has a unique representation of the form $f = \sum c_n f_n$ with suitable coefficients c_n.

Let $x_0, x_1, \ldots$ be a dense sequence of distinct elements in a compact interval $I = [a, b]$ such that $x_0 = a$ and $x_1 = b$. Set $f_0(x) = 1$ and $f_1(x) = (x - a)/(b - a)$. Furthermore, for $n \geq 2$ set

$$a_n := \max \{ x_j : j < n \quad \text{and} \quad x_j < x_n \},$$
$$b_n := \min \{ x_j : j < n \quad \text{and} \quad x_j > x_n \},$$
$$f_n(x) := \text{med} \{ (x - a_n)/(x_n - a_n), (b_n - x)/(b_n - x_n), 0 \}.$$

Draw a figure.

Finally, for $f \in C(I)$ and $n \geq 1$ we denote by $L_n f \in C(I)$ the polygonal approximation of f consisting of n linear segments and coinciding with f in $x_0, x_1, \ldots, x_n$. Set also $L_0 f := f(a)$.

Prove the following statements:

(i) $\| f - L_n f \|_\infty \to 0$.
(ii) We have

$$L_n f = L_{n-1} f + (f - L_{n-1} f)(x_n) f_n, \quad n = 1, 2, \ldots .$$

---

[82]Schauder [412].

(iii) We have

$$L_n f = \sum_{j=0}^{n} c_j f_j$$

with

$$c_0 = f(x_0) \quad \text{and} \quad c_j = (f - L_{j-1}f)(x_j) \quad \text{for} \quad j = 1, \ldots, n.$$

(iv) If $\sum c_n f_n \equiv 0$, then all coefficients c_n vanish.

Chapter 9
Spaces of Integrable Functions

Beauty is the first test: there is no permanent place in the world for ugly mathematics.—
G. Hardy

The function spaces introduced in this chapter play an important role in many branches of mathematics, including the theory of probability and partial differential equations. They are based on the Lebesgue integral.

We consider an arbitrary *measure space* $(X, \mathcal{M}, \mu)$, i.e., μ is a σ-finite, complete measure on a σ-ring $\mathcal{M}$ in X.

If $X = I$ is an interval of $\mathbb{R}$, then we usually consider the ordinary Lebesgue measure on I.[1]

As usual, we identify two functions if they are equal almost everywhere.

9.1 L^p Spaces, $1 \le p \le \infty$

Definitions Given a measurable function f on X, we set[2]

$$\|f\|_p := \left(\int_X |f|^p \, d\mu \right)^{1/p}, \quad 1 \le p < \infty$$

---

[1]We consider only real-valued functions. See, e.g., Dunford–Schwartz [117], Edwards [119] or Yosida [488] for the study of spaces of Banach space-valued *Bochner-integrable* functions.

[2]Riesz [377]. More general spaces were introduced by Orlicz [347, 348]; see Krasnoselskii–Rutickii [267].

© Springer-Verlag London 2016
V. Komornik, *Lectures on Functional Analysis and the Lebesgue Integral*,
Universitext, DOI 10.1007/978-1-4471-6811-9_9

and[3]

$$\|f\|_\infty := \inf \{M \geq 0 \, : \, |f| \leq M \quad \text{p.p.}\}.$$

Furthermore, we denote by $L^p(X, \mathcal{M}, \mu)$ or shortly by L^p the set of measurable functions satisfying $\|f\|_p < \infty$.

We will soon justify the notation by showing that $\|\cdot\|_p$ is a norm on L^p for each p.

Remarks • The norm $\|f\|_2$ is associated with the scalar product

$$(f, g) := \int_X fg \, d\mu.$$

• The notation $\|f\|_\infty$ is motivated by the relation

$$\|f\|_\infty = \lim_{p \to \infty} \|f\|_p,$$

valid for all $f \in L^\infty$ if $\mu(X) < \infty$.[4]
• If we consider the counting measure on the set X of natural numbers, then the spaces L^p reduce to the spaces ℓ^p investigated in Part I of this book.

First we generalize Proposition 2.14 and Theorem 5.12 (pp. 70 and 184).

Proposition 9.1 *Let $p, q \in [1, \infty]$ be conjugate exponents.*

(a) *(Hölder's inequality)[5] If $f \in L^p$ and $g \in L^q$, then $fg \in L^1$ and*

$$\|fg\|_1 \leq \|f\|_p \cdot \|g\|_q.$$

(b) *(Minkowski's inequality)[6] If $f, g \in L^p$, then $f + g \in L^p$ and*

$$\|f + g\|_p \leq \|f\|_p + \|g\|_p.$$

(c) *(Riesz–Fischer)[7] L^p is a Banach space. L^2 is a Hilbert space.*

---

[3]This is in fact a *minimum* by an elementary argument.
[4]Private communication of E. Fischer to F. Riesz, see [379, 380].
[5]Riesz [379, 384].
[6]Riesz [379, 384].
[7]See Footnote 17 on p. 184.

For the proof we first generalize Lemma 5.13:

Lemma 9.2 (Riesz)[8] *Let (f_n) be a Cauchy sequence in L^p, $1 \le p \le \infty$. There exists a subsequence (f_{n_k}) and two functions $f, g \in L^p$ such that $|f_{n_k}| \le g$ for all k, and $f_{n_k} \to f$ a.e.*

Remark For $p = \infty$ we do not need subsequences, and the following property holds: f_n *converges uniformly to some* $f \in L^\infty$ *outside a null set.*

Indeed, if (f_n) is a Cauchy sequence in L^∞, then there exist a null set $A \subset X$ and a sequence (h_n) of bounded functions on $K := X \setminus A$ such that $f_n \equiv h_n$ on K for each n, and (h_n) is a Cauchy sequence in $\mathcal{B}(K)$.

Since $\mathcal{B}(K)$ is complete, (h_n) converges uniformly to some $h \in \mathcal{B}(K)$. Setting $f := h$ on K and $f := 0$ on A, we obtain a bounded, measurable function f, satisfying $f_n \to f$ in L^∞.

Example (Fréchet)[9] The sequence of functions

$$f_{2^k+i}(t) := \begin{cases} 1 & \text{if } \frac{i}{2^k} \le t \le \frac{i+1}{2^k}, \\ 0 & \text{otherwise,} \end{cases} \quad k = 0, 1, \ldots, \quad i = 0, 1, \ldots, 2^k - 1$$

converges to zero in $l^p(0,1)$ for each $1 \le p < \infty$, but the numerical sequence $(f_n(t))$ is divergent for each fixed $t \in [0,1]$.

The use of subsequences is therefore necessary in the lemma.

Proof The case $p = 1$ has already been proved in Lemma 5.13 (p. 184). Let $1 < p < \infty$, and choose a subsequence (f_{n_k}) satisfying

$$\|f_n - f_{n_k}\|_p \le 2^{-k} \quad \text{for all} \quad n \ge n_k, \quad k = 1, 2, \ldots.$$

Next, using Lemma 7.5 (p. 220) choose a sequence (A_m) of sets of finite measure such that each f_{n_k} vanishes outside $A := \cup A_m$.

Applying the Hölder inequality we obtain for each m the inequalities

$$\sum_{k=1}^{\infty} \int_{A_m} |f_{n_{k+1}} - f_{n_k}| \, d\mu \le \sum_{k=1}^{\infty} \mu(A_m)^{1/q} \cdot \|f_{n_{k+1}} - f_{n_k}\|_p$$

$$\le \mu(A_m)^{1/q} < \infty,$$

where q stands for the conjugate exponent of p.

---

[8]Riesz [377].
[9]Fréchet [160].

Applying Corollary 5.9 (p. 180), it follows that the series

$$|f_{n_1}| + \sum_{k=1}^{\infty} |f_{n_{k+1}} - f_{n_k}| \quad \text{and} \quad f_{n_1} + \sum_{k=1}^{\infty} (f_{n_{k+1}} - f_{n_k})$$

converge a.e. on $A = \cup A_m$ to some limit functions g and f.

Comparing their partial sums g_k and f_{n_k} we have $f_{n_k} \to f$ a.e., and $|f_{n_k}| \le g_k \le g$ for all k by the triangle inequality. Hence $|f| \le g$.

Extending f and g by zero outside A, these relations hold on the whole X. Since $\|g_k\|_p \le \|f_{n_1}\|_p + 1$ by the choice of the subsequence (f_{n_k}), we have $g \in L^p$ by the Fatou lemma (p. 183), and then $f \in L^p$, because $|f| \le g$. $\qquad\qquad \square$

Proof of Proposition 9.1

(a) If $f \in L^p$ and $g \in L^q$, then f, g are measurable and hence fg is also measurable. If $p = 1$, then $q = \infty$, and the inequality follows by a straightforward computation:

$$\|fg\|_1 = \int |fg| \, dt \le \int |f| \cdot \|g\|_\infty \, dt = \|f\|_1 \|g\|_\infty .$$

The case $p = \infty$ is analogous.

If $1 < p < \infty$ and $1 < q < \infty$, then we may assume by homogeneity that $\|f\|_p = \|g\|_q = 1$. Using Young's inequality (p. 70) we obtain that

$$\|fg\|_1 = \int_I |f| \cdot |g| \, dt \le \int_I \frac{|f|^p}{p} + \frac{|g|^q}{q} \, dt = \frac{1}{p} + \frac{1}{q} = 1 = \|f\|_p \cdot \|g\|_q .$$

(b) If $f, g \in L^p$, then f, g are measurable and hence $f + g$ is also measurable. The case $p = 1$ is easy:

$$\|f + g\|_1 = \int |f + g| \, dt \le \int |f| + |g| \, dt = \|f\|_1 + \|g\|_1 .$$

The case $p = \infty$ is also simple: we have

$$|f + g| \le |f| + |g| \le \|f\|_\infty + \|g\|_\infty ,$$

and hence

$$\|f + g\|_\infty \le \|f\|_\infty + \|g\|_\infty$$

by definition.

Now let $1 < p < \infty$ and $1 < q < \infty$. Since $|f + g| \le |f| + |g|$, we may assume that both f and g are nonnegative.

Since f, g are measurable, there exists a non-decreasing sequence $A_1 \subset A_2 \subset \cdots$ of sets of finite measure such that $f = g = 0$ a.e. outside $\cup A_n$. Let us introduce the nonnegative functions

$$f_n := \chi_{A_n} \min\{f, n\} \quad \text{and} \quad g_n := \chi_{A_n} \min\{g, n\},$$

then

$$f_n \nearrow f \quad \text{and} \quad g_n \nearrow g \quad \text{a.e.,}$$

and

$$\int (f_n + g_n)^p \, dt \le (2n)^p \mu(A_n) < \infty \quad \text{for each} \quad n.$$

Applying (a) we have for each n the following estimate:

$$\|f_n + g_n\|_p^p = \int_I (f_n + g_n)^p \, dt$$

$$\le \int_I f_n (f_n + g_n)^{p-1} \, dt + \int_I g_n (f_n + g_n)^{p-1} \, dt$$

$$\le \|f_n\|_p \cdot \left\| (f_n + g_n)^{p-1} \right\|_q + \|g_n\|_p \cdot \left\| (f_n + g_n)^{p-1} \right\|_q$$

$$= \left(\|f_n\|_p + \|g_n\|_p \right) \|f_n + g_n\|_{q(p-1)}^{p-1}$$

$$= \left(\|f_n\|_p + \|g_n\|_p \right) \|f_n + g_n\|_p^{p-1},$$

whence[10]

$$\|f_n + g_n\|_p \le \|f_n\|_p + \|g_n\|_p.$$

Applying the generalized Beppo Levi theorem we have

$$\int_I f_n^p \, dt \to \int_I f^p \, dt,$$

i.e., $\|f_n\|_p \to \|f\|_p$. We have similarly $\|g_n\|_p \to \|g\|_p$ and $\|f_n + g_n\|_p \to \|f + g\|_p$. Therefore, letting $n \to \infty$ in the preceding inequality we conclude that $\|f + g\|_p \le \|f\|_p + \|g\|_p$. Finally, since the right-hand side of the last inequality if finite by our assumption $f, g \in L^p$, the left-hand side is also finite, so that $f + g \in L^p$.

---

[10]Here we use the finiteness of $\|f_n + g_n\|_p$.

(c) The case $p = 1$ has already been proved in Theorem 5.12 (p. 184). For $1 < p <$ ∞ we adapt that proof as follows.

Let (f_n) be a Cauchy sequence in L^p. By Lemma 9.2 there exist $f \in L^p$ and a subsequence (f_{n_k}) such that $f_{n_k} \to f$ a.e.

For any given $\varepsilon > 0$ there exists an N such that

$$\int |f_m - f_n|^p \, d\mu < \varepsilon$$

for all $m, n \geq N$. Choosing $n = n_k$ and letting $k \to \infty$, an application of the Fatou lemma yields the inequalities

$$\int |f_m - f|^p \, d\mu \leq \varepsilon$$

for all $m \geq N$.   □

Next we study the density of step functions in L^p spaces.

Proposition 9.3 (a) *Let* $f \in L^p$, $1 \leq p \leq \infty$. *There exist step functions* φ_n *and* $h \in L^p$ *such that*

$$|\varphi_n| \leq h \quad \text{for all} \quad n, \quad \text{and} \quad \varphi_n \to f \quad \text{a.e.} \tag{9.1}$$

(b) *If* $1 \leq p < \infty$, *then the step functions are dense in* L^p.
(c) *The characteristic functions of measurable sets generate* L^∞.

Remark The step functions are *not* dense L^∞ in general, but they are dense in the weaker locally convex topology $\sigma(L^\infty, L^1)$, defined by the family of seminorms[11]

$$p_g(f) := \left| \int fg \, d\mu \right|, \quad g \in L^1.$$

Indeed, for any given $f, g \in L^1$ we have $\|(\varphi_n - f)g\|_1 \to 0$ by Lebesgue's dominated convergence theorem (p. 181), with the sequence (φ_n) defined in (a).

Proof (a) If $f \in L^\infty$, then f is measurable by definition, and hence there exists a sequence of step functions satisfying $\psi_n \to f$ a.e. Furthermore, all functions ψ_n and f vanish outside some measurable set A. Then the functions

$$h := \|f\|_\infty \chi_A \quad \text{and} \quad \varphi_n := \text{med} \{- \|f\|_\infty, \psi_n, \|f\|_\infty\}$$

have the required properties.

---

[11] See Sect. 9.7 (p. 336) for the study of this topology.

If $f \in L^p$ for some $1 \leq p < \infty$, then separating the positive and negative parts of f we may assume that $f \geq 0$.

Since $f^p \in L^1$ by Proposition 5.14 (p. 185), there exists a sequence (ψ_n) of step functions satisfying $\|f^p - \psi_n\|_1 \to 0$.

Applying Lemma 5.13 (or the preceding lemma), by taking a subsequence we may also assume that there exist two functions $\tilde{h}, \tilde{f} \in L^1$ satisfying $\psi_n \to \tilde{f}$ a.e., and $|\psi_n| \leq \tilde{h}$ for all n. Then we have $\|\tilde{f} - \psi_n\|_1 \to 0$ by the dominated convergence theorem, and hence necessarily $\tilde{f} = f^p$ a.e. We conclude that (9.1) is satisfied with $h := \tilde{h}^{1/p} \in L^p$ and $\varphi_n := |\psi_n|^{1/p}$.

(b) Given any $f \in L^p$, the step functions of (a) satisfy $\int |\varphi_n - f|^p \, dx \to 0$ by the dominated convergence theorem.

(c) Given any $g \in L^\infty$ and $\varepsilon > 0$,

$$h(t) := \left[\frac{g(t)}{\varepsilon} \right] \varepsilon$$

is a *finite* linear combination of characteristic functions of measurable sets, satisfying the inequality $\|g - h\|_\infty \leq \varepsilon$. $\qquad\square$

Now we prove the L^2 version of the Hilbert–Schmidt theorem (p. 38). Similarly to Sect. 7.3 (p. 224) we consider a product measure $\mu \times \mu$ on $X \times X$.

Proposition 9.4 (Hilbert–Schmidt)[12] *If $a \in L^2(X \times X)$, then the formula*

$$(Af)(t) := \int_X a(t,s)f(s) \, ds, \quad t \in X$$

defines a completely continuous operator in $L^2(X)$.

Proof Using the Cauchy–Schwarz inequality and applying Tonelli's theorem (p. 228), the following estimate holds for all $f \in L^2(X)$:

$$\int_X \left| \int_X a(t,s)f(s) \, ds \right|^2 dt \leq \int_X \left(\int_X |a(t,s)|^2 \, ds \right) \cdot \left(\int_X |f(s)|^2 \, ds \right) dt$$
$$= \|a\|_2^2 \cdot \|f\|_2^2 .$$

Hence A is a continuous operator on $L^2(X)$, and $\|A\| \leq \|a\|_2$.[13]

To prove the compactness, in view of Proposition 2.37 (p. 101), it is sufficient construct a sequence (A_n) of continuous operators of finite rank on $L^2(X)$, satisfying $\|A - A_n\| \to 0$.

---

[12]Hilbert [209], Schmidt [415].
[13]We even have equality here.

Applying Proposition 9.3 we choose a sequence (a_n) of step functions satisfying $a_n \to a$ in $L^2(X \times X)$, and we define

$$(A_n f)(t) := \int_X a_n(t, s) f(s) \, ds, \quad f \in L^2(X), \quad t \in X.$$

Repeating the above estimates with a_n and $a - a_n$ instead of a, we obtain that the operators A_n are continuous in $L^2(X)$, and that

$$\|A - A_n\| \le \|a - a_n\|_2 \to 0.$$

It remains to show that each A_n has a finite rank. For this we observe that, by the definition of the product measure, each step function a_n on $X \times X$ is of the form

$$a_n(t, s) = \sum_{i=1}^{N} \chi_{J_i}(t) \cdot \chi_{K_i}(s)$$

with some sets $J_i, K_i \in \mathcal{M}$ of finite measure, and hence the range of A_n is generated by the N functions $\chi_{K_1}, \dots, \chi_{K_N}$. $\qquad\square$

The rest of this section is devoted to the study of some important special cases. Let I be an *open* interval and $w : I \to \mathbb{R}$ a nonnegative *measurable* function with respect to the usual Lebesgue measure. Assume that w is integrable on every *compact* subinterval of I,[14] and denote by $\mathcal{P}$ the semiring of bounded intervals whose closures are in I. Then the formula $\mu(J) := \int_J w \, dt$ defines a finite measure on $\mathcal{P}$. Consider the corresponding integral, and denote by L_w^p the corresponding L^p spaces.

For $w = 1$ this reduces to the usual $L^p(I)$ spaces.

We denote by $C_c(I)$ the vector space of continuous functions $g : I \to \mathbb{R}$ that vanish outside some *compact* subinterval of I, i.e., vanish in some neighborhood of the endpoints of I.[15]

Proposition 9.5 *Let* $1 \le p < \infty$.

(a) L_w^p *is separable.*
(b) $C_c(I)$ *is dense in* L_w^p.[16]
(c) *If* I *is bounded and* w *is integrable on* I, *then the algebraic polynomials are dense in* L_w^p.
(d) *If* $|I| \le 2\pi$ *and* w *is integrable in* I, *then the trigonometric polynomials are dense in* L_w^p.

---

[14] We say in such cases that w is *locally* integrable.
[15] The compact subinterval may depend on g.
[16] Moreover, the proof will show that for each $f \in L_w^p$ there exists a function $h \in L_w^p$ and a sequence $(\varphi_n) \subset C_c(I)$ satisfying the relations (9.1) of Proposition 9.3.

Fig. 9.1 Graph of g_n

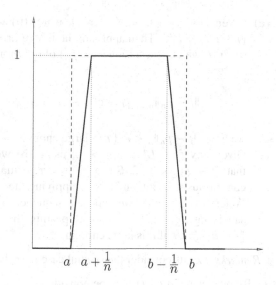

Proof We denote by $\|\cdot\|_p$ the norm of L^p_w.

(a) By Proposition 9.3 the characteristic functions of the intervals in $\mathcal{P}$ generate L^p_w. If we consider only the intervals with rational endpoints, then we obtain countably many functions that still generates L^p_w.

(b) By Proposition 9.3 it is sufficient to find for each fixed compact interval $J = [a, b] \subset I$ a sequence of functions $(g_n) \subset C_c(I)$ converging to χ_J in L^p_w. The formulas

$$g_n(t) := \begin{cases} 0 & \text{if } t \le a, \\ n(t-a) & \text{if } a \le t \le a+n^{-1}, \\ 1 & \text{if } a+n^{-1} \le t \le b-n^{-1}, \\ n(b-t) & \text{if } b-n^{-1} \le t \le b, \\ 0 & \text{if } t \ge b \end{cases}$$

for $n > 2/(b-a)$ yield such a sequence (see Fig. 9.1). Indeed,

$$\|\chi_J - g_n\|_p^p = \int_a^b |1 - g_n(t)|^p w(t)\, dt \to 0$$

by the dominated convergence theorem, because $g_n \to 1$ a.e. in $[a, b]$,

$$0 \le |1 - g_n|^p w \le w$$

for all n, and w is integrable.

(c) Given any $f \in L^p_w$ and $\varepsilon > 0$, using (b) we choose $g \in C_c(I)$ such that $\|f - g\|_p < \varepsilon/2$. Then applying the first approximation theorem of Weierstrass (p. 260) we choose a sequence (p_n) of polynomials satisfying $\|g - p_n\|_\infty \to 0$. Since

$$\|f - p_n\|_p \leq \|f - g\|_p + \|g - p_n\|_p < \frac{\varepsilon}{2} + \|g - p_n\|_\infty \cdot \|1\|_p,$$

we have $\|f - p_n\|_p < \varepsilon$ if n is large enough.

(d) Given any $f \in L^p_w$ and $\varepsilon > 0$, using (b) we choose $g \in C_c(I)$ again such that $\|f - g\|_p < \varepsilon/2$. Since $|I| \leq 2\pi$, g may be extended to a 2π-periodic, continuous function on $\mathbb{R}$. Now applying the second approximation theorem of Weierstrass (p. 264) we choose a sequence (h_n) of trigonometric polynomials satisfying $\|g - h_n\|_\infty \to 0$. Repeating the reasoning in (c) we obtain that $\|f - h_n\|_p < \varepsilon$ if n is large enough.   □

Remarks Let us consider the special case $w = 1$.

- By property (b) $L^p(I)$ may be considered as a completion of $C_c(I)$ with respect to the norm $\|\cdot\|_p$.
- None of the four properties holds for $L^\infty(I)$ in general. Indeed, each of (b), (c), (d) would imply (a), i.e., the separability of $L^\infty(I)$.

  But $L^\infty(I)$ is not separable, because it contains uncountably many pairwise disjoint non-empty open sets. Indeed, the $2^{\aleph_0}$ open balls $B_{1/2}(\chi_J)$, where J runs over the compact subintervals of I, are pairwise disjoint.[17]
- On the other hand, the four properties remain valid if we consider in $L^\infty(I)$ the weak star topology $\sigma(L^\infty, L^1)$.[18]

Now we prove the completeness of several classical orthonormal sequences introduced in Chap. 1. We recall the importance of this property for the corresponding Fourier series.[19]

Consider the Hilbert space L^2_w with the scalar product $(f, g) := \int fgw \, dt$. If the functions $t \mapsto t^k w(t)$ are integrable for all $k = 0, 1, \ldots$, then all algebraic polynomials belong to L^2_w.[20] Applying the Gram–Schmidt method (Proposition 1.15, p. 28) to the sequence 1, id, id², ... we obtain an orthonormal sequence (P_k) of polynomials in L^2_w such that $\deg p_k = k$ for every k.

Corollary 9.6

(a) *If I is bounded and w is integrable on I, then (P_k) is an orthonormal basis of L^2_w.*

---

[17] Compare this proof with that of the non-separability of ℓ^∞, p. 74.

[18] Use the remark following the statement of Proposition 9.3.

[19] See Proposition 1.13, p. 25.

[20] This happens, for example, if I is bounded and w is integrable on I.

(b) *If I is an interval of length 2π, then the* trigonometric system:

$$e_0 = \frac{1}{\sqrt{2\pi}} \quad and \quad e_{2k-1} = \frac{\sin kt}{\sqrt{\pi}}, \quad e_{2k} = \frac{\cos kt}{\sqrt{\pi}}, \quad k = 1, 2, \ldots$$

is an orthonormal basis of $L^2(I)$.
(c) *The functions*

$$\sqrt{\frac{2}{\pi}} \sin kx, \quad k = 1, 2, \ldots$$

form an orthonormal basis of $L^2(0, \pi)$.
(d) *The functions*

$$\sqrt{\frac{1}{\pi}} \quad and \quad \sqrt{\frac{2}{\pi}} \cos kx, \quad k = 1, 2, \ldots$$

form an orthonormal basis of $L^2(0, \pi)$.

Proof The orthonormality of the functions in (b), (c), (d) may be verified by a straightforward computation.[21]

(a) and (b) follow from parts (c), (d) of the preceding proposition and from Proposition 1.14 (p. 27).

(c) It suffices to show that if $h \in L^2(0, \pi)$ is orthogonal to the functions $\sin kt$ for all $k = 1, 2, \ldots$, then $h = 0$. Extending h to an *odd* function on $(-\pi, \pi)$, we obtain a function $H \in L^2(-\pi, \pi)$ that is orthogonal to the whole trigonometric system. Using (b) we conclude that $H = 0$ on $(-\pi, \pi)$, and hence $h = 0$ on $(0, \pi)$.

(d) It suffices to show that if $h \in L^2(0, \pi)$ is orthogonal to the functions $\cos kt$ for all $k = 0, 1, \ldots$, then $h = 0$. Extending h to an *even* function on $(-\pi, \pi)$, we obtain a function $H \in L^2(-\pi, \pi)$ that is orthogonal to the whole trigonometric system. Using (b) we conclude that $H = 0$ on $(-\pi, \pi)$, and hence $h = 0$ on $(0, \pi)$. $\square$

Remarks

- Without the additional hypotheses in (a) the orthonormal sequence (P_k) may be incomplete.[22]

  However, the Laguerre and Hermite polynomials, that occur in many applications, are complete, although they are defined on the unbounded intervals $(0, \infty)$ and $\mathbb{R}$.[23]

---

[21] This has already been noted on p. 24.
[22] A counterexample was given by Stieltjes [435].
[23] Steklov [434]. See Kolmogorov–Fomin 1981. See also a proof of von Neumann in: Courant–Hilbert [91] or Szegő [446].

- Since convergence in L^2 spaces does not imply a.e. convergence in general, Proposition 1.14 (p. 27) does not imply the a.e. convergence of Fourier series.

  Nevertheless, Carleson proved that the trigonometric Fourier series of every function $f \in L^2(I)$ converges to f a.e.[24]
- Applying an equiconvergence theorem of Haar,[25] the a.e. convergence also holds for the Fourier series associated with the Legendre polynomials.[26]

9.2 * Compact Sets

In this section we characterize the compact sets in L^p for the usual Lebesgue measure in $\mathbb{R}$. As in Proposition 8.7 (p. 268), it is sufficient to characterize the totally bounded sets.

Proposition 9.7 (Kolmogorov–Riesz)[27] *Let* $1 \le p < \infty$. *A bounded set* $\mathcal{F} \subset L^p(\mathbb{R})$ *is totally bounded* $\Longleftrightarrow$ *the following two conditions are satisfied:*

$$\sup_{f \in \mathcal{F}} \int_{|t|>R} |f(t)|^p \, dt \to 0 \quad as \quad R \to \infty,$$

and

$$\sup_{f \in \mathcal{F}} \int_{-\infty}^{\infty} |f(t) - f(t+h)|^p \, dt \to 0 \quad as \quad h \to 0.$$

We introduce for commodity the translated functions $f_h(t) := f(t+h)$, and we rewrite the conditions in the equivalent forms

$$\sup_{f \in \mathcal{F}} \|f\|_{L^p(\mathbb{R} \setminus [-R,R])} \to 0 \quad as \quad R \to \infty \tag{9.2}$$

and

$$\sup_{f \in \mathcal{F}} \|f - f_h\|_p \to 0 \quad as \quad h \to 0. \tag{9.3}$$

---

[24]Carleson [78]. His theorem was generalized to $f \in L^p(I)$ with $p > 1$ by Hunt [220].

[25]Haar [177]. The result remains valid for all classical orthogonal polynomials: see Joó–Komornik [228], Komornik [255–257]. Other equiconvergence theorems have already been obtained by Liouville in [306].

[26]The Legendre polynomials are the orthogonal polynomials associated with the constant weight function $w = 1$.

[27]Kolmogorov [251], Riesz [388]. See also Hanche-Olsen and Holden [189] for a survey and historical comments.

Proof of the Necessity

First step. If $f \in L^p$ and $1 \leq p < \infty$, then

$$\|f\|_{L^p(\mathbb{R}\setminus[-R,R])} \to 0 \quad \text{as} \quad R \to \infty,$$

and

$$\|f - f_h\|_p \to 0 \quad \text{as} \quad h \to 0.$$

For the proof we set

$$g_n(t) := \begin{cases} |f(t)|^p & \text{if } |t| > n, \\ 0 & \text{if } |t| \leq n. \end{cases}$$

These functions are integrable, $|g_n| \leq |f|^p$, and $g_n \to 0$ a.e. Applying the dominated convergence theorem it follows that

$$0 \leq \int_{|t|>R} |f(t)|^p \, dt \leq \int_{-\infty}^{\infty} g_{[R]}(t) \, dt \to 0$$

as $R \to \infty$. (Here $[R]$ stands for the integer part of R.) This proves the first relation.

The second relation is obvious if f is the characteristic function of some bounded interval. By the triangle inequality the relation holds for all step functions as well. Finally, given any $f \in L^p$ and $\varepsilon > 0$, we choose a step function φ satisfying $\|f - \varphi\|_p < \varepsilon$. Then we also have $\|f_h - \varphi_h\|_p < \varepsilon$ for all h. If h is sufficiently close to zero, then $\|\varphi - \varphi_h\|_p < \varepsilon$, and therefore

$$\|f - f_h\|_p \leq \|f - \varphi\|_p + \|\varphi - \varphi_h\|_p + \|\varphi_h - f_h\|_p < 3\varepsilon.$$

Second step. If $\mathcal{F}$ is totally bounded, then for each fixed $\varepsilon > 0$ it can be covered by finitely many balls of radius ε. Let us denote by $f_1, \ldots, f_m$ the centers of these balls.

By the first step there exists $R > 0$ and $\delta > 0$ such that

$$\|f_i\|_{L^p(\mathbb{R}\setminus[-R,R])} < \varepsilon$$

and

$$\|f_i - f_{i,h}\|_p < \varepsilon \quad \text{if} \quad |h| < \delta$$

for $i = 1, \ldots, m$.

Each $f \in \mathcal{F}$ belongs to one of the balls $B_\varepsilon(f_i)$, so that

$$\|f\|_{L^p(\mathbb{R}\setminus[-R,R])} \leq \|f - f_i\|_{L^p(\mathbb{R}\setminus[-R,R])} + \|f_i\|_{L^p(\mathbb{R}\setminus[-R,R])} < 2\varepsilon$$

and

$$\|f - f_h\|_p \leq \|f - f_i\|_p + \|f_i - f_{i,h}\|_p + \|f_{i,h} - f_h\|_p < 3\varepsilon$$

if $|h| < \delta$.   □

Proof of the Sufficiency

First step. Applying Steklov's regularization method[28] we reduce the problem to the case of continuous functions. Setting

$$(S_r f)(t) := \frac{1}{r} \int_0^r f(t + s)\, ds, \quad f \in L^p,\ r > 0,$$

first we establish the following estimates:

$$\|S_r f\|_\infty \leq r^{-1/p} \|f\|_p; \tag{9.4}$$

$$|(S_r f)(t) - (S_r f)(t + h)| \leq r^{-1/p} \|f - f_h\|_p \tag{9.5}$$

for all $t \in \mathbb{R}$;

$$\|f - S_r f\|_p \leq \sup_{0 < h \leq r} \|f - f_h\|_p. \tag{9.6}$$

The first estimate is obtained by applying Hölder's inequality:

$$|(S_r f)(t)| \leq r^{-1} \int_0^r |f(t + s)|\, ds \leq r^{-1/p} \|f\|_{L^p(t,t+r)} \leq r^{-1/p} \|f\|_p$$

for all $t \in \mathbb{R}$.
Applying (9.4) to $f - f_h$ instead of f we get (9.5).
Finally, we have

$$|(f - S_r f)(t)| = \left| r^{-1} \int_0^r f(t) - f(t + s)\, ds \right|$$

$$\leq r^{-1/p} \left(\int_0^r |f(t) - f(t + s)|^p\, ds \right)^{1/p}$$

---

[28]Steklov [433].

for each t, and hence (9.6) follows:

$$\int_{-\infty}^{\infty} |(f - S_r f)(t)|^p \, dt \leq r^{-1} \int_{-\infty}^{\infty} \int_0^r |f(t) - f(t + s)|^p \, ds \, dt$$

$$= r^{-1} \int_0^r \int_{-\infty}^{\infty} |f(t) - f(t + s)|^p \, dt \, ds$$

$$\leq \sup_{0 < h \leq r} \|f - f_h\|_p^p .$$

Second step. For any fixed $\varepsilon > 0$, we will cover $\mathcal{F}$ with finitely many balls of radius $\leq 3\varepsilon$.
Applying (9.2) we choose $R > 0$ such that

$$\|f\|_{L^p(\mathbb{R}\setminus[-R,R])} < \varepsilon \quad \text{for all} \quad f \in \mathcal{F}.$$

Furthermore, using (9.3) and (9.6) we choose $r > 0$ such that

$$\|f - S_r f\|_p < \varepsilon \quad \text{for all} \quad f \in \mathcal{F}.$$

Since $\mathcal{F}$ is bounded, by (9.4) and (9.5) the function system $\{S_r f : f \in \mathcal{F}\}$ is uniformly bounded and equicontinuous. Applying the Arzelà–Ascoli theorem (p. 268) on the interval $[-R, R]$, we obtain a finite number of continuous functions $g_1, \ldots, g_m$ such that each $f \in \mathcal{F}$ satisfies for some index i the inequalities

$$|S_r f - g_i| \leq (2R)^{-1/p} \varepsilon \quad \text{in} \quad [-R, R]. \tag{9.7}$$

Extending the functions g_i by zero to $\mathbb{R}$, we obtain $f_1, \ldots, f_m \in L^p$. To conclude we show that $\|f - f_i\|_p < 3\varepsilon$ for every $f \in \mathcal{F}$, where the index i is the same as in (9.7).
For the proof we use the triangle inequality, the definition of R and r, and finally the choice of i:

$$\|f - f_i\|_p = \|f\|_{L^p(\mathbb{R}\setminus[-R,R])} + \|f - g_i\|_{L^p(-R,R)}$$

$$< \varepsilon + \|f - S_r f\|_{L^p(-R,R)} + \|S_r f - g_i\|_{L^p(-R,R)}$$

$$< 2\varepsilon + (2R)^{1/p} \|S_r f - g_i\|_{L^\infty(-R,R)}$$

$$\leq 3\varepsilon. \qquad \square$$

9.3   * Convolution

We have encountered integrals of the form

$$\int f(s)g(t-s)\,ds$$

many times: in the methods of Landau and de la Vallée-Poussin, in the closed forms of the Dirichlet and Fejér kernels in the preceding chapter, and in the Steklov functions in the preceding section.

Such integrals often occur in the theory of partial differential equations and in harmonic analysis to prove density theorems.[29]

In this section we give only one basic result.[30]

Proposition 9.8 *Let* $1 \leq p, q, r \leq \infty$ *satisfy the equality*

$$\frac{1}{p} + \frac{1}{q} = \frac{1}{r} + 1,$$

and let $f \in L^p(\mathbb{R}^N)$, $g \in L^q(\mathbb{R}^N)$.
The formula

$$(f * g)(x) := \int f(x-y)g(y)\,dy$$

defines a function $f * g \in L^r(\mathbb{R}^N)$, *and*

$$\|f * g\|_r \leq \|f\|_p \cdot \|g\|_q.$$

If f *vanishes outside* A *and* g *vanishes outside* B, *then* $f * g$ *vanishes outside*

$$A + B := \{a + b \in \mathbb{R}^N : a \in A \quad and \quad b \in B\}.$$

Definition The function $f * g$ is called the *convolution* of f and g.[31]

Remarks

- The definition shows that the convolution is *commutative*: $f * g = g * f$.

---

[29]The latter applications are based on the celebrated *Haar measure* (Haar [178]), a natural generalization of the usual Lebesgue measure to *topological groups*.

[30]There are many more results and applications in Brezis [65], Hörmander [218, 219], Katznelson [245], Pontryagin [364], Rudin [402, 405, 406], Schwartz [420], Weil [485].

[31]Fourier [148].

- It follows by induction on k that if

$$f_1 \in L^{p_1}(\mathbb{R}^N), \ldots, f_k \in L^{p_k}(\mathbb{R}^N)$$

for some $k \geq 2$, where $1 \leq p_1, \ldots, p_k, r \leq \infty$ satisfy the equality

$$\frac{1}{p_1} + \cdots + \frac{1}{p_k} = \frac{1}{r} + k - 1,$$

then

$$g := f_1 * (\cdots * f_k) \cdots) \in L^r(\mathbb{R}^N)$$

and

$$\|g\|_r \leq \|f_1\|_{p_1} \cdots \|f_k\|_{p_k}.$$

Moreover, the *associativity* relation $(f * g) * h = f * (g * h)$ holds, so that we may remove the parentheses in the definition of g.

The condition on the exponents is equivalent to the simpler relation

$$\frac{1}{p_1'} + \cdots + \frac{1}{p_k'} = \frac{1}{r'}$$

where we use the conjugate exponents.

Proof We proceed in several steps.

(i) If the step functions φ_n, ψ_n converge a.e. to f and g, respectively in $\mathbb{R}$, then the *step functions* $\varphi_n(x-y)\psi_n(y)$ converge a.e. to $f(x-y)g(y)$ in $\mathbb{R}^2$; the verification is left to the reader. Hence the function $(x, y) \mapsto f(x - y)g(y)$ is measurable.

(ii) The case $r = \infty$ of the theorem readily follows from Hölder's inequality. Henceforth we assume that $r < \infty$. Since $p \leq r$ and $q \leq r$, then p and q are also finite.

(iii) If f and g are nonnegative and integrable, then applying Tonelli's theorem we obtain that

$$\int (f * g)(x)\, dx = \int \left(\int f(x-y)g(y)\, dy \right) dx$$

$$= \int \left(\int f(x-y)g(y)\, dx \right) dy$$

$$= \int \left(\int f(x-y)\, dx \right) g(y)\, dy$$

$$= \|f\|_1 \cdot \|g\|_1 < \infty.$$

Hence $f * g \in L^1(\mathbb{R}^N)$, and

$$\|f * g\|_1 = \|f\|_1 \cdot \|g\|_1 . \tag{9.8}$$

(iv) Turning to the general case ($f \in L^p$, $g \in L^q$, $r < \infty$), first we prove the following inequality:

$$(|f| * |g|)^r \le \|f\|_p^{r-p} \cdot \|g\|_q^{r-q} \cdot (|f|^p * |g|^q) \quad \text{a.e.} \tag{9.9}$$

Introducing the conjugates p' and q' of p and q, we have

$$\frac{1}{p'} + \frac{1}{q'} + \frac{1}{r} = 1.$$

Since

$$1 - \frac{p}{r} = p\left(\frac{1}{p} - \frac{1}{r}\right) = p\left(1 - \frac{1}{q}\right) = \frac{p}{q'}$$

and

$$1 - \frac{q}{r} = q\left(\frac{1}{q} - \frac{1}{r}\right) = q\left(1 - \frac{1}{p}\right) = \frac{q}{p'},$$

the following equality holds a.e.:

$$|f(x-y)g(y)| = \left(|f(x-y)|^p\right)^{1/q'}\left(|g(y)|^q\right)^{1/p'}\left(|f(x-y)|^p|g(y)|^q\right)^{1/r}.$$

Integrating with respect to y, applying Hölder's inequality and using (iii) we obtain

$$(|f| * |g|)(x) \le \|f\|_p^{p/q'} \cdot \|g\|_q^{q/p'} \cdot \left|(|f|^p * |g|^q)(x)\right|^{1/r},$$

or equivalently

$$\left|(|f| * |g|)(x)\right|^r \le \|f\|_p^{rp/q'} \cdot \|g\|_q^{rq/p'} \cdot (|f|^p * |g|^q)(x).$$

We conclude by observing that $rp/q' = r - p$ and $rq/p' = r - q$.

(v) The right-hand side of (9.9) is integrable by (iii). Hence $|f| * |g| \in L^r(\mathbb{R}^N)$, i.e.,

$$\int\left(\int |f(x-y)g(y)| \, dy\right)^r dx < \infty.$$

Applying this to the positive and negative parts of f and g we conclude that the four functions

$$y \mapsto f_\pm(x-y)g_\pm(y)$$

are integrable for a.e. x. Hence their linear combination

$$y \mapsto f(x-y)g(y)$$

is also (measurable and) integrable for a.e. x. Therefore $f * g$ is well defined a.e.

Next, applying (9.8) and (9.9) we obtain the following estimate:

$$\int |(f*g)(x)|^r \, dx = \int \left| \int f(x-y)g(y) \, dy \right|^r dx$$

$$\leq \int \left(\int |f(x-y)g(y)| \, dy \right)^r dx$$

$$= \||f| * |g|\|_r^r$$

$$\leq \|f\|_p^{r-p} \cdot \|g\|_q^{r-q} \cdot \||f|^p * |g|^q\|_1$$

$$= \|f\|_p^r \cdot \|g\|_q^r.$$

Hence $f * g \in L^r(\mathbb{R}^N)$ and $\|f * g\|_r \leq \|f\|_p \cdot \|g\|_q$.

(vi) If $(f*g)(x)$ is defined for some $x \notin A + B$, then $x - y \notin A$ for all $y \in B$. Consequently, $f(x-y)g(y) = 0$ for a.e. $y \in \mathbb{R}^N$, whence $(f*g)(x) = 0$.   □

9.4 Uniformly Convex Spaces

The parallelogram identity is an important property of Euclidean spaces. For $1 < p < \infty$ the L^p spaces have a weaker, but still useful property:

Definition A normed space X is *uniformly convex*[32] if for each $\varepsilon > 0$ there exists a $\delta > 0$ such that if two vectors $x, y \in X$ satisfy the inequalities

$$\|x\| \leq 1, \ \|y\| \leq 1 \quad \text{and} \quad \|x+y\| > 2 - \delta,$$

---

[32]Clarkson [89].

Fig. 9.2 Uniform convexity

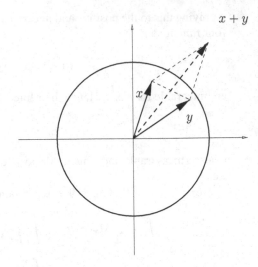

then

$$\|x - y\| < \varepsilon.$$

(See Fig. 9.2.)

It follows from the definition that every uniformly convex space is strictly convex (see p. 67).

Examples • Every Euclidean space is uniformly convex. Indeed, since

$$\|x - y\|^2 = 2\,\|x\|^2 + 2\,\|y\|^2 - \|x + y\|^2 < 4 - (2 - \delta)^2 < 4\delta,$$

we may choose $\delta := \varepsilon^2/4$ for each ε.
• The space ℓ^1 is not uniformly convex, because

$$\|e_1\| = \|e_2\| = 1 \quad \text{and} \quad \|e_1 + e_2\| = \|e_1 - e_2\| = 2,$$

so that for $\varepsilon < 2$ there is no suitable $\delta > 0$.
• The space ℓ^∞ is not uniformly convex either, because the vectors $x := e_1 + e_2$ and $y := e_1 - e_2$ satisfy

$$\|x\| = \|y\| = 1 \quad \text{and} \quad \|x + y\| = \|x - y\| = 2,$$

so that for $\varepsilon < 2$ there is no suitable $\delta > 0$.

On the other hand, ℓ^p is uniformly convex if $1 < p < \infty$. More generally:

Proposition 9.9 *Let $(X, \mathcal{M}, \mu)$ be an arbitrary measure space and $1 < p < \infty$.
Then $L^p(X, \mathcal{M}, \mu)$ is uniformly convex.*[33]

Proof

First step. If x and y are distinct real numbers, then

$$\left|\frac{x+y}{2}\right|^p < \frac{|x|^p + |y|^p}{2}$$

by the strict convexity of the function $t \mapsto |t|^p$.
Second step. For each $\varepsilon \in (0, 2^{1-p}]$ we denote by $\varrho = \varrho(\varepsilon)$ the minimum of the
function

$$\frac{|x|^p + |y|^p}{2} - \left|\frac{x+y}{2}\right|^p$$

on the non-empty[34] compact set

$$\left\{(x, y) \in \mathbb{R}^2 \ : \ |x|^p + |y|^p = 2 \quad \text{and} \quad \left|\frac{x-y}{2}\right|^p \geq \varepsilon\right\}.$$

By the preceding step we have $\varrho > 0$. By homogeneity it follows that if $x, y \in \mathbb{R}$
satisfy the inequality

$$\left|\frac{x-y}{2}\right|^p \geq \varepsilon \frac{|x|^p + |y|^p}{2},$$

then

$$\varrho \frac{|x|^p + |y|^p}{2} \leq \frac{|x|^p + |y|^p}{2} - \left|\frac{x+y}{2}\right|^p.$$

Third step. For any given $\varepsilon > 0$ we have to find $\delta > 0$ such that if two functions
$f, g \in L^p$ satisfy the inequalities

$$\int |f|^p \, dx \leq 1, \quad \int |g|^p \, dx \leq 1 \quad \text{and} \quad \int \left|\frac{f+g}{2}\right|^p \, dx > 1 - \delta,$$

[33]Clarkson [89]. The proof given here is due to McShane [320].
[34]$(2^{1/p}, 0)$ belongs to the set.

then

$$\int \left| \frac{f-g}{2} \right|^p dx < 2\varepsilon.$$

We may assume that $\varepsilon \in (0, 2^{1-p}]$. Setting

$$M := \left\{ \left| \frac{f-g}{2} \right|^p \geq \varepsilon \frac{|f|^p + |g|^p}{2} \right\},$$

applying the convexity of the function $t \mapsto |t|^p$, and using the preceding step we obtain the following estimate:

$$\int_X \left| \frac{f-g}{2} \right|^p dx$$

$$= \int_{X \setminus M} \left| \frac{f-g}{2} \right|^p dx + \int_M \left| \frac{f-g}{2} \right|^p dx$$

$$\leq \varepsilon \int_{X \setminus M} \frac{|f|^p + |g|^p}{2} dx + \int_M \frac{|f|^p + |g|^p}{2} dx$$

$$\leq \varepsilon \int_{X \setminus M} \frac{|f|^p + |g|^p}{2} dx + \frac{1}{\varrho} \int_M \left(\frac{|f|^p + |g|^p}{2} - \left| \frac{f+g}{2} \right|^p \right) dx$$

$$\leq \varepsilon \int_X \frac{|f|^p + |g|^p}{2} dx + \frac{1}{\varrho} \int_X \left(\frac{|f|^p + |g|^p}{2} - \left| \frac{f+g}{2} \right|^p \right) dx$$

$$\leq \varepsilon + \frac{1}{\varrho} - \frac{1-\delta}{\varrho}$$

$$= \varepsilon + \frac{\delta}{\varrho}.$$

We conclude by choosing $\delta < \varepsilon \varrho$. $\square$

The following variant of the orthogonal projection (p. 12) is valid in all uniformly convex Banach spaces:

Proposition 9.10 (Sz.-Nagy)[35] *Let K be a non-empty convex closed set in a uniformly convex Banach space X. For each $x \in X$ there exists in K a unique closest point y to x.*

---

[35]Sz.-Nagy [447].

Proof

Existence. The result is obvious if $x \in K$. Henceforth we assume that $x \notin K$, and we choose a *minimizing sequence*: $(y_n) \subset K$, and

$$\|x - y_n\| \to d := \mathrm{dist}(x, K).$$

Setting

$$t_n := 1 / \|x - y_n\| \quad \text{and} \quad z_n := t_n(x - y_n),$$

we have $\|z_n\| = 1$ for every n. Furthermore, applying the convexity of K and the definition of d we obtain the following relation:

$$
\begin{aligned}
\|z_n + z_m\| &= \|t_n(x - y_n) + t_m(x - y_m)\| \\
&= (t_n + t_m) \left\| x - \left(\frac{t_n}{t_n + t_m} y_n + \frac{t_m}{t_n + t_m} y_m \right) \right\| \\
&\geq (t_n + t_m)d \\
&\to 2.
\end{aligned}
$$

By the uniform convexity this implies that (z_n) is a Cauchy sequence; since, moreover, X is complete, it converges to some point $z \in X$. Consequently,

$$y_n = x - \frac{z_n}{t_n} \to x - dz =: y.$$

Hence $y \in K$ because K is closed, and $\|x - y\| = \lim \|x - y_n\| = d$.

Uniqueness. If $y, y' \in K$ and $\|x - y\| = \|x - y'\| = d$, then the formulas $y_{2n-1} := y$ and $y_{2n} := y'$, $n = 1, 2, \ldots$ define a minimizing sequence. This sequence is convergent by the preceding step, but this is possible only if $y = y'$. □

Examples The spaces L^1 and L^∞ do not always have the property of the last proposition, so they are not uniformly convex.

- Consider in $X = L^1(-1, 1)$ the closed subspace M formed by the functions having integral zero, and the constant function $g = 1$. If $f \in M$, then

$$\|g - f\|_1 = \int_{-1}^{1} |1 - f(t)| \, dt \geq \int_{-1}^{1} 1 - f(t) \, dt = 2,$$

with equality for all $f \in M$ satisfying $f \leq 1$. Therefore the distance $\mathrm{dist}(g, M) = 2$ is attained at infinitely many points.

• Consider in $X = L^\infty(-1, 1)$ the closed subspace M formed by the functions vanishing a.e. on $[-1, 0]$, and the constant function $g = 1$. We have

$$\|g - f\|_\infty \geq \|g - f\|_{L^\infty(-1,0)} = 1$$

for all $f \in M$, with equality whenever $0 \leq f \leq 2$. Therefore the distance $\text{dist}(g, M) = 2$ is attained at infinitely many points.

In uniformly convex spaces we may complete Proposition 2.22 (p. 80) on the relation between strong and weak convergence:

*Proposition 9.11 (Radon–Riesz)[36] *In uniformly convex spaces we have*

$$x_n \to x \iff x_n \rightharpoonup x \quad and \quad \|x_n\| \to \|x\| .$$

Proof The implication $\implies$ holds in all normed spaces by Proposition 2.22 (p. 80).

The converse implication is obvious if $x = 0$. Assume henceforth that $\|x\| > 0$, then $\|x_n\| > 0$ for all sufficiently large n. The assumptions $x_n \rightharpoonup x$ and $\|x_n\| \to \|x\|$ imply that

$$\frac{x_n}{\|x_n\|} + \frac{x}{\|x\|} \rightharpoonup 2\frac{x}{\|x\|}.$$

Since the norm of the limit is equal to 2,

$$\liminf \left\| \frac{x_n}{\|x_n\|} + \frac{x}{\|x\|} \right\| \geq 2$$

by Proposition 2.22 (f).[37]

By the definition of uniform convexity this implies that

$$\left\| \frac{x_n}{\|x_n\|} - \frac{x}{\|x\|} \right\| \to 0.$$

Consequently,

$$x_n = \|x_n\| \cdot \frac{x_n}{\|x_n\|} \to \|x\| \cdot \frac{x}{\|x\|} = x. \qquad \square$$

---

[36]Hildebrandt [210] (ℓ^p), Radon [366] (p. 1358: L^p), Riesz [382] (pp. 58–59: ℓ^p), Riesz [385] (simple proof for L^p).
[37]In fact, the left-hand norm converges to 2.

*Remarks
- We recall (p. 83) that the equivalence fails, for example, in c_0 and ℓ^∞.
- We also recall that ℓ^1, although not uniformly convex, has the Radon–Riesz property: see Proposition 2.26, p. 84.
- The preceding example is an exception: we will soon show (p. 338) that $L^1(-\pi, \pi)$ does not have the Radon–Riesz property.
- By a theorem of Kadec[38] every *separable* Banach space has an equivalent norm having the Radon–Riesz property.

9.5 Reflexivity

Unlike the spaces $C(K)$, most L^p spaces are reflexive:

Proposition 9.12 (Clarkson)[39] *For any given measure space* $(X, \mathcal{M}, \mu)$, $L^p(X, \mathcal{M}, \mu)$ *is reflexive for all* $1 < p < \infty$.

In view of Proposition 9.9 it suffices to establish the following result:

Proposition 9.13 (Milman–Pettis)[40] *Every uniformly convex Banach space is reflexive.*

Remark This result clarifies the relationship between Proposition 2.31 (c) and Proposition 9.10 (pp. 91 and 326) on the distance from closed convex sets.

Proof [41] Consider the canonical isometry $J : X \to X''$ of Proposition 2.28 (p. 87). Since J is homogeneous, it is sufficient to show that if $\Phi \in X''$ and $\|\Phi\| = 1$, then there exists an $x \in X$ satisfying $Jx = \Phi$.

Denote the closed unit balls of X and X'' by B and B''. By Goldstein's theorem (p. 139) there exists a net (x_n) in B such that $J(x_n) \to \Phi$ in the topology $\sigma(X'', X')$. It follows that the "doubled" net converges to 2Φ:

$$J(x_m + x_n) = J(x_m) + J(x_n) \to 2\Phi.$$

Consequently,

$$\|x_m + x_n\| \to \|2\Phi\| = 2.$$

---

[38] Kadec [234–236]. See also Bessaga–Pelczyński [40].

[39] Clarkson [89].

[40] Milman [322], Pettis [358].

[41] We follow Lindenstrauss–Tzafriri [303, p. 61]. See, e.g., Brezis [65] for a proof without using nets.

Indeed, in the contrary case there would exist a subnet belonging to the ball $\alpha B''$ for some $0 < \alpha < 2$. This ball would be compact by the Banach–Alaoglu theorem (p. 139), and hence closed in the Hausdorff topology $\sigma(X'', X')$. This would imply $\|2\Phi\| \leq \alpha < 2$, contradicting the choice of Φ.

Since X is uniformly convex, the relation $\|x_m + x_n\| \to 2$ implies that (x_n) is a Cauchy net in X. Since X is complete, it converges to some point $x \in X$. Then $J(x_n) \to J(x)$ in $\sigma(X'', X')$ by the definition of this topology. But we also have $J(x_n) \to \Phi$, so that $\Phi = J(x)$ by the uniqueness of the limit.   □

The spaces L^1 and L^∞ are not reflexive in general:

Examples • We have seen several proofs of the non-reflexivity of $C([0, 1])$ in the preceding chapter. Since it is a closed subspace of $L^\infty(0, 1)$, by Proposition 3.23 (p. 143) $L^\infty(0, 1)$ cannot be reflexive either.

• The space $L^1(0, 1)$ is not reflexive, because there exist linear functionals $\varphi \in (L^1(0, 1))'$ whose norms are not attained.[42] For example, let

$$\varphi(f) := \int_0^1 tf(t)\, dt, \quad f \in L^1(0, 1).$$

The inequalities

$$|\varphi(f)| \leq \int_0^1 t\,|f(t)|\, dt \leq \int_0^1 |f(t)|\, dt = \|f\|_1 \tag{9.10}$$

imply that $\|\varphi\| \leq 1$. Furthermore, the functions (see Fig. 9.3)

$$f_n := n\chi_{[1-n^{-1},1]}$$

have unit norm in $L^1(0, 1)$, and $|\varphi(f_n)| \to 1$, so that $\|\varphi\| = 1$.

But this norm is not attained, because the second inequality in (9.10) is strict for every non-zero function.

• The non-reflexivity of $L^1(X, \mathcal{M}, \mu)$ for most measure spaces also follows from the existence of bounded sequences with no weakly converging subsequences. (See Theorem 2.30, p. 90.)

More precisely, if there exists a disjoint set sequence (A_n) such that $0 < \mu(A_n) < \infty$ for all n, then the functions $f_n := \mu(A_n)^{-1}\chi_{A_n}$ form a bounded sequence having no weakly converging subsequences.

---

[42]See Proposition 2.1, p. 55.

Fig. 9.3 Graph of $n\chi_{[1-n^{-1},1]}$

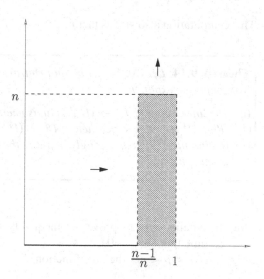

Indeed, for any given subsequence (f_{n_k}) consider the linear functional defined by the formula

$$\varphi(f) := \sum_{k=1}^{\infty}(-1)^k \int_{A_{n_k}} f\, d\mu.$$

Then the numerical sequence $(\varphi(f_{n_k})) = ((-1)^k)$ is divergent.
We return to the question of reflexivity at the end of the next section.

9.6 Duals of L^p Spaces

In this section we generalize the relations $(\ell^p)' = \ell^q$ of Proposition 2.15, p. 73). If $p, q \in [1, \infty]$ are conjugate exponents, then the formula

$$(jg)(f) := \int_X fg\, d\mu$$

defines a continuous linear functional on L^p for each $g \in L^q$.
    Indeed, the integrals are well defined by Hölder's inequality, and

$$|(jg)(f)| \le \|g\|_q \cdot \|f\|_p .$$

Since jg is clearly linear, hence

$$jg \in (L^p)' \quad \text{and} \quad \|jg\| \le \|g\|_q .$$

This computation also shows that $j : L^q \to (L^p)'$ is a continuous linear map of norm ≤ 1.

Theorem 9.14 *Let $(X, \mathcal{M}, \mu)$ be an arbitrary measure space, and $p, q \in [1, \infty]$ two conjugate exponents.*

(a) *The linear map $j : L^q \to (L^p)'$ is an isometry.*[43]
(b) *(Riesz)[44] If $1 < p < \infty$, then $j : L^q \to (L^p)'$ is an isometric isomorphism.*
(c) *(Steinhaus)[45] If μ is strongly σ-finite, then $j : L^\infty \to (L^1)'$ is an isometric isomorphism.*

Proof

(a) It remains only to prove the inequality $\|jg\| \geq \|g\|_q$.[46] We may therefore assume that $\|g\|_q > 0$.

If $1 < p < \infty$, then the function

$$ f := |g|^{q-1} \operatorname{sign} g $$

satisfies the equalities

$$ \|f\|_p^p = \int |f|^p \, d\mu = \int |g|^{p(q-1)} \, d\mu = \int |g|^q \, d\mu = \|g\|_q^q = \|g\|_q^{p(q-1)} . $$

Hence

$$ f \in L^p, \quad \|f\|_p = \|g\|_q^{q-1} > 0, $$

and

$$ (jg)(f) = \int |g|^q \, d\mu = \|g\|_q^q = \|g\|_q \cdot \|f\|_p . $$

Since $\|f\|_p > 0$, we conclude that $\|jg\| \geq \|g\|_q$.

If $p = \infty$, then setting $f := \operatorname{sign} g \in L^\infty$ we have

$$ \|g\|_1 = \int |g| \, d\mu = (jg)(f) \leq \|jg\| \cdot \|f\|_\infty = \|jg\| . $$

---

[43]In the case $p = 1$ it is essential for the existence of B that the functions in L^∞ are *measurable* by our definition, and not only locally measurable. It is instructive to consider on an uncountable set X the measure μ that is equal to zero on countable sets, and equal to ∞ otherwise. This is another reason in favour of the constructive measurability definition adopted in this book.

[44]Riesz [380] for $X = [0, 1]$, Nikodým [343], McShane [320].

[45]Steinhaus [432] for $X = [0, 1]$, Dunford [116].

[46]See also a direct proof for $X = \mathbb{R}$ in Riesz and Sz.-Nagy [394].

Finally, if $p = 1$, then for any fixed number $0 < c < \|g\|_\infty$ the set

$$A := \{x \in X : |g(x)| \geq c\}$$

has a positive measure. Applying Lemma 7.5 (p. 220) there exists a $B \subset A$ satisfying $0 < \mu(B) < \infty$. Then $f := \chi_B \operatorname{sign} g \in L^1$, and

$$c\mu(B) \leq \int fg \, d\mu = (jg)(f) \leq \|jg\| \cdot \|f\|_1 = \|jg\| \cdot \mu(B).$$

Hence $c \leq \|jg\|$ for all $c < \|g\|_\infty$, so that $\|g\|_\infty \leq \|jg\|$.

(b) We have to prove that j is onto. Since j is an isometry and L^q is complete, the range $R(j)$ of j is a closed subspace of $(L^p)'$. It remains to show that it is dense in $(L^p)'$.

By Corollary 2.9 (p. 64) it suffices to show that if $\Phi \in (L^p)''$ is orthogonal to $R(j) \subset (L^p)'$, then $\Phi = 0$. Since L^p is reflexive, identifying $(L^p)''$ with L^p this is equivalent to the following property: *if $f \in L^p$ and $\int fg \, d\mu = 0$ every $g \in L^q$, then $f = 0$.*

Setting

$$g := |f|^{p-1} \operatorname{sign} f$$

and repeating the computation of (a), reversing the role of p and q, we obtain that

$$g \in L^q \quad \text{and} \quad 0 = \int fg \, d\mu = \int |f|^p \, d\mu.$$

Hence $f = 0$ a.e.

* (c) Given $\varphi \in (L^1)'$ we have to find $g \in L^\infty$ satisfying

$$\varphi(f) = \int_X fg \, d\mu \tag{9.11}$$

for all $f \in L^1$.[47]

First we assume that $\mu(X) < \infty$. Then the formula

$$\nu(A) := \varphi(\chi_A)$$

defines a set function on $\mathcal{M}$. It is finitely additive by the linearity of φ. Moreover, it is σ-additive. Indeed, if $A = \bigcup^* A_n$ with $A, A_n \in \mathcal{M}$, then $\sum \chi_{A_n} = \chi_A$ in L^1 by Corollary 5.9 (p. 180). Using the continuity of $\varphi \in (L^1)'$

---

[47] The following reasoning may be adapted for $1 < p < \infty$ as well: see Dunford–Schwartz [117].

we conclude that

$$\nu(A) = \varphi(\chi_A) = \sum \varphi(\chi_{A_n}) = \sum \nu(A_n).$$

Observe that $\nu \ll \mu$. Indeed, if $\mu(A) = 0$, then $\chi_A = 0$ a.e., and hence

$$\nu(A) = \varphi(\chi_A) = 0.$$

Applying the Radon–Nikodým theorem (p. 240) there exists a measurable function g such that

$$\nu(A) = \int_A g \, d\mu \tag{9.12}$$

for every set A of finite measure.

We show that $g \in L^\infty$. Given any number $0 < c < \|g\|_\infty$, at least one of the two sets

$$\{x \in X : g(x) \geq c\} \quad \text{and} \quad \{x \in X : -g(x) \geq c\}$$

has a positive measure, and then (as in the proof of (a)) it contains a set B of finite positive measure. If for example $g \geq c$ on B (the other case is analogous), then

$$c\mu(B) \leq \int_B g \, d\mu = \nu(B) = \varphi(\chi_B) \leq \|\varphi\| \cdot \|\chi_B\|_1 = \|\varphi\| \cdot \mu(B).$$

Hence $c \leq \|\varphi\|$ for all $c < \|g\|_\infty$, so that $\|g\|_\infty \leq \|\varphi\| (< \infty)$.

We deduce from (9.12) by linearity that (9.11) is satisfied for all step functions f. Since they are dense in L^1 by Proposition 5.14 (p. 185), by continuity (9.11) holds for all $f \in L^1$, too.

In the general case there exists a finite or countable disjoint sequence (P_n) such that $0 < \mu(P_n) < \infty$ for all n, and $\mu(A) = 0$ for all $A \in \mathcal{M}$ satisfying $A \subset X \setminus \cup^* P_n$.[48]

Applying the preceding result for each P_n we obtain a function $g \in L^\infty$ vanishing outside $\cup^* P_n$ and satisfying (9.11) for the functions $f = h\chi_{P_n}$, $h \in L^1$, $n = 1, 2, \ldots$.

Using the dominated convergence theorem, the linearity and the continuity of φ, (9.11) follows again:

$$\varphi(h) = \sum \varphi(h\chi_{P_n}) = \sum \int_X h\chi_{P_n} g \, d\mu = \int_X hg \, d\mu. \qquad \square$$

---

[48] We use the strong σ-additivity assumption.

Remarks

- Hildebrandt and Fichtenholz–Kantorovich characterized $(L^\infty)'$.[49]
- The map $j : L^1 \to (L^\infty)'$ is onto only in degenerate cases, for example when μ is the counting measure on a finite set.

  We have already seen (p. 79) that $j : \ell^1 \to (\ell^\infty)'$ is not onto.

  The map $j : L^1(\mathbb{R}) \to L^\infty(\mathbb{R})'$ is not onto either because $L^1(\mathbb{R})$ is not even a dual space.[50] This follows (similarly to the analogous result on c_0 on p. 140) from the theorems of Banach–Alaoglu and Krein–Milman, because the closed unit ball of $L^1(\mathbb{R})$ has no extremal points.

  For the last property we show that if $\int |f| \, dx = 1$, then there exists a non-zero function $g \in L^1(\mathbb{R})$ satisfying $\int |f + tg| \, dx = 1$ for all $t \in [-1, 1]$.

  For this we first choose a set A of finite positive measure and a number $\varepsilon > 0$ such that $f > \varepsilon$ or $f < -\varepsilon$ on A. Then we choose any non-zero function g such that $\int g \, dx = 0$, $g = 0$ outside A, and $|g| < \varepsilon$ on A.

- Let us also give a direct proof of the non-surjectivity of the map $j : L^1(\mathbb{R}) \to L^\infty(\mathbb{R})'$. The *Dirac functional*, defined by the formula

$$\delta(g) := g(0), \quad g \in C_b(\mathbb{R})$$

is a continuous linear functional of norm one on $C_b(\mathbb{R})$. Applying the Helly–Hahn–Banach theorem (p. 65) it can be extended to a continuous linear functional on $L^\infty(\mathbb{R})$. We claim that no function $f \in L^1(\mathbb{R})$ satisfies the equality

$$\int fg \, dt = g(0) \tag{9.13}$$

for all $g \in C_b(\mathbb{R})$.[51]

Assume on the contrary that there exists such a function f. The formula $g_n(x) := \min\{n|x|, 1\}$ defines a sequence of functions in $C_b(\mathbb{R})$ satisfying $g_n(0) = 0$, $fg_n \to f$ a.e., and $|fg_n| \le |f|$ for all n. Applying the dominated convergence theorem it follows that

$$\int f \, dt = \lim \int fg_n \, dt = \lim g_n(0) = 0.$$

But this is impossible because choosing $g = 1$ in (9.13) we get $\int f \, dt = 1$.

---

[49]Hildebrandt [213, p. 875], Fichtenholz–Kantorovich [145, p. 76]. See also Dunford–Schwartz [117], Kantorovich–Akilov [243].

[50]This property and the following proof remain valid for all measure spaces where each set A of positive measure has a subset B satisfying $0 < \mu(B) < \mu(A)$.

[51]This is an important theorem in the *theory of distributions*, asserting that the Dirac functional is not a *regular distribution*. See Schwartz [420].

- In the preceding remark we have found a linear functional in $L^\infty(\mathbb{R})'$ not represented by any $f \in L^1(\mathbb{R})$. Since $L^\infty(\mathbb{R})' = L^1(\mathbb{R})''$, this proves directly the non-reflexivity of $L^1(\mathbb{R})$.
- Since $L^1(\mathbb{R})' = L^\infty(\mathbb{R})$, by Proposition 3.23 (p. 143) $L^\infty(\mathbb{R})$ is not reflexive either.

Example We show that the strong σ-finiteness assumption cannot be omitted in Part (c).[52]

Consider the measure space $(X, \mathcal{M}, \mu)$ and the measure ν of the counterexample on page 243.

Since $\nu \le \mu$, we have

$$\int |f|\, d\nu \le \int |f|\, d\mu = \|f\|_1$$

for all $f \in L^1$, so that the formula

$$\varphi(f) := \int f\, d\nu$$

defines an element φ of $(L^1)'$.

We claim that φ is not represented by any (measurable or locally measurable) function $g \in L^\infty$. Indeed, if we had

$$\int f\, d\nu = \int gf\, d\mu$$

for all $f \in L^1$, then (taking $f = \chi_A$ for $A \in \mathcal{M}$) g would be a (measurable or locally measurable) Radon–Nikodým derivative of ν with respect to μ, contradicting our results on pp. 243 and 251.

9.7 Weak and Weak Star Convergence

The purpose of this section is to characterize the weak and weak star convergence of L^p spaces. Since all weakly convergent and weak star convergent sequences are bounded by Propositions 2.24 and 3.18 (pp. 82 and 138), it is sufficient to consider bounded sequences.

---

[52]See Schwartz [419] and Ellis–Snow [123] for the characterization of $(L^1)'$ in the general case.

Let $p, q \in [1, \infty]$ be conjugate exponents, and let us denote by $\sigma(L^p, L^q)$ the locally convex topology on L^p, defined by the family of seminorms

$$p_g(f) := \left| \int fg \, d\mu \right|, \quad g \in L^q.$$

If $1 < p < \infty$, then this is the weak topology of L^p. If our measure space is strongly σ-finite, then $\sigma(L^1, L^\infty)$ is the weak topology of L^1, and $\sigma(L^\infty, L^1)$ is the weak star topology of L^∞.

Proposition 9.15 *Let (f_n) be a bounded sequence in L^p, and $f \in L^p$.*

(a) *(Riesz)[53] If $1 < p \leq \infty$, then $f_n \to f$ in $\sigma(L^p, L^q) \iff$*

$$\int_A f_n \, d\mu \to \int_A f \, d\mu \tag{9.14}$$

for each set A of finite measure.
(b) *If $p = 1$, then $f_n \to f$ in $\sigma(L^1, L^\infty) \iff$ (9.14) holds for all measurable sets A.*

Remarks

- If $1 < p \leq \infty$, then using Proposition 9.3 (p. 310) the proof below shows that it suffices to consider in (9.14) the sets A of the semiring at the origin of the definition of the integral.
  Consequently, for the usual Lebesgue measure on an interval $I \subset \mathbb{R}$ the condition (9.14) is equivalent to the *pointwise* convergence $F_n \to F$, where F_n and F are some primitives of f_n and f that coincide at some fixed point of I.
- Let (I_n) be a sequence of disjoint subintervals of an interval $I = [a, b]$ such that $|I_n| > 0$ and $I_n \subset (a, a + 2^{-n})$ for every n. The formula

$$f_n := |I_{2n-1}|^{-1} \chi_{I_{2n-1}} - |I_{2n}|^{-1} \chi_{I_{2n}}$$

  defines a bounded sequence in $L^1(I)$ satisfying the relation $F_n \to F$ of the preceding remark with $F = f = 0$.
  But f_n does not converge to f in $\sigma(L^1, L^\infty)$ because (9.14) fails for $A := \cup I_{2n}$.
- The functions $f_n := \chi_{[n,n+1]}$ in $\mathbb{R}$ show that it is not sufficient to consider sets of finite measure in (9.14) when $p = 1$.

Proof of Proposition 9.15 Let us rewrite (9.14) in the form

$$\int \chi_A f_n \, d\mu \to \int \chi_A f \, d\mu. \tag{9.15}$$

---

[53]Riesz [380] (for finite p).

If $f_n \to f$ in $\sigma(L^p, L^q)$, then (9.15) is satisfied for all sets A with the indicated properties because $\chi_A \in L^q$.

The converse implications hold because the characteristic functions χ_A of the indicated sets A generate L^q in all cases by Proposition 9.3 (b), (c) (p. 310), and because the functions $g \in L^q$ satisfying $\int gf_n \, d\mu \to \int gf \, d\mu$ form a *closed* subspace of L^q by the boundedness of the sequence (f_n) (see Lemma 2.25, p. 83).   □

We end this section by presenting a basic example of weak convergence. Given a sequence (λ_n) of real numbers, tending to infinity, we consider the functions

$$f_n(t) := \sin \lambda_n t \quad \text{and} \quad g_n(t) := \cos \lambda_n t.$$

*Proposition 9.16 (Riemann–Lebesgue)[54] *Given any conjugate exponents $p, q \in [1, \infty]$, we have $f_n \to 0$ and $g_n \to 0$ in $\sigma(L^p, L^q)$ on each bounded interval I.*

Proof The sequences (f_n), (g_n) are bounded in L^∞ and hence in all spaces $L^p(I)$. Since $L^q \subset L^1$, it is sufficient to prove the convergences in the topology $\sigma(L^\infty, L^1)$.

For any fixed point $a \in I$, the primitives of the functions f_n, g_n vanishing at a converge pointwise to zero, because

$$\left| \int_a^x \sin \lambda_n t \, dt \right| = \left| \frac{\cos \lambda_n a - \cos \lambda_n x}{\lambda_n} \right| \le \frac{2}{|\lambda_n|} \to 0,$$

and a similar estimate holds for $\cos \lambda_n t$ as well. We conclude by applying the first remark on the preceding page.   □

*Remark In the special case where $|I| = 2\pi$, $p = 2$ and $\lambda_n = n$, the proposition follows from the Bessel inequality for the trigonometric system[55]: the Fourier coefficients of each $f \in L^2(I)$ converge to zero.

*Example We recall[56] that ℓ^1 has the Radon–Riesz property.

On the other hand, $L^1(-\pi, \pi)$ does not have this property. Indeed, the functions $h_n(t) := 1 + \sin nt$ converge weakly to $h(t) := 1$ in $L^1(-\pi, \pi)$ by the Riemann–Lebesgue lemma. Furthermore,

$$\|h_n\|_1 = \int_{-\pi}^{\pi} 1 + \sin nt \, dt = 2\pi = \|h\|_1$$

---

[54]Riemann [371], Lebesgue [289, p. 473] and [293, p. 61]. See an interesting application of Poincaré [363] to the distribution of small planets.

[55]Halphén [188].

[56]See the example preceding Proposition 2.26, p. 84.

for every n. Nevertheless, h_n does not converge strongly to h because

$$\|h_n - h\|_1 = \int_{-\pi}^{\pi} |\sin nt| \, dt = 4$$

for all n.

9.8 Exercises

In the first seven exercises we consider the Hilbert space $H = L^2(0, 1)$ with the scalar product $(f, g) := \int_0^1 fg \, dt$.

Exercise 9.1

(i) Show that every uniformly convergent sequence $(x_n) \subset H$ also converges in H.
(ii) Set $x_n(t) := n^2 te^{-nt}$. Show that (x_n) converges pointwise to 0 but it does not converge in H.
(iii) Construct a sequence of continuous functions converging in H but diverging at each point.

Exercise 9.2 Consider the following sets in H:

(i) The set of functions $x \in H$ vanishing a.e. on some neighborhood of $t = 1/2$.[57]
(ii) The set of functions $x \in H$ with values in $[-1, 1]$.

Are they convex? Are they closed?

Exercise 9.3

(i) For each $\lambda \in \mathbb{R}$ we denote by M_λ the set of all continuous functions $x \in H$ satisfying $x(0) - \lambda$. Show that the sets are convex, dense and disjoint.
(ii) Show that the set of polynomials P vanishing at 1 is convex and dense in H.

Exercise 9.4 Show that

$$M := \left\{ f \in L^2(0, 1) \; : \; \int_0^1 f(t) \, dt = 0 \right\}$$

is a closed subspace of $L^2(0, 1)$. Determine $M^\perp$.

Exercise 9.5 The formula $(Af)(t) := tf(t)$ defines a continuous self-adjoint operator on the Hilbert space $H = L^2(0, 1)$ which has no eigenvalues.

Exercise 9.6 There is no translation invariant measure in $L^2(0, 1)$ such that $0 < \mu(A) < \infty$ for all open balls.

---

[57]The neighborhood may depend on x.

Exercise 9.7 There exists a continuous, injective function $f : [0, 1] \to L^2(0, 1)$ such that the vectors $f(b) - f(a)$ and $f(d) - f(c)$ are orthogonal whenever $0 \leq a < b < c < d \leq 1$. What is the geometric meaning of this property of the "curve" f?

Exercise 9.8 (Haar System)[58] Set

$$\psi(x) := \begin{cases} 1 & \text{if } 0 \leq x < 1/2, \\ -1 & \text{if } 1/2 \leq x < 1, \\ 0 & \text{otherwise}, \end{cases}$$

and introduce the functions

$$\psi_{n,k}(x) := 2^{n/2}\psi(2^n x - k), \quad x \in \mathbb{R}, \quad n, k \in \mathbb{Z}.$$

Prove the following:

(i) The functions $\psi_{n,k}$ form an orthonormal basis in $L^2(\mathbb{R})$.
(ii) The functions 1 and $\psi_{n,k}$ for $n \geq 0$ and $0 \leq k < 2^n$ form an orthonormal basis in $L^2(0, 1)$.
(iii) Consider the orthonormal basis of (ii) by starting with 1 and then ordered according to the lexicographic ordering of the pairs (n, k). If $f \in C([0, 1])$, then its Fourier series converges *uniformly* to f.

Exercise 9.9 Consider the spaces L^p corresponding to a probability measure.

(i) Show that if $1 \leq p < q \leq \infty$, then $L^q \subset L^p$.
(ii) Show that if $1 \leq p < q \leq \infty$ and $x^k \to x$ in L^q, then $x^k \to x$ in L^p.
(iii) Investigate the validity of the equalities

$$\bigcup_{q > p} L^q = L^p \quad \text{and} \quad \bigcap_{p < q} L^p = L^q.$$

Exercise 9.10 Consider the L^p spaces on a measure space.

(i) If there are no sets of arbitrarily small positive measure, then $p < q \implies L^p \subset L^q$.
(ii) If there are no sets of arbitrarily large measure, then $p < q \implies L^p \supset L^q$.
(iii) Are the above conditions also necessary?

---

[58] Haar [176]. This is the first wavelet, in modern terminology; see, e.g., Strichartz [442].

Chapter 10
Almost Everywhere Convergence

A youth who had begun to read geometry with Euclid, when he had learnt the first proposition, inquired, "What do I get by learning these things?" So Euclid called a slave and said "Give him threepence, since he must make a gain out of what he learns".—Stobaeus

There is no royal road to geometry.—Menaechmus to Alexander the Great[1]

We have seen in Part II the importance of *a.e. convergence* in integration theory. The purpose of this last chapter of our book is to clarify its relationship to other convergence notions.

As usual, we consider a measure space $(X, \mathcal{M}, \mu)$, and we identify two functions if they are equal a.e.

10.1 L^p Spaces, $1 \le p \le \infty$

First we compare the strong and a.e. convergences. We may generalize the theorems of Lebesgue and Fatou (pp. 181 and 183):

Proposition 10.1 *Let* (f_n) *be a* bounded *sequence in* L^p, $p \in [1, \infty)$, *and assume that* $f_n \to f$ *a.e.*

(a) $f \in L^p$, *and* $\|f\|_p \le \liminf \|f_n\|_p$.
(b) *If there exists a* $g \in L^p$ *such that* $|f_n| \le g$ *for all n, then* $\|f_n - f\|_p \to 0$.
(c) *If* $\|f_n\|_p \to \|f\|_p$, *then* $\|f_n - f\|_p \to 0$.[2]

Proof (a) We apply the Fatou lemma to the sequence of functions $|f_n|^p$.

---

[1] By other sources, Euclid to King Ptolemy.
[2] Radon [366, p. 1358], Riesz [385].

© Springer-Verlag London 2016
V. Komornik, *Lectures on Functional Analysis and the Lebesgue Integral*,
Universitext, DOI 10.1007/978-1-4471-6811-9_10

(b) We apply Lebesgue's convergence theorem to the sequence of functions $|f_n - f|^p$. This is justified because

$$|f_n - f|^p \le (|f_n| + |f|)^p \le 2^p g^p,$$

and the function $2^p g^p$ is integrable by our assumption.

(c) Following Novinger[3] we apply the Fatou lemma to the sequence of functions

$$\frac{|f_n|^p + |f|^p}{2} - \left|\frac{f_n - f}{2}\right|^p,$$

converging a.e. to $|f|^p$. (They are nonnegative by the convexity of the function $t \mapsto |t|^p$.) We obtain that

$$\int |f|^p \, d\mu \le \liminf \int \frac{|f_n|^p + |f|^p}{2} - \left|\frac{f_n - f}{2}\right|^p \, d\mu$$

$$= \int |f|^p \, d\mu - \limsup \int \left|\frac{f_n - f}{2}\right|^p \, d\mu.$$

Hence $\limsup \|f_n - f\|_p^p \le 0$, and therefore $\|f_n - f\|_p \to 0$.

$\square$

*Remarks

- Part (a) remains valid for $p = \infty$ with a simple proof.
- The characteristic functions of the intervals $[n^{-1}, 1]$ show that (b) and (c) fail in $L^\infty(0, 1)$.

Next we investigate the relations between the weak and a.e. convergences. As usual we denote by q the conjugate exponent of p.

Proposition 10.2 *Let (f_n) be a bounded sequence in L^p, $p \in (1, \infty]$. If $f_n \to f$ a.e., then $f_n \to f$ in $\sigma(L^p, L^q)$ as well.*

Proof Since $f \in L^p$ by Proposition 10.1 (a), changing (f_n) to $(f_n - f)$ we may assume that $f = 0$.

Let us introduce[4] for $N = 1, 2, \ldots$ the sets

$$E_N := \{x \in X : |f_n(x)| \le 1 \quad \text{for all} \quad n \ge N\}$$

and

$$G_N := \{g \in L^q : g = 0 \quad \text{a.e. outside} \quad E_N\}.$$

---

[3]Novinger [345].
[4]See Lions [304].

Since $f_n \to 0$ a.e., almost every $x \in X$ belongs to $\cup E_N$. Since, moreover, the set sequence (E_N) is non-decreasing, $\cup G_N$ is dense L^q. Indeed, each $g \in L^q$ is the limit in L^q of the sequence of functions $\chi_{E_N} g \in G_N$ by Proposition 10.1 (b).

Now assume first that $\mu(X) < \infty$. Since (f_n) is bounded in L^p, and $L^q \subset (L^p)'$, by Lemma 2.25 (p. 83) it is sufficient to show that

$$\int f_n g \, d\mu \to 0 \quad \text{for each} \quad g \in \cup G_N.$$

The last relation follows by applying the dominated convergence theorem. Indeed, if $g \in G_N$, then $f_n g \to 0$ a.e., $|f_n g| \leq |g|$ for all $n \geq N$, and $g \in L^q \subset L^1$, because $\mu(X) < \infty$.

In the general case we change G_N to $G_N \cap L^1$ in the above proof. We have to show that $G_N \cap L^1$ is dense in G_N with respect to the topology of L^q. For this we approximate each $g \in G_N$ by a suitable sequence (φ_n) of step functions (this is possible by Proposition 9.3 (a), p. 310), and then we change φ_n to $\varphi_n \chi_{E_N}$.   □

Remark The case $p = 1$ is different: if $f_n \to f$ a.e., then the weak and strong convergences are the same: see Theorem 10.10 of Vitali–Hahn–Saks below, p. 357.

Examples Consider the usual Lebesgue measure on $X = [0, 1]$.

- The nonnegative functions

$$f_n(t) := ne^{-nt}$$

converge a.e. to zero, and

$$0 \leq \int f_n(t) \, dt \; \to 1 \neq 0 = \int 0 \, dt.$$

Hence (f_n) is bounded in L^1, but does not converge weakly to zero in L^1. (See Fig. 10.1.) Hence the proposition fails for $p = 1$.
- For any fixed $p \in (1, \infty)$ the functions

$$f_n(t) := n^{1/p} e^{-nt}$$

converge to $f := 0$ a.e. Furthermore,

$$\|f_n\|_p^p = \frac{1 - e^{-np}}{p} \to \frac{1}{p},$$

so that (f_n) is bounded in L^p, but does not converge strongly to f in L^p.

The same conclusion holds in L^∞ for the limit functions $f_n(t) := e^{-nt}$: $f_n \to 0$ a.e., and $\|f_n\|_\infty = 1$ for all n.

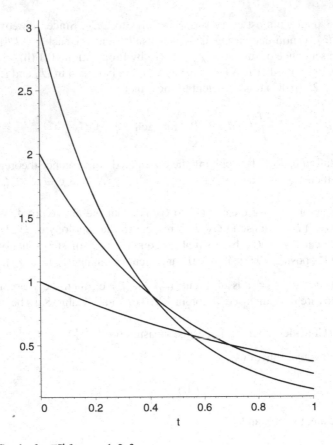

Fig. 10.1 Graph of ne^{-nt} for $n = 1, 2, 3$

Thus we cannot replace weak convergence by strong convergence in the proposition.

10.2 L^p Spaces, $0 < p \leq 1$

The definition of the *sets L^p* remains meaningful for all $0 < p < \infty$: a measurable function f belongs to L^p if

$$\int |f|^p \, d\mu < \infty.$$

But (except for some degenerate cases) the usual formula

$$\|f\|_p := \left(\int |f|^p \, d\mu \right)^{1/p}$$

does not define a norm if $0 < p < 1$: the inequality sign in the triangle inequality and in the other usual inequalities is reversed[5]:

Proposition 10.3 *Let* $0 < p < 1$ *and* $q = p/(p-1) < 0$ *be two conjugate exponents.*[6]

(a) *(Reverse Young inequality) If* x, y *are nonnegative numbers, then*[7]

$$xy \ge \frac{x^p}{p} + \frac{y^q}{q}.$$

(b) *(Reverse Hölder inequality) If* f *and* g *are measurable functions, then*

$$\|fg\|_1 \ge \|f\|_p \cdot \|g\|_q.$$

(c) *(Reverse Minkowski inequality) If* f *and* g *are nonnegative, measurable functions, then*

$$\|f + g\|_p \ge \|f\|_p + \|g\|_p.$$

Proof

(a) We may assume that $x, y > 0$. Consider the graph of the convex function given by the equivalent equations $y = x^{p-1}$ and $x = y^{q-1}$. The shaded region in Fig. 10.2 belongs to the rectangle of sides x and y, hence its area is at most xy. Furthermore, it is the difference of two unbounded regions, limited by the coordinate axes, the sides of the rectangle and the graph of our function. Consequently,

$$xy \ge \int_0^x s^{p-1} \, ds - \int_y^\infty t^{q-1} \, dt = \frac{x^p}{p} + \frac{y^q}{q}.$$

---

[5]Compare with Proposition 9.1, p. 306.

[6]The usual relation $p^{-1} + q^{-1} = 1$ still holds. See Hardy–Littlewood–Pólya [191] or Sobolev [428].

[7]If $y = 0$, then the last fraction is replaced by its limit: $-\infty$.

Fig. 10.2 Reverse Young
inequality

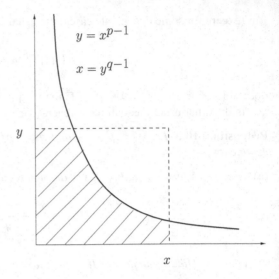

$$y = x^{p-1}$$

$$x = y^{q-1}$$

(b) The cases $\|f\|_p = 0$ and $\|g\|_q = 0$ being obvious, we may assume by
homogeneity that $\|f\|_p = \|g\|_q = 1$. Applying the reverse Young inequality
we obtain that

$$\|fg\|_1 = \int |f| \cdot |g| \, d\mu \geq \int \frac{|f|^p}{p} + \frac{|g|^q}{q} \, d\mu = \frac{1}{p} + \frac{1}{q} = 1 = \|f\|_p \cdot \|g\|_q.$$

(c) We may assume by homogeneity that $\|f + g\|_p = 1$. Applying the reverse
Hölder inequality we obtain that

$$\|f + g\|_p = \|f + g\|_p^p$$

$$= \int (f + g)^p \, d\mu$$

$$= \int f(f + g)^{p-1} + g(f + g)^{p-1} \, d\mu$$

$$\geq \|f\|_p \cdot \left\| (f + g)^{p-1} \right\|_q + \|g\|_p \cdot \left\| (f + g)^{p-1} \right\|_q$$

$$= \|f\|_p \cdot \|f + g\|_{q(p-1)}^{p-1} + \|g\|_p \cdot \|f + g\|_{q(p-1)}^{p-1}$$

$$= \|f\|_p + \|g\|_p.$$

In the last step we have used the relation $(p - 1)q = p$.

$\square$

Despite the last proposition we may introduce a natural *metric* on *L^p* for $0 < p < 1$. For this we first generalize the notion of the norm:

Definition Let X be a vector space. A function $N : X \to \mathbb{R}$ is a *pseudonorm* if the following conditions are satisfied for all $x, y \in X$:

- $N(x) \geq 0$;
- $N(x) = 0 \iff x = 0$;
- $N(x + y) \leq N(x) + N(y)$;
- $N(cx) \leq N(x)$ for all $-1 \leq c \leq 1$;
- $N(n^{-1}x) \to 0$ as $n \to \infty$.

Note that every norm is also a pseudonorm.

Proposition 10.4 *If N is a pseudonorm on X, then the formula*

$$d(x, y) := N(x - y)$$

defines a metric on X, and X is a separated topological vector space with respect to the corresponding topology.

Proof The only non-trivial property is the continuity of the multiplication. For any given $x_0 \in X$, $\lambda_0 \in \mathbb{R}$ and $\varepsilon > 0$ we choose a large integer satisfying $n \geq 1 + |\lambda_0|$ and $N(n^{-1}x_0) < \varepsilon$. If

$$|\lambda - \lambda_0| < 1/n \quad \text{and} \quad N(x - x_0) < \varepsilon/n,$$

then $|\lambda| < |\lambda_0| + 1/n \leq n$, and therefore

$$N(\lambda x - \lambda_0 x_0) \leq N(\lambda(x - x_0)) + N((\lambda - \lambda_0)x_0)$$
$$\leq N(n(x - x_0)) + N(n^{-1}x_0)$$
$$< nN(x - x_0) + \varepsilon < 2\varepsilon.$$

$\square$

Remark If N does not satisfy the last condition of the definition of the pseudonorm, then we still get a metric space, but not a topological vector space.

Now we endow the spaces L^p with a natural pseudonorm:

Proposition 10.5 *Let* $0 < p \leq 1$.

(a) L^p *is a vector space.*
(b) *The formula*

$$N_p(f) := \|f\|_p^p = \int_X |f|^p \, d\mu$$

defines a pseudonorm on L^p.
      Henceforth we consider this metric in L^p.
(c) *For each Cauchy sequence* (f_n) *in* L^p *there exist two functions* $f, g \in L^p$ *and a subsequence* (f_{n_k}) *such that* $|f_{n_k}| \leq g$ *for all* k, *and* $f_{n_k} \to f$ *a.e.*
(d) L^p *is a complete metric space.*

Remark For $p = 1$ the pseudonorm N_1 is equal to the norm $\|\cdot\|_1$.

Proof

(a) If $f \in L^p$ and $c \in \mathbb{R}$, then $N_p(cf) = |c|^p N_p(f) < \infty$, and hence $cf \in L^p$. It remains to show that if $f, g \in L^p$, then $f + g \in L^p$. This follows by applying the elementary inequality of Lemma 3.24 (p. 144):

$$N_p(f + g) = \int |f + g|^p \, d\mu \leq \int |f|^p + |g|^p \, d\mu = N_p(f) + N_p(g) < \infty.$$

(b) The first two properties of the pseudonorms are obvious, while the last two follow from the equality $N_p(cf) = |c|^p N_p(f)$. Finally, the triangle inequality has been proved in (a).
(c) Following the proofs of Lemmas 5.13 and 9.2 (pp. 184 and 307) we may choose a subsequence (f_{n_k}) satisfying

$$\sum_{k=1}^{\infty} \int |f_{n_{k+1}} - f_{n_k}|^p \, d\mu \leq 1.$$

Hence

$$\sum_{k=1}^{\infty} |f_{n_{k+1}} - f_{n_k}|^p < \infty$$

a.e. by Corollary 5.9 (p. 180) of the Beppo Levi theorem. This implies the inequality

$$\sum_{k=1}^{\infty} |f_{n_{k+1}} - f_{n_k}| < \infty$$

a.e., because in almost every fixed $t \in X$ we have $\left|f_{n_{k+1}} - f_{n_k}\right|^p \le 1$ if k is sufficiently large, and this implies

$$\left|f_{n_{k+1}} - f_{n_k}\right| \le \left|f_{n_{k+1}} - f_{n_k}\right|^p$$

because $0 < p \le 1$.

It follows that the series

$$|f_{n_1}| + \sum_{k=1}^{\infty} \left|f_{n_{k+1}} - f_{n_k}\right| \quad \text{and} \quad f_{n_1} + \sum_{k=1}^{\infty} (f_{n_{k+1}} - f_{n_k})$$

converge a.e. to some functions g, f. The partial sums g_k and f_{n_k} satisfy the inequality $|f_{n_k}| \le g_k$; letting $k \to \infty$ this yields $|f| \le g$.

It remains to show that $g \in L^p$. Thanks to the choice of (f_{n_k}) we have

$$\int |g_k|^p \, d\mu = N_p(g_k) \le N_p(f_{n_1}) + \sum_{k=1}^{\infty} N_p(f_{n_{k+1}} - f_{n_k}) \le N_p(f_{n_1}) + 1$$

for all k. Since $|g_k|^p \to |g|^p$ a.e., $|g|^p$ is integrable by the Fatou lemma (p. 183), i.e., $g \in L^p$.

(d) We may repeat the proof of Proposition 9.1 (p. 307), by using property (c) above instead of Lemma 9.2 (p. 307).

$\square$

Proposition 10.1 (p. 341) remains valid for all $0 < p \le 1$:

Proposition 10.6 *Let (f_n) be a* bounded *sequence in L^p for some $p \in (0, 1]$, converging a.e. to some function f.*

(a) $f \in L^p$ and $N_p(f) \le \liminf N_p(f_n)$.
(b) *If there exists a $g \in L^p$ such that $|f_n| \le g$ for all n, then $N_p(f_n - f) \to 0$.*
(c) *If $N_p(f_n) \to N_p(f)$, then $N_p(f_n - f) \to 0$.*

Proof

(a) We apply the Fatou lemma to the functions $|f_n|^p$.
(b) Since

$$|f_n - f|^p \le |f_n|^p + |f|^p \le 2g^p$$

a.e., we may again apply the dominated convergence theorem.
(c) The functions

$$|f_n|^p + |f|^p - |f_n - f|^p$$

are nonnegative by Lemma 3.24 (p. 144), and converge to $2\,|f|^p$ a.e. Applying the Fatou lemma we obtain that

$$\int 2\,|f|^p \; d\mu \le \liminf \int |f_n|^p + |f|^p - |f_n - f|^p \; d\mu$$

$$= \int 2\,|f|^p \; d\mu - \limsup \int |f_n - f|^p \; d\mu.$$

Hence $\limsup N_p(f_n - f) \le 0$, i.e., $N_p(f_n - f) \to 0$.

$\square$

Example The spaces $L^p([0,1])$ are not locally convex for $0 < p < 1$, because their only convex open subsets are $\varnothing$ and L^p.[8]

By translation invariance it suffices to show that if K is a convex open set containing 0, then $K = L^p$.

Fix $r > 0$ such that $B_r(0) \subset K$, and fix $x \in L^p$ arbitrarily. For each natural number n there exists a finite subdivision $0 = t_0 < \cdots < t_n = 1$ such that

$$\int_{t_{i-1}}^{t_i} |x(t)|^p \; dt = n^{-1} \int_0^1 |x(t)|^p \; dt, \quad i = 1, \dots, n.$$

Setting

$$x_i := n \chi_{[t_{i-1}, t_i]} x, \quad i = 1, \dots, n$$

we have

$$N_p(x_i) = n^{p-1} \int_0^1 |x(t)|^p \; dt, \quad i = 1, \dots, n.$$

Consequently, choosing a sufficiently large n we have $x_1, \dots, x_n \in B_r(0)$. Since $B_r(0) \subset K$ and K is convex, we conclude that

$$x = (x_1 + \cdots + x_n)/n \in K.$$

It follows from this result that $(L^p)' = \{0\}$, so that no two points of $L^p([0,1])$ may be separated by a closed affine hyperplane.[9]

---

[8]This property and the following proof remains valid in much more general measure spaces.
[9]Day [96].

As the preceding example indicates, for $0 < p < 1$ the spaces L^p are not normable, and not even locally convex in general. Therefore they are much less useful than the spaces for $p \geq 1$.[10]

10.3 L^0 Spaces

The spaces to be studied in this section provide a better understanding of a.e. convergence.

We denote by L^0 the set of measurable, *a.e. finite-valued* functions satisfying

$$N_0(f) := \int \frac{|f|}{1+|f|}\, d\mu < \infty.$$

Proposition 10.7

(a) L^0 *is a vector space.*
(b) N_0 *is a pseudonorm on L^0.*
  Henceforth we endow L^0 with the corresponding metric.
(c) *(Riesz)[11] For every Cauchy sequence (f_n) of L^0 there exist two functions $f, g \in L^0$ and a subsequence (f_{n_k}) such that $|f_{n_k}| \leq g$ for all k, and $f_{n_k} \to f$ a.e.*
(d) L^0 *is a complete metric space.*

Proof (a) If $f \in L^0$ and $c \in \mathbb{R}$, then $cf \in L^0$. This follows from the estimate

$$N_0(cf) = \int \frac{|cf|}{1+|cf|}\, d\mu \leq \int \frac{|cf|}{|c|+|cf|}\, d\mu = N_0(f) < \infty$$

if $|c| \leq 1$, and from

$$N_0(cf) = \int \frac{|cf|}{1+|cf|}\, d\mu \leq \int \frac{|cf|}{1+|f|}\, d\mu = |c|\, N_0(f) < \infty$$

if $|c| \geq 1$.

---

[10]Except for the construction of counterexamples. For example, Roberts [395, 396] constructed non-empty, compact, convex sets in $L^p([0,1])$ with $0 < p < 1$ that have no extremal points. Hence the Krein–Milman theorem (p. 129) does not hold in these spaces. See also Kalton [241], Kalton and Peck [242], Narici–Beckenstein [331].
[11]Riesz [377].

If $f, g \in L^0$, then $f + g \in L^0$. Indeed, using the monotonicity of the function $t \mapsto t/(1 + t)$ in $[0, \infty)$, we have

$$N_0(f + g) = \int \frac{|f + g|}{1 + |f + g|}\, d\mu$$

$$\leq \int \frac{|f| + |g|}{1 + |f| + |g|}\, d\mu$$

$$\leq \int \frac{|f|}{1 + |f|}\, d\mu + \int \frac{|g|}{1 + |g|}\, d\mu$$

$$= N_0(f) + N_0(g)$$

$$< \infty.$$

(b) It follows from the definition that $N_0(0) = 0$, and $N_0(f) > 0$ if $f \neq 0$. The properties

$$N_0(f + g) \leq N_0(f) + N_0(g)$$

and

$$N_0(cf) \leq N_0(f) \quad \text{if} \quad -1 \leq c \leq 1$$

have been shown in (a). Finally,[12] for $f \in L^0$ and $n \to \infty$ we have

$$\int \frac{|n^{-1}f|}{1 + |n^{-1}f|}\, d\mu \to 0$$

by the dominated convergence theorem (p. 181), i.e., $N_0(n^{-1}f) \to 0$.
(c) Adapting the proof of Proposition 10.5 there exists a subsequence (f_{n_k}) satisfying

$$\sum_{k=1}^{\infty} \int \frac{|f_{n_k} - f_{n_{k+1}}|}{1 + |f_{n_k} - f_{n_{k+1}}|}\, d\mu \leq 1;$$

hence

$$\sum_{k=1}^{\infty} \frac{|f_{n_k} - f_{n_{k+1}}|}{1 + |f_{n_k} - f_{n_{k+1}}|} < \infty$$

---

[12]It is important here that f is finite-valued a.e. by assumption.

a.e. by the Beppo Levi theorem (p. 178). Since

$$\frac{|x|}{1+|x|} \le \frac{1}{2} \implies |x| \le 1 \implies |x| \le 2\frac{|x|}{1+|x|},$$

this implies that

$$\sum_{k=1}^{\infty} |f_{n_k} - f_{n_{k+1}}| < \infty$$

a.e. Consequently, the partial sums g_k, f_{n_k} of the series

$$|f_{n_1}| + \sum_{k=1}^{\infty} |f_{n_{k+1}} - f_{n_k}| \quad \text{and} \quad f_{n_1} + \sum_{k=1}^{\infty}(f_{n_{k+1}} - f_{n_k})$$

converge a.e. to some finite-valued functions g, f, satisfying $|f| \le g$.
It remains to show that $g \in L^0$. By the choice of (f_{n_k}) we have

$$\int \frac{|g_k|}{1+|g_k|} \, d\mu = N_0(g_k) \le N_0(f_{n_1}) + \sum_{k=1}^{\infty} N_0(f_{n_{k+1}} - f_{n_k}) \le N_0(f_{n_1}) + 1$$

for all k. Since

$$\frac{|g_k|}{1+|g_k|} \to \frac{|g|}{1+|g|}$$

a.e., the limit function is integrable by the Fatou lemma, i.e., $g \in L^0$.

(d) Using (c) we fix a subsequence (f_{n_k}) converging a.e. to some $f \in L^0$. Next for any given $\varepsilon > 0$ we choose an integer M such that

$$\int \frac{|f_m - f_n|}{1+|f_m - f_n|} \, d\mu = N_0(f_m - f_n) < \varepsilon$$

for all $m, n \ge M$. Taking $n = n_k$ and letting $k \to \infty$, an application of the Fatou lemma yields

$$N_0(f_m - f) = \int \frac{|f_m - f|}{1+|f_m - f|} \, d\mu \le \varepsilon$$

for all $m \ge M$.

$\square$

Now we extend Propositions 10.1 and 10.6 (pp. 341 and 349) to L^0:

Proposition 10.8 *Let* (f_n) *be a* bounded *sequence in* L^0, *converging a.e. to some finite-valued function* f.

(a) $f \in L^0$ *and* $N_0(f) \leq \liminf N_0(f_n)$;
(b) *If there exists a* $g \in L^0$ *such that* $|f_n| \leq g$ *for all* n, *then* $N_0(f_n - f) \to 0$;
(c) *If* $N_0(f_n) \to N_0(f)$, *then* $N_0(f_n - f) \to 0$.

Remark If $\mu(X) < \infty$, then $N_0(f) \leq \mu(X)$ for all $f \in L^0$, so that every sequence (f_n) is bounded in L^0.

Proof of Proposition 10.8

(a) We apply the Fatou lemma to the functions $\dfrac{|f_n|}{1 + |f_n|}$.

(b) Since

$$\frac{|f_n - f|}{1 + |f_n - f|} \leq \frac{|f_n|}{1 + |f_n|} + \frac{|f|}{1 + |f|} \leq 2\frac{|g|}{1 + |g|},$$

we may apply the dominated convergence theorem.

(c) The functions

$$\frac{|f_n|}{1 + |f_n|} + \frac{|f|}{1 + |f|} - \frac{|f_n - f|}{1 + |f_n - f|}$$

are nonnegative[13] and converge a.e. to $2|f|/(1 + |f|)$. Applying the Fatou lemma we get

$$2 \int \frac{|f|}{1 + |f|} \, d\mu \leq \liminf \int \frac{|f_n|}{1 + |f_n|} + \frac{|f|}{1 + |f|} - \frac{|f_n - f|}{1 + |f_n - f|} \, d\mu$$

$$= 2 \int \frac{|f|}{1 + |f|} \, d\mu - \limsup \int \frac{|f_n - f|}{1 + |f_n - f|} \, d\mu$$

$$= 2 \int \frac{|f|}{1 + |f|} \, d\mu - \limsup N_0(f_n - f).$$

Hence $\limsup N_0(f_n - f) \leq 0$, i.e., $N_0(f_n - f) \to 0$.

$\square$

Remark The theorems of Lebesgue and Fatou (pp. 181, 183) and Propositions 10.1, 10.2, 10.6 and 10.8 (pp. 341, 342, 349) remain valid if we assume the convergence in L^0 instead of a.e. convergence.

For example, if (f_n) is *bounded* in L^p for some $1 < p \leq \infty$ and $f_n \to f$ in L^0, then $f_n \to f$ in $\sigma(L^p, L^q)$, where q denotes the conjugate exponent.

---

[13]See the proof of Proposition 10.7 (a).

Indeed, by Cantor's lemma (p. 36) it is sufficient to show that every subsequence (f_{n_k}) of (f_n) has a subsequence $(f_{n_{k_\ell}})$ converging to f in the topology $\sigma(L^p, L^q)$. Since $f_{n_k} \to f$ in L^0, by the Riesz lemma [Proposition 10.9 (c)] there exists a subsequence $(f_{n_{k_\ell}})$ converging to f a.e. We conclude by applying Proposition 10.2.

The other proofs are analogous.

Like L^p for $0 < p < 1$, the L^0 spaces are not locally convex in general:

Example The only convex open sets in $L^0([0, 1])$ are $\varnothing$ and L^0. Hence $(L^0)' = \{0\}$, and no two points of $L^0([0, 1])$ may be separated by a closed affine hyperplane.[14]

As before, it is sufficient to show that if a convex open set K contains the point 0, then $K = L^0$. Fix $r > 0$ such that $B_r(0) \subset K$ and a positive integer $n > 1/r$.

For any given $x \in L^0$ we consider the functions

$$x_i(t) := \begin{cases} nx(t) & \text{if } (i-1)/n \le t \le i/n, \\ 0 & \text{otherwise,} \end{cases} \quad i = 1, \ldots, n.$$

We have

$$N_0(x_i) = \int_{(i-1)/n}^{i/n} \frac{|nx(t)|}{1 + |nx(t)|} \, dt \le \frac{1}{n} < r,$$

so that $x_1, \ldots, x_n$ belong to the ball $B_r(0)$. Since $B_r(0) \subset K$ and K is convex, we conclude that

$$x = (x_1 + \cdots + x_n)/n \in K.$$

10.4   Convergence in Measure

In view of the usefulness of a.e. convergence we might try to associate it with some norm, metric or topology.

As in the preceding section, we consider only measurable, *a.e. finite-valued* functions. In *finite* measure spaces we have a simple characterization of the convergence in L^0. The following notion is frequently used in the theory of probability:

Definition A sequence of functions (f_n) *converges in measure*[15] or *stochastically* to f if for each fixed $\varepsilon > 0$ we have

$$\mu(\{t \in X : |f_n(t) - f(t)| \ge \varepsilon\}) \to 0$$

as $n \to \infty$.

---

[14]Nikodým [343].
[15]Lebesgue [293].

Proposition 10.9 *Assume that $\mu(X) < \infty$.*

(a) *If $0 \leq p \leq q \leq \infty$, then $L^q \subset L^p$, and the embedding $i : L^q \to L^p$ is continuous.*
(b) *L^0 is the set of all measurable and a.e. finite-valued functions.*
(c) *(Fréchet)[16] The convergence of L^0 is convergence in measure.*
(d) *(Riesz)[17] If $f_n \to f$ in measure, then there exists a subsequence (f_{n_k}) converging to f a.e.*

Proof

(a) The case $p = q$ is obvious. If $0 < p < q < \infty$, then applying the Hölder inequality to the product $1 \cdot |f|^p$ we obtain the estimate

$$\int |f|^p \, d\mu \leq \|1\|_{q/(q-p)} \cdot \||f|^p\|_{q/p} = \mu(X)^{1-\frac{p}{q}} \cdot \|f\|_q^p,$$

whence

$$\|f\|_p \leq \mu(X)^{\frac{1}{p}-\frac{1}{q}} \cdot \|f\|_q.$$

The last inequality holds for $q = \infty$ as well, by a direct computation. This proves the continuity of the embedding $i : L^q \to L^p$ for all $0 < p < q \leq \infty$.

It remains to show that the embedding $i : L^q \to L^0$ is continuous for all $0 < q < 1$. There exists a constant $c_q > 0$ such that

$$\frac{|f|}{1 + |f|} \leq c_q |f|^q$$

for all $f \in L^q$, because the function $t \mapsto |t|^{1-q}/(1 + |t|)$ is continuous and bounded on $\mathbb{R}$. This implies the inequality $N_0(f) \leq c_q N_q(f)$ and thus the required continuity.

(b) This follows from the definition of L^0 because $|f|/(1 + |f|)$ is bounded.
(c) We may assume that $f = 0$. If $N_0(f_n) \to 0$ and $\varepsilon > 0$, then using the non-decreasingness of the function $t \mapsto t/(1 + t)$ on $[0, \infty)$ we obtain that

$$0 \leq \mu(\{|f_n| \geq \varepsilon\}) = \int_{\{|f_n| \geq \varepsilon\}} 1 \, d\mu$$

$$\leq \frac{1+\varepsilon}{\varepsilon} \int_{\{|f_n| \geq \varepsilon\}} \frac{|f_n|}{1 + |f_n|} \, d\mu \leq \frac{1+\varepsilon}{\varepsilon} N_0(f_n) \to 0.$$

---

[16]Fréchet [159, 160]. He used an equivalent metric.
[17]Riesz [377].

Conversely, if $f_n \to 0$ in measure, then

$$N_0(f_n) = \int_{\{|f_n| < \varepsilon\}} \frac{|f_n|}{1 + |f_n|}\, d\mu + \int_{\{|f_n| \geq \varepsilon\}} \frac{|f_n|}{1 + |f_n|}\, d\mu$$

$$\leq \frac{\varepsilon}{1 + \varepsilon}\mu(X) + \mu(\{|f_n| \geq \varepsilon\})$$

for each $\varepsilon > 0$. Hence $N_0(f_n) < \varepsilon\mu(X)$ if n is large enough.
(d) We combine (c) with Proposition 10.7 (c) (p. 351).

□

Remark If $\mu(X) < \infty$, then the theorems of Lebesgue and Fatou (pp. 181, 183) and Propositions 10.1, 10.2, 10.6 and 10.8 (pp. 341, 342, 349) remain valid if we assume convergence in measure instead of a.e. convergence.

This follows from the last remark of the preceding section by applying Proposition 10.9 (d) instead of Proposition 10.9 (c).

The assumption $\mu(X) < \infty$ may be omitted if we assume convergence in measure on every set of finite measure.

Using convergence in measure we may characterize strong convergence in L^p, and we may clarify the relationship between weak and strong convergence:

Proposition 10.10 *Assume that $\mu(X) < \infty$.*

(a) *(Vitali)*[18] *Let $0 < p < \infty$. We have $\|f_n - f\|_p \to 0 \iff f_n \to f$ in measure, and*

$$\sup_n \int_A |f_n|^p\, d\mu \to 0 \quad as \quad \mu(A) \to 0.$$

(b) *Let $0 < r < p \leq \infty$. If (f_n) is bounded in L^p and $f_n \to f$ in measure, then $\|f_n - f\|_r \to 0$.*

(c) *(Vitali–Hahn–Saks)*[19] *The following equivalence holds in L^1:*

$$\|f_n - f\|_1 \to 0 \iff f_n \rightharpoonup f, \quad and \quad f_n \to f \quad in\ measure.$$

Example The functions $f_n(t) := n^{1/p}e^{-nt}$ defined on $[0, 1]$ have the following properties[20]:

- they are bounded in L^p for all $p \in (0, \infty]$;
- they converge to zero in measure;

---

[18]Vitali [471, p. 147]. This strengthens the dominated convergence theorem.
[19]Vitali [471, p. 147]; Hahn [181]; Saks [408]. This contains Schur's theorem (p. 84) as a special case.
[20]We have already studied them in Sect. 10.1.

- they do not converge strongly in L^p.

This shows the optimality of (b) in the proposition.

Remark Convergence in measure is not necessary for weak convergence.[21] For example, the functions $\sin nt$ do not converge to zero in measure on any bounded interval I, but $\sin nt \to 0$ in $\sigma(L^p, L^q)$ for all $p \in [1, \infty]$ by the Riemann–Lebesgue lemma (p. 338).

Proof

(a) First we assume that $\|f_n - f\|_p \to 0$. Then $f_n \to f$ in measure by parts (a) and (c) of the preceding proposition. Furthermore, for any fixed $\varepsilon > 0$ we may fix an integer N such that $\|f_n - f\|_p < \varepsilon$ for all $n \geq N$. By Lemma 7.14 (p. 235) there exists a $\delta > 0$ such that

$$\int_A |f|^p \, d\mu < \varepsilon^p, \quad \text{and} \quad \int_A |f_n|^p \, d\mu < \varepsilon^p, \quad n = 1, \ldots, N-1,$$

whenever $\mu(A) < \delta$. Then the following conditions are satisfied for all $n \geq N$:

$$\int_A |f_n|^p \, d\mu \leq \int_A |f|^p \, d\mu + \int_A |f_n - f|^p \, d\mu < 2\varepsilon^p$$

if $0 < p \leq 1$, and

$$\left(\int_A |f_n|^p \, d\mu \right)^{1/p} \leq \left(\int_A |f|^p \, d\mu \right)^{1/p} + \left(\int_A |f_n - f|^p \, d\mu \right)^{1/p} < 2\varepsilon$$

if $1 \leq p < \infty$.

For the converse direction it suffices to show that (f_n) is a Cauchy sequence in L^p. Indeed, then f_n converges to some $g \in L^p$ by the completeness of L^p, and then also in measure by parts (a) and (c) of the preceding Proposition. Since $f_n \to f$ in measure by assumption, we conclude that $f = g$ a.e. by the uniqueness of the limit.

For the proof of the Cauchy property we fix $\varepsilon > 0$ arbitrarily, and we choose $\delta > 0$ such that

$$\mu(A) < \delta \implies \int_A |f_n|^p \, d\mu < \varepsilon^p \quad \text{for all} \quad n.$$

Since $f_n \to f$ in measure, we may choose a large N such that

$$\mu\left(\left\{ |f_n - f| \geq \frac{\varepsilon}{2} \right\} \right) < \frac{\delta}{2} \quad \text{for all} \quad n \geq N.$$

---

[21]Except some special spaces like ℓ^1 by Schur's theorem.

Applying the triangle inequality this yields

$$\mu\big(\{|f_m - f_n| \geq \varepsilon\}\big) < \delta \quad \text{for all} \quad m, n \geq N.$$

Consequently, using the elementary inequality[22]

$$|x - y|^p \leq \max\{1, 2^{p-1}\}\,(|x|^p + |y|^p)$$

for $x, y \in \mathbb{R}$, we obtain following estimate for all $m, n \geq N$:

$$\int_X |f_m - f_n|^p \, d\mu$$

$$= \int_{\{|f_m - f_n| \geq \varepsilon\}} |f_m - f_n|^p \, d\mu + \int_{\{|f_m - f_n| < \varepsilon\}} |f_m - f_n|^p \, d\mu$$

$$\leq \max\{1, 2^{p-1}\} \int_{\{|f_m - f_n| \geq \varepsilon\}} |f_m|^p + |f_n|^p \, d\mu + \mu(X)\varepsilon^p$$

$$\leq \big(\max\{2, 2^p\} + \mu(X)\big)\varepsilon^p.$$

We conclude by observing that the right-hand side tends to zero as $\varepsilon \to 0$.

(b) Applying Hölder's inequality we have the following estimate for each measurable set A:

$$\int_A |f_n|^r \, d\mu = \int_X \chi_A |f_n|^r \, d\mu \leq \|\chi_A\|_{p/(p-r)} \cdot \||f_n|^r\|_{p/r}$$

$$= \mu(A)^{(p-r)/p} \cdot \|f_n\|_p^r.$$

Since (f_n) is bounded in L^p, the right-hand side tends to zero uniformly in n as $\mu(A) \to 0$. We conclude by applying (a).

(c) If $\|f_n - f\|_1 \to 0$, then $f_n \rightharpoonup f$ by a general property of weak convergence, and $f_n \to f$ in measure by part (a) above.

For the converse direction, by (a) it suffices to show that

$$\sup_n \int_A |f_n| \, d\mu \to 0 \quad \text{as} \quad \mu(A) \to 0.$$

Using the decomposition

$$\int_A |f_n| \, d\mu = \int_{A \cap \{f_n > 0\}} f_n \, d\mu - \int_{A \cap \{f_n < 0\}} f_n \, d\mu$$

---

[22] The inequality was proved in Lemma 3.24 (p. 144) for $0 < p \leq 1$. For $p \geq 1$ it follows from the convexity of the function $t \mapsto |t|^p$.

it suffices to show that

$$\int_A f_n \, d\mu \to 0 \quad \text{as} \quad \mu(A) \to 0, \tag{10.1}$$

uniformly in n.

(10.1) holds for each n by Lemma 7.14 (p. 235). In order to prove the uniformity in n we denote by $\tilde{L}^1$ the set of characteristic functions of measurable sets.

$\tilde{L}^1$ is *closed* in L^1, and hence a complete metric space. Indeed, if $\chi_{A_m} \to g \, L^1$, then by the Riesz lemma (p. 184) there exists an a.e. convergent subsequence $\chi_{A_{m_k}} \to g$. Then $g(t) \in \{0,1\}$ for a.e. t, i.e., g is the characteristic function of some measurable set.

Fix $\varepsilon > 0$ arbitrarily. Since (f_n) is weakly convergent, the sets

$$F_N := \left\{\chi_A \in \tilde{L}^1 \ : \ \left|\int_A (f_m - f_n) \, d\mu\right| \le \varepsilon \quad \text{for all} \quad m, n \ge N\right\}$$

$(N = 1, 2, \ldots)$ cover $\tilde{L}^1$. These sets are closed. Indeed, if $\chi_{A_k} \to \chi_A$ in L^1, then

$$\mu(A \setminus A_k) + \mu(A_k \setminus A) = \|\chi_{A_k} - \chi_A\|_1 \to 0.$$

Applying (10.1) for any fixed m, n this yields the estimate

$$\left|\int_A (f_m - f_n) \, d\mu\right| = \lim_{k \to \infty} \left|\int_{A_k} (f_m - f_n) \, d\mu\right| \le \varepsilon.$$

Applying Baire's lemma (p. 32) at least one of these sets contains a ball, say $B_r(\chi_A) \subset F_N$. This implies the implication

$$\mu(B) < r \implies \left|\int_B (f_m - f_n) \, d\mu\right| \le 2\varepsilon \quad \text{for all} \quad m, n \ge N. \tag{10.2}$$

Indeed, using the relations

$$\chi_B = \chi_{A \cup B} - \chi_{A \setminus B} \quad \text{and} \quad \chi_{A \cup B}, \chi_{A \setminus B} \in B_r(\chi_A)$$

we obtain the inequalities

$$\left|\int_B f_m - f_n \, d\mu\right| \le \left|\int_{A \cup B} f_m - f_n \, d\mu\right| + \left|\int_{A \setminus B} f_m - f_n \, d\mu\right| \le 2\varepsilon.$$

Applying (10.1) for $n = 1, \ldots N$, there exist $r_1, \ldots r_N > 0$ such that

$$\mu(B) < r_i \implies \left|\int_B f_i \, d\mu\right| \le \varepsilon, \quad i = 1, \ldots N. \tag{10.3}$$

Setting $\delta := \min\{r, r_1 \ldots, r_N\}$ we deduce from (10.2) and (10.3) that

$$\mu(B) < \delta \implies \left| \int_B f_n \, d\mu \right| \leq 3\varepsilon \quad \text{for all} \quad n.$$

□

We have seen in classical analysis the difference between pointwise and uniform convergence. This difference is smaller than expected from the point of view of measure theory. As a byproduct we find a close connection between pointwise convergence and convergence in measure.

Definition The sequence (f_n) converges *quasi-uniformly* to f if for each $\delta > 0$ there exists a set B of measure $< \delta$ such that f_n converges uniformly to f on $X \setminus B$.

Quasi-uniform convergence implies a.e. convergence. Indeed, if $f_n \to f$ quasi-uniformly, then for each $k = 1, 2, \ldots$ there exists a set B_k of measure $< 1/k$ such that $f_n \to f$ uniformly in $X \setminus B_k$. Then $B := \cap B_k$ is a null set, and $f_n \to f$ in $X \setminus B$.

By a surprising discovery of Egorov the quasi-uniform and a.e. convergences are in fact *equivalent*:

Proposition 10.11 *Assume that $\mu(X) < \infty$, and let f_n, f be measurable, a.e. finite-valued functions.*

(a) *(Egorov)[23] If $f_n \to f$ a.e., then $f_n \to f$ quasi-uniformly.*
(b) *(Lebesgue)[24] If $f_n \to f$ a.e., then $f_n \to f$ in measure.*

Remark The functions $f_n = \chi_{[n,n+1]}$ on $X = \mathbb{R}$ show the necessity of the assumption $\mu(X) < \infty$.

Proof

(a) Fix $\delta > 0$ arbitrarily. Since $f_n \to f$ a.e., for each fixed positive integer k, a.e. $x \in X$ belongs to the union of the sets

$$B_{k,m} := \{x \in X : |f_n - f| \leq 1/k \quad \text{for all} \quad n \geq m\}, \quad m = 1, 2, \ldots.$$

Using the assumption $\mu(X) < \infty$, and applying Proposition 7.3 (c) (p. 216) to the non-decreasing set sequence $(B_{k,m})$, there exists a sufficiently large index m_k such that $\mu(X \setminus B_{k,m_k}) < 2^{-k}\delta$. Then $f_n \to f$ uniformly in $B := \cap_{k=1}^{\infty} B_{k,m_k}$, and $\mu(X \setminus B) < \delta$.

(b) Given $\delta > 0$ and $\varepsilon > 0$ arbitrarily, we seek N such that

$$\mu(\{|f_n - f| > \delta\}) < \varepsilon \quad \text{for all} \quad n \geq N.$$

---

[23]Egorov [121].
[24]Lebesgue [293].

By Egorov's theorem there exists a set of measure $< \varepsilon$ such that $f_n \to f$ uniformly in $X \setminus A$. It remains to choose a sufficiently large N such that $|f_n - f| < \delta$ in $X \setminus A$ for all $n \geq N$.

□

We end this section (and the book) by proving that a.e. convergence is *not* a topological convergence in general. This explains some of the difficulties when dealing with this notion.

Corollary 10.12 (Fréchet)[25] *In $L^0([0, 1])$ a.e. convergence is not topologizable.*

Proof Consider the sequence of functions (f_n) introduced on p. 307. Since it converges to zero in measure, by the Riesz lemma (p. 354) every subsequence of (f_n) has a subsequence converging a.e. to zero.

If a.e. convergence were topologizable, then by Cantor's lemma (p. 36) we could conclude that (f_n) itself converges a.e. to zero. But this is false: the numerical sequence $(f_n(t))$ is divergent for every $t \in [0, 1]$.   □

Remark Combining Propositions 10.9 (c), (d) and 10.11 (b) we conclude that among the topological convergences, convergence in measure is the closest to a.e. convergence.

---

[25]Fréchet [160].

Hints and Solutions to Some Exercises

Exercise 1.3. The vectors $e_1 + \cdots + e_n$ form a divergent Cauchy sequence.

Exercise 1.4. Consider the identities

$$(c_1 x_1 + \cdots + c_k x_k, c_1 y_1 + \cdots + c_k y_k) = |c_1|^2 + \cdots + |c_k|^2, \quad k = 1, 2, \ldots.$$

If $c_1 x_1 + \cdots + c_k x_k = 0$ or $c_1 y_1 + \cdots + c_k y_k = 0$, then we conclude that $|c_1|^2 + \cdots + |c_k|^2 = 0$ and hence $c_1 = \cdots = c_k = 0$.

Exercise 1.6. If $(f_n) \subset M$ and $f_n \to f$, then $f \in M$. Indeed, we deduce from the relations

$$\int_0^1 f^2 \, dt = \int_0^1 |f - f_n|^2 \, dt \le \int_{-1}^1 |f - f_n|^2 \, dt \to 0$$

and the continuity of f that $f = 0$ in $[0, 1]$.

If $g \in M^\perp$, then $g - 0$ on $[-1, 0]$. Indeed, the formula

$$f(t) := \begin{cases} t^2 g(t) & \text{if } t \le 0; \\ 0 & \text{if } t \ge 0 \end{cases}$$

defines a function $f \in M$, so that

$$0 = \int_{-1}^1 fg \, dt = \int_{-1}^0 t^2 |g(t)|^2 \, dt.$$

Since g is continuous, we conclude that $g = 0$ in $[-1, 0]$. Hence

$$M \oplus M^\perp \subset \{ f \in X \ : \ f(0) = 0 \}.$$

The converse inclusion is obvious.

© Springer-Verlag London 2016
V. Komornik, *Lectures on Functional Analysis and the Lebesgue Integral*,
Universitext, DOI 10.1007/978-1-4471-6811-9

Notice that X is not complete.

Exercise 1.8. Consider the sets $H = \mathbb{R}$, $M = \mathbb{Z}$ and $N = [0, 1)$.

Exercise 1.10. It suffices to choose an orthonormal basis in G: the proof of its existence, given in the text, does not use completeness.

Exercise 1.11. The density has already been proved on pp. 7–8.

Second solution. The vectors $e_1 - e_2, e_1 - e_3, \ldots$ belong to M, and they generate ℓ^2. Indeed, if $x \in \ell^2$ is orthogonal to them, then $(x, e_n) = (x, e_1)$ for all n. Since $\sum (x, e_n)^2 < \infty$, $(x, e_n) = 0$ for all n, and therefore $x = 0$.

The sequence $(e_1 - e_n)$ is linearly independent; by orthogonalization we obtain an orthonormal basis of ℓ^2.

Exercise 1.12. The orthonormal sequence $e_2, e_3, \ldots$ does not satisfy (a) because f_1 is not the sum of its Fourier series:

$$\sum_{n=2}^{\infty} (f_1, e_n) e_n = \sum_{n=2}^{\infty} \frac{e_n}{n} = f_1 - e_1.$$

Nevertheless, it satisfies (d). Indeed, let $x = c_1 f_1 + c_2 e_2 + \cdots + c_m e_m$ be a *finite* linear combination satisfying $(x, e_n) = 0$ for all $n \geq 2$. Writing them explicitly we have the equations

$$\frac{c_1}{n} + c_n = 0, \quad n = 2, \ldots, m$$

and

$$\frac{c_1}{n} = 0, \quad n = m+1, m+2, \ldots.$$

Hence we first deduce that $c_1 = 0$, and then that $c_n = 0$ for $n = 2, \ldots, m$. Thus $x = 0$.

Exercise 1.14.

(ii) Let $F_n = [n, \infty) \subset \mathbb{R}$, $n = 1, 2, \ldots$.

(iii) Let (e_n) be an orthonormal sequence, and

$$F_n := \{e_k : k > n\}, \quad n = 1, 2, \ldots$$

or

$$F_n = \{x \in H : \|x\| = 1 \text{ and } x \perp e_1, \ldots, x \perp e_n\}, \quad n = 1, 2, \ldots.$$

Exercise 1.24.[1] If $Tx = x$, then using $\|T^*\| = \|T\| \leq 1$ we get

$$\|x\|^2 = (Tx, x) = (x, T^*x) \leq \|x\| \cdot \|T^*x\| \leq \|x\|^2;$$

---

[1] We follow Riesz and Sz. Nagy [393].

hence $(x, T^*x) = \|x\| \cdot \|T^*x\|$ and $\|T^*x\| = \|x\|$. Using these equalities we obtain that

$$\|x - T^*x\|^2 = \|x\|^2 - (x, T^*x) - (T^*x, x) + \|T^*x\|^2 = 0,$$

i.e., $T^*x = x$. Exchanging the role of T and T^* we conclude that $N(I - T) = N(I - T^*)$.

Exercise 2.2.

(i) Consider the sequences $x_n := n^{-1/p}$ and $y_n := n^{-1/q}(\ln n)^{-2/q}$.
(ii) The sequence

$$x^k = (1^{-1/p}, 2^{-1/p}, \dots, k^{-1/p}, 0, 0, \dots), \quad k = 1, 2, \dots$$

converges in $\ell^q \iff q > p$.

Exercise 2.4. Both sequences converge pointwise to zero. Since

$$\sup x_n = x_n \left(\frac{n}{n+1} \right) = \left(1 - \frac{1}{n+1} \right)^n - \left(1 - \frac{1}{n+1} \right)^{n+1} \to 0,$$

the first sequence is uniformly convergent.
   Since

$$\sup y_n = y_n(2^{-1/n}) = \frac{1}{4} \not\to 0,$$

the second convergence is not uniform.
Exercise 2.5.

(i) Since $|x(1)| \le \|x\|_\infty$ for all $x \in A$, the linear functional is continuous, of norm ≤ 1.
(ii) *First solution.* For $x_n(t) = t^n$ we have $x_n(1) = 1$ and $\|x_n\|_2^2 = 1/(2n+1)$, $n = 1, 2, \dots$. Since

$$\sup_{x \in A, x \ne 0} \frac{|x(1)|}{\|x\|_2} \ge \sup_n \frac{|x_n(1)|}{\|x_n\|_2} = \infty,$$

the linear functional is not continuous.

   Second solution. Define $y_n \in A$ by $y_n = 0$ in $[0, 1 - 1/n]$ and $y_n(1 - t) = nt$ in $[1 - 1/n, 1]$. Then $y_n(1) = 1$ and $\|x_n\|_2^2 = 1/(3n)$.
Exercise 2.6.

(i) The bilinear map $g(x, y) := xy$ is continuous from $A_\infty \times A_\infty$ into A_∞ because

$$\|xy\|_\infty \le \|x\|_\infty \cdot \|y\|_\infty.$$

for all $x, y \in A$.

The linear map $h(x) := (x, x)$ of A_∞ into $A_\infty \times A_\infty$ is obviously continuous, hence $f = g \circ h$ is continuous, too.

(ii) The functions

$$z_n(t) := \min\left\{n, x^{-1/4}\right\}, \quad n = 1, 2, \ldots$$

satisfy

$$\|z_n\|_2^2 \leq \int_0^1 x^{-1/2}\, dx = [2\sqrt{x}]_0^1 = 2$$

for all n, and

$$\|z_n^2\|_2^2 \to \int_0^1 x^{-1}\, dx = \infty.$$

Hence our map is not continuous.

(iii) The continuity of f follows from (i) because we have weakened the topology of the space of arrival.

Exercise 2.10. Write $[f] := f + L$ for brevity. If $([f_n])$ is a Cauchy sequence in X/L, then there exists a subsequence satisfying

$$\left\|[f_{n_{k+1}}] - [f_{n_k}]\right\| < 2^{-k}, \quad k = 1, 2, \ldots.$$

Choose $h_k \in [f_{n_{k+1}}] - [f_{n_k}]$ such that $\|h_k\| < 2^{-k}$, then $h := \sum h_k$ is a well-defined element of X. Since

$$[f_{n_k}] - [f_{n_1}] = \sum_{i=1}^{k-1}[f_{n_{i+1}} - f_{n_i}] = \sum_{i=1}^{k-1}[h_i],$$

we have $[f_{n_k}] - [f_{n_1}] \to [h]$ and therefore $[f_{n_k}] \to [h + f_{n_1}]$ in X/L.

Exercise 2.11.

(i) *First solution.* If $\overline{B}_{r_1}(x_1) \supset \overline{B}_{r_2}(x_2) \supset \cdots$, then the sequence (r_k) is non-increasing, hence converges to some $r \geq 0$. Then we have $\overline{B}_{r_1-r}(x_1) \supset$

$\overline{B}_{r_2-r}(x_2) \supset \cdots$ because

$$\overline{B}_{r_1}(x_1) \supset \overline{B}_{r_2}(x_2) \iff r_1 \geq r_2 + \|x_1 - x_2\| \iff \overline{B}_{r_1-r}(x_1) \supset \overline{B}_{r_2-r}(x_2).$$

We conclude by applying Cantor's theorem.

Second solution.[2] If $n > m$, then $\|x_n - x_m\| \leq r_m - r_n$. Since (r_n) is a bounded and non-increasing sequence, it is a Cauchy sequence. Its limit belongs to each closed ball.

(ii) *First solution.*[3] We consider the linear subspace $X := \text{Vect}\,\{e_1, e_2, \ldots\}$ of ℓ^1 with the restriction of the norm. Choose a sequence $y = (y_n) \in \ell^1$ with $y_n > 0$ for all n, and consider the closed balls $\overline{B}_{r_n}(x_n)$ with

$$x_n = (y_1, \ldots, y_n, 0, 0, \ldots) \quad \text{and} \quad r_n = y_{n+1} + y_{n+2} + \cdots, \quad n = 1, 2, \ldots.$$

Second solution.[4] Let Y be the completion of a non-complete normed space X, and $y \in Y \setminus X$. Starting with an arbitrary point $x_1 \in X$, we construct a sequence $(x_n) \subset X$ satisfying $\|y - x_{n+1}\| < \|y - x_n\|/3$, and we consider in Y the closed balls $F_n = \overline{B}_{r_n}(x_n)$ of radius $r_n := 2\|y - x_n\|$.

If $x \in F_{n+1}$ for some $n \geq 1$, then

$$\begin{aligned}
\|x - x_n\| &\leq \|x - x_{n+1}\| + \|x_{n+1} - y\| + \|y - x_n\| \\
&\leq 2\|y - x_{n+1}\| + \|x_{n+1} - y\| + \|y - x_n\| \\
&< 2\|y - x_n\|,
\end{aligned}$$

and hence $x \in F_n$.

Finally, since $y \in F_n$ for all n and $\text{diam}\, F_n \to 0$, $\cap F_n$ does not meet X.

Exercise 2.12.

(ii) Let $K_1 \supset K_2 \supset \cdots$ be a decreasing sequence of non-empty bounded closed convex sets in a reflexive space. Choosing a point $x_n \in K_n$ for each n we obtain a bounded sequence. There exists a weakly convergent subsequence $x_{n_k} \rightharpoonup x$. Each K_m contains all but finitely many elements of (x_{n_k}), so that $x \in K_m$.

(ii) *First solution.* Consider in $X = c_0$ the sets

$$K_n := \{x = (x_i) \in c_0 \;:\; x_1 = \cdots = x_n = \|x\| = 1\}, \quad n = 1, 2, \ldots.$$

Second solution. If X is not reflexive, then there exists a non-empty closed convex set $K \subset X$ and a point $x \in X$ such that the distance $d := \text{dist}(x, K)$ is not attained. Set $K_n := K \cap \overline{B}_{d+n-1}(x), 1, 2, \ldots.$

---

[2] F. Alabau-Boussouira, private communication.
[3] M. Ounaies, private communication.
[4] With Á. Besenyei.

Exercise 2.13.

(i) In finite dimensions the bounded closed sets are compact, and we may apply Cantor's intersection theorem.
(ii) In infinite dimensions there exists a sequence (x_n) of unit vectors satisfying $\|x_n - x_k\| \geq 1$ for all $n \neq k$.[5] Set $F_n := \{x_n, x_{n+1}, \ldots\}$, $n = 1, 2, \ldots$.

Exercise 2.17.

(iii) If X is reflexive, then there is a weakly convergent subsequence $x_{n_k} \rightharpoonup x$ of (x_n). Therefore $\varphi(x_{n_k}) \to \varphi(x)$ for each $\varphi \in X'$. Since a (numerical) Cauchy sequence converges to its accumulation points, $\varphi(x_n) \to \varphi(x)$ for each $\varphi \in X'$, i.e., $x_n \rightharpoonup x$.
(ii) follows from (iii) because the Hilbert spaces are reflexive.
(i) follows from (iii) because the finite-dimensional normed spaces are reflexive, and the weak and strong convergences are the same.
(iv) See Dunford and Schwartz [117].
(v) Setting $x_n := e_1 + \cdots + e_n$ we get a weak Cauchy sequence because each $\varphi \in c_0'$ is represented by some $(y_k) \in \ell^1$, and hence

$$\varphi(x_n) - \varphi(x_m) = y_{m+1} + \cdots + y_n \to 0$$

as $n > m \to \infty$. Considering the linear functionals $\varphi \in c_0'$ associated with the sequences e_j we obtain that the only possible weak limit of (x_n) is the constant sequence $(1, 1, \ldots)$. Since it does not belong to c_0, (x_n) does not converge weakly.
(vi) Argue as in the last example of Sect. 2.5, p. 79.

Exercise 2.18. The linearly independent subsets of X satisfy the assumptions of Zorn's lemma, hence there exists a maximal linearly independent subset B. This is necessarily a basis of the *vector space* X. Choose an infinite sequence $(f_n) \subset B$, define $\varphi(f_n) := n\,|f_n\|$ for $n = 1, 2, \ldots$, and define $\varphi(x)$ arbitrarily for $x \in B \setminus \{f_1, f_2, \ldots\}$. Then φ extends to a unique linear functional $\psi : X \to \mathbb{R}$, and ψ is not continuous.

Exercise 2.19. If a normed space X has a countably infinite Hamel basis $f_1, f_2, \ldots$, then X is the union of the (finite-dimensional and hence) closed subspaces $\mathrm{Vect}\{f_1, \ldots, f_n\}$, $n = 1, 2, \ldots$. Since none of them has interior points, by Baire's theorem X cannot be complete.

Exercise 2.20.[6]

(i) For each $\theta \in [0, \pi)$ let S_θ be the intersection of $\mathbb{Z}^2$ with an infinite strip of inclination θ and width greater than one. Each S_θ is infinite, but the intersection of two such sets belongs to a bounded parallelogram and hence is finite. Since

---

[5]This was an application of the Helly–Hahn–Banach theorem in the course.
[6]We present the proofs of Buddenhagen [67] and Lacey [276], respectively.

$(0, 1) \subset [0, \pi)$ and since there is a bijection between $\mathbb{N}$ and $\mathbb{Z}^2$, the desired result follows.

(ii) By the Helly–Hahn–Banach theorem there exist two sequences $(x_n) \subset X$ and $(\varphi_n) \subset X'$ satisfying $\varphi_n(x_k) \neq 0 \Longleftrightarrow n = k$. Then (x_n) is linearly independent; moreover, no x_n belongs to the closed linear span of the remaining vectors x_m. We may assume by normalization that the sequence (x_n) is bounded. Then the vectors

$$\sum_{n \in N_t} \frac{x_n}{2^n}, \quad t \in (0, 1)$$

form a linearly independent set of vectors, having $2^{\aleph_0}$ elements.

Exercise 2.21.

(i) Consider the sets N_t of the preceding exercise. Setting

$$x_n^t = \begin{cases} 1 & \text{if } n \in N_t, \\ 0 & \text{otherwise} \end{cases}$$

we obtain $2^{\aleph_0}$ linearly independent functions $x^t \in \ell^\infty$.

Since ℓ^∞ itself has $2^{\aleph_0}$ elements, its Hamel dimension is $2^{\aleph_0}$.

(ii) Fix a sequence of vectors $x_1, x_2, \ldots$ satisfying

$$\|x_n\| = \text{dist}\,(x_n, \text{Vect}\,\{x_1, \ldots, x_{n-1}\}) = 3^{-n}, \quad n = 1, 2, \ldots,$$

and define

$$Ac := \sum_{n=1}^{\infty} c_n x_n \in X$$

for all $c \in \ell^\infty$.

These vectors are well defined because X is complete and

$$\sum_{n=1}^{\infty} \|c_n x_n\| \leq \|c\|_\infty \sum_{n=1}^{\infty} \|x_n\| < \infty.$$

It remains to show that $Ac = 0$ implies $c = 0$.

We have for each positive integer N the following estimate:

$$\|Ac\| \geq \left\| \sum_{n=1}^{N} c_n x_n \right\| - \left\| \sum_{n=N+1}^{\infty} c_n x_n \right\|$$

$$\geq |c_N| 3^{-N} - \sum_{n=N+1}^{\infty} |c_n| 3^{-n}$$

$$\geq |c_N| 3^{-N} - \|c\|_\infty \sum_{n=N+1}^{\infty} 3^{-n}.$$

If $Ac = 0$, then

$$|c_N| \leq \|c\|_\infty \sum_{n=1}^{\infty} 3^{-n} = \frac{1}{2} \|c\|_\infty$$

for all N; therefore $\|c\|_\infty \leq \frac{1}{2} \|c\|_\infty$ and thus $c = 0$.

Exercise 4.1. The set of continuous functions $f : \mathbb{R} \to \mathbb{R}$ has the power $2^{\aleph_0}$ of $\mathbb{R}$ because it is determined by its values at rational points. The set of jump functions also has the power $2^{\aleph_0}$. Consequently, the set of monotone functions has the power $2^{\aleph_0}$.

On the other hand, the set of null sets has the power of $2^{2^{\aleph_0}} > 2^{\aleph_0}$.

Exercise 4.2. It suffices to prove that the line $y = x + \alpha$ meets $C \times C$ for each $\alpha \in [-1, 1]$. We recall that $C = \cap C_n$ where each C_n is the disjoint union of 2^n intervals of length 3^{-n}. Hence each $C_n \times C_n$ is the disjoint union of 4^n squares of side 3^{-n}.

Prove that the line $y = x + \alpha$ meets at least one of the squares in $C_1 \times C_1$, say S_1.

Next prove that $y = x + \alpha$ meets at least one of the squares in $C_1 \times C_1$, lying in S_1, say S_2.

Construct recursively a decreasing sequence of squares $S_1, S_2, \ldots$, each meeting the line $y = x + \alpha$.

Exercise 4.7. $\alpha > \beta$ or $\alpha = \beta \leq 0$.

Exercise 4.11. Apply Jordan's theorem in (i), Cantor's diagonal method in (ii) and (v), and use Proposition 4.2 (a), p. 153.

Exercise 5.6. (i) There is a compact subset of positive measure. Apply the Cantor–Bendixson theorem. (ii) All subsets of Cantor's ternary set are measurable. (iii) For otherwise A is countable. (iv) Apply Vitali's method modulo 1.

Exercise 5.7. See Rudin [404].

Exercise 6.1. (i) f is continuous and strictly monotone. (ii) The image of its complement is a union of intervals of total length one. (iii) Consider the inverse image of a non-measurable subset of $f(C)$.

Exercise 6.2. (i) For $\alpha = 0$ we can take Cantor's ternary set. For $\alpha \in (0, 1)$ modify the construction by changing the length of the removed open intervals. (ii) Take $A = \cup C_{\alpha_n}$ with a sequence $\alpha_n \to 1$. (iii) Take the complement of A.

Exercise 7.2. Let $\mu(A) = 0$ if A is finite, and $\mu(A) = \infty$ otherwise.

Exercise 7.3. If $A \subset \mathbb{R}$ is a non-measurable set, then

$$\{(x,x) \in \mathbb{R}^2 \ : \ x \in A\} \tag{10.1}$$

is a two-dimensional null set.

Exercise 7.5. See, e.g., Riesz and Sz.-Nagy [394] and Sz.-Nagy [448] for detailed proofs and applications to Fourier series and to the Riesz representation theorem 8.23 (p. 291).

Exercise 7.6. $\alpha > 0$.

Exercise 7.7. Consider in $\mathbb{R}$ the measure generated by the length of bounded subintervals of $[0, \infty)$.

Exercise 7.8. For example, let

$$f_1(x,y) := \begin{cases} 1 & \text{if } x < y < x+1, \\ -1 & \text{if } x-1 < y < x, \\ 0 & \text{otherwise,} \end{cases}$$

$$f_2(x,y) := \begin{cases} 1 & \text{if } 0 < x < y < 2x, \\ -1 & \text{if } 0 < 2x < y < 3x, \\ 0 & \text{otherwise,} \end{cases}$$

$$f_3(x,y) := \begin{cases} 1 - 2^{-n-1} & \text{if } x, y \in (n, n+1), \\ 2^{-n-1} - 1 & \text{if } x, y-1 \in (n, n+1), \\ 0 & \text{otherwise} \end{cases}$$

for $n = 0, 1, 2, \ldots,$

$$f_4(x,y) = -f_4(-x,y) := \begin{cases} 1 & \text{if } 0 < y < x, \\ -1 & \text{if } x < y < 2x, \\ 0 & \text{otherwise.} \end{cases}$$

Exercise 7.9.

(iii) If (I_i) is a δ-cover with $0 < \delta < 1$ and $t > s$, then

$$\sum_{i=1}^{\infty} |I_i|^t \leq \delta^{t-s} \sum_{i=1}^{\infty} |I_i|^s.$$

Hence

$$H^t_\delta(A) \le \delta^{t-s} H^s_\delta(A).$$

If $H^s(A) < \infty$, then

$$\delta^{t-s} H^s_\delta(A) \le \delta^{t-s} H^s(A) \to 0$$

as $\delta \to 0$, and therefore $H^t(A) = 0$.

Exercise 8.1. Use Dini's theorem.

Exercise 8.2. If $c_1 |x - x_1| + \cdots + c_n |x - x_n| \equiv 0$ in I, then each term on the left-hand side is differentiable everywhere.

Exercise 8.4. (We follow Natanson [333].)

(ii) The case $d = 0$ is trivial. In the case $d > 0$ prove the following assertions:

- There exists a subdivision $a = x_0 < \cdots < x_n = b$ such that the oscillation of $f - p$ is less than d on each subinterval.
- Let us denote, numbering from left to right, by $I_1, \ldots, I_m$ those closed subintervals where $\max |f - p| = d$. Choose a point x_k between I_k and I_{k+1} whenever the sign of $f - p$ is different on I_k and I_{k+1}. If property (ii) fails, then the product ω of the corresponding factors $x - x_k$ belongs to $\mathcal{P}_n$.
- Changing ω to $-\omega$ if necessary, ω and $f - p$ have the same signs on each subinterval $I_1, \ldots, I_m$.
- If $c > 0$ is sufficiently small, then $|f - p - c\omega| < d$ on $[a, b]$.

(iii) Assume that both $p, q \in \mathcal{P}_n$ are closest polynomials to f. Prove the following assertions:

- $r := (p + q)/2$ also satisfies $|f - r| \le d$ on $[a, b]$.
- There exist $n + 2$ consecutive values $a \le x_1 < \cdots < x_{n+2} \le b$ at which $f(x_i) - r(x_i) = \pm d$, with alternating signs.
- $(f - p)(x_i) = (f - q)(x_i) = (f - r)(x_i)$ for each i.
- $p - q$ vanishes at more than $n + 1$ points, and hence $p = q$.

Exercise 8.5.

(i) follows from Bessel's inequality (Proposition 1.16, p. 29).

Exercise 8.8.

(ii) If

$$t = 2\left(\frac{t_1}{3} + \frac{t_2}{3^2} + \cdots + \frac{t_n}{3^n} + \cdots\right)$$

and

$$t' = 2\left(\frac{t'_1}{3} + \frac{t'_2}{3^2} + \cdots + \frac{t'_n}{3^n} + \cdots\right)$$

are two points of C such that $t_n \neq t'_n$, then $|t - t'| \geq 1/3^n$. Therefore, if $|t - t'| < 1/3^{2n}$, then $t_k = t'_k$ for $k = 1, 2, \ldots, 2n$ and therefore

$$\left|f_i(t) - f_i(t')\right| \leq 1/2^n, \quad i = 1, 2.$$

(iii) Since $[0,1] \setminus C$ is a union of pairwise disjoint open intervals, and since f_i is defined at the endpoints of these intervals, we may extend f_i linearly to each open interval.

(iv) Define $\alpha \in (0,1)$ by $9^\alpha = 2$. If

$$\frac{1}{9^{n+1}} \leq |t - t'| < \frac{1}{9^n}$$

for some integer n, then the above computation shows that

$$\left|f_i(t) - f_i(t')\right| \leq \frac{1}{2^n} = \frac{1}{9^{n\alpha}} \leq 9^\alpha \left|t - t'\right|^\alpha.$$

Hence f is Hölder continuous with the exponent α.

Exercise 8.10. Using the complexification method (2.16) of Murray (p. 112) we may assume that L_m is complex linear.

If $k > m$ and $h_k(x) := e^{ikx}$, then $(T_s h_k)(x) = e^{iks} h_k(x)$, and therefore

$$\int_{-\pi}^{\pi} (T_{-s} L_m T_s h_k)(x) \, ds = \int_{-\pi}^{\pi} e^{iks} (L_m h_k)(x - s) \, ds = 0$$

because $L_m h_k$ has order $< k$ and thus is orthogonal to h_k.

Exercise 8.11.

(iv) If c_m is the first non-zero coefficient in $\sum c_n f_n$, then $f_n(x_m) = 0$ for all $n > m$, and hence $\sum c_n f_n(x_m) = c_m f_m(x_m) = c_m \neq 0$.

Exercise 9.1.

(iii) Modify Fréchet's example (p. 307) by making the functions continuous.

Exercise 9.3.

(i) For each $n = 1, 2, \ldots$ we define $f_n \in M_\lambda$ such that $f_n = f$ in $[1/n, 1]$, and f_n is affine in $[0, 1/n]$ with $f_n(0) = \lambda$. Then

$$\|f - f_n\|_2 \leq \frac{|\lambda| + \|f\|_\infty}{\sqrt{n}}.$$

(ii) *First solution.* Given $f \in H$ and $\varepsilon > 0$ arbitrarily, first we choose $g \in H$ satisfying $\|f - g\| < \varepsilon$ and vanishing in a neighborhood of 1, and then we choose a polynomial p such that $\|g - p\|_\infty < \varepsilon$. Then $|p(1)| < \varepsilon$, and hence the polynomial $P := p - p(1)$ satisfies $P(1) = 0$ and

$$\|f - P\| \le \|f - g\| + \|g - p\| + \|p - P\| \le \|f - g\| + \|g - p\|_\infty + |p(1)| < 3\varepsilon.$$

Second solution. The linear functional $\varphi(P) := P(1)$, defined on the linear subspace $\mathcal{P}$ of the polynomials is not continuous, because $\mathrm{id}^n \to 0$ for the norm of X, but $\varphi(\mathrm{id}^n) = 1$ does not converge to $\varphi(0) = 0$. Therefore its kernel $N(\varphi)$ is dense in $\mathcal{P}$. Since $\mathcal{P}$ is dense in X by the Weierstrass approximation theorem, $N(\varphi)$ is dense in X.

Exercise 9.4. We have $M = 1^\perp$ and hence $M^\perp = 1^{\perp\perp} = \mathrm{Vect}\{1\}$ is the linear subspace of constant functions.

Exercise 9.6. If (e_k) is an orthonormal sequence and $0 < r \le \sqrt{2}/2$, then the pairwise disjoint balls $B_r(e_k)$ belong to the ball $B_{1+r}(0)$.

Exercise 9.7. Set $f(t) = \chi_{(0,t)}$.

Exercise 9.9.

(iii) Consider the functions

$$x(t) := t^{-1/p} \quad \text{and} \quad x(t) := t^{-1/q} \, |\ln t|^{-2/q} .$$

Teaching Remarks

Functional Analysis

- Most results of functional analysis and their optimality may be and are illustrated by the small ℓ^p spaces.
- Although we assume that the reader is familiar with the basic notions of topology, we could not resist presenting a little-known beautiful short proof of the classical Bolzano–Weierstrass theorem, based on an elementary lemma of a combinatorial nature, perhaps due to Kürschák (p. 6).
- We have included in the English edition a transparent elementary proof of the Farkas–Minkowski lemma, of fundamental importance in linear programming (p. 133), the Taylor–Foguel theorem on the uniqueness of Hahn–Banach extensions, and the Eberlein–Šmulian characterization of reflexive spaces.
- The simple proof of Lemma 3.24 (p. 144) may be new.
- Chapter 1 and the first seven sections of Chap. 2 may be covered in a one-semester course if we omit the material marked by $*$. Chapter 3 may be treated later, in a course devoted to the theory of distributions.
- It seems to be a good idea to treat the ℓ^p spaces only for $1 < p < \infty$ in the lectures, and to consider ℓ^1, ℓ^∞, c_0 later as exercises.

The Lebesgue Integral

- For didactic reasons Chap. 5 is devoted to the case of functions $f : \mathbb{R} \to \mathbb{R}$. However, it is shown subsequently in Chap. 7 that all results and almost all proofs remain valid word for word in arbitrary measure spaces. This approach may lead to a better understanding of the theory without loss of time.

© Springer-Verlag London 2016
V. Komornik, *Lectures on Functional Analysis and the Lebesgue Integral*,
Universitext, DOI 10.1007/978-1-4471-6811-9

- Applying Riesz's constructive definition of measurable functions we quickly arrive at essentially the most general forms of the Fubini–Tonelli and Radon–Nikodým theorems. For strongly σ-finite measures this is equivalent to the familiar inverse image definition. Otherwise the latter definition is weaker (in this book it is called *local measurability*), and, as we explain at the end of Sect. 7.7, the usual unpleasant counterexamples to some important theorems appear because of this weaker measurability notion.
- A one-semester course could start with the definition of null sets and with Proposition 4.3 (p. 155), followed by Chaps. 5 and 7, except Sect. 7.7. We suggest, however, to state without proof two further classical theorems of Lebesgue on the differentiability of monotone functions and on the generalized Newton–Leibniz formula (pp. 157, 204), and to treat briefly the L^p spaces by following Sect. 9.1 (p. 305) in *Function spaces*.

Function Spaces

- In order to make our exposition of *functional analysis* more accessible, we have avoided the spaces of continuous and Lebesgue integrable functions. This was anachronistic, because it was precisely the investigation of these spaces that led to the first great discoveries of functional analysis. Since they continue to play an important role in mathematics and its applications, we devote the last part of the book to these spaces.
- Contrary to the preceding parts, we give several different proofs of various important theorems, in order to stress the multiple interconnections among different branches of analysis.
- We present a large number of important examples that are not easy to localize in the literature.

Bibliography

1. N.I. Achieser, *Theory of Approximation* (Dover, New York, 1992)
2. N.I. Akhieser, I.M. Glazman, *Theory of Linear Operators in Hilbert Space I-II* (Dover, New York, 1993)
3. L. Alaoglu, Weak topologies of normed linear spaces, Ann. Math. (2) **41**, 252–267 (1940)
4. P.S. Alexandroff, Einführung in die Mengenlehre und in die allgemeine Topologie (German) [Introduction to Set Theory and to General Topology]. Translated from the Russian by Manfred Peschel, Wolfgang Richter and Horst Antelmann. *Hochschulbücher für Mathematik.* University Books for Mathematics, vol. 85 (VEB Deutscher Verlag der Wissenschaften, Berlin, 1984), 336 pp
5. F. Altomare, M. Campiti, *Korovkin-Type Approximation Theory and its Applications* (De Gruyter, Berlin, 1994)
6. Archimedes, *Quadrature of the parabola*; [199], 235–252
7. C. Arzelà, Sulla integrazione per serie. Rend. Accad. Lincei Roma **1**, 532–537, 566–569 (1885)
8. C. Arzelà, Funzioni di linee. Atti Accad. Lincei Rend. Cl. Sci. Fis. Mat. Nat. (4) **5I**, 342–348 (1889)
9. C. Arzelà, Sulle funzioni di linee. Mem. Accad. Sci. Ist. Bologna Cl. Sci. Fis. Mat. (5) **5**, 55–74 (1895)
10. C. Arzelà, Sulle serie di funzioni. Memorie Accad. Sci. Bologna **8**, 131–186 (1900), 701–744
11. G. Ascoli, Sul concetto di integrale definite. Atti Acc. Lincei (2) **2**, 862–872 (1875)
12. G. Ascoli, Le curve limiti di una varietà data di curve. Mem. Accad. dei Lincei (3) **18**, 521–586 (1883)
13. G. Ascoli, Sugli spazi lineari metrici e le loro varietà lineari. Ann. Mat. Pura Appl. (4) **10**, 33–81, 203–232 (1932)
14. D. Austin, A geometric proof of the Lebesgue differentiation theorem. Proc. Am. Math. Soc. **16**, 220–221 (1965)
15. V. Avanissian, Initiation à l'analyse fonctionnelle (Presses Universitaires de France, Paris, 1996)
16. R. Baire, Sur les fonctions discontinues qui se rattachent aux fonctions continues. C. R. Acad. Sci. Paris **126**, 1621–1623 (1898). See in [18]
17. R. Baire, Sur les fonctions à variables réelles. Ann. di Mat. (3) **3**, 1–123 (1899). See in [18]
18. R. Baire, *Oeuvres Scientifiques*, (Gauthier-Villars, Paris, 1990)
19. S. Banach, Sur les opérations dans les ensembles abstraits et leur application aux équations intégrales. Fund. Math. **3**, 133–181 (1922); [26] II, 305–348

© Springer-Verlag London 2016

V. Komornik, *Lectures on Functional Analysis and the Lebesgue Integral*,

Universitext, DOI 10.1007/978-1-4471-6811-9

20. S. Banach, An example of an orthogonal development whose sum is everywhere different from the developed function. Proc. Lond. Math. Soc. (2) **21**, 95–97 (1923)

21. S. Banach, Sur le problème de la mesure. Fund. Math. **4**, 7–33 (1923)

22. S. Banach, Sur les fonctionnelles linéaires I-II. Stud. Math. **1**, 211–216, 223–239 (1929); [26] II, 375–395

23. S. Banach, *Théorèmes sur les ensembles de première catégorie.* Fund. Math. **16**, 395–398 (1930); [26] I, 204–206

24. S. Banach, *Théorie des opérations linéaires* (Monografje Matematyczne, Warszawa, 1932); [26] II, 13–219

25. S. Banach, *The Lebesgue Integral in Abstract Spaces*; Jegyzet a [409] könyvben (1937)

26. S. Banach, *Oeuvres I-II* (Państwowe Wydawnictwo Naukowe, Warszawa, 1967, 1979)

27. S. Banach, S. Mazur, Zur Theorie der linearen Dimension. Stud. Math. **4**, 100–112 (1933); [26] II, 420–430

28. S. Banach, H. Steinhaus, Sur le principe de la condensation de singularités. Fund. Math. **9**, 50–61 (1927); [26] II, 365–374

29. S. Banach, A. Tarski, *Sur la décomposition des ensembles de points en parties respectivement congruentes.* Fund. Math. **6**, 244–277 (1924)

30. R.G. Bartle, *A Modern Theory of Integration.* Graduate Studies in Mathematics, vol. 32 (American Mathematical Society, Providence, RI, 2001)

31. W.R. Bauer, R.H. Brenner, The non-existence of a banach space of countably infinite hamel dimension. Am. Math. Mon. **78**, 895–896 (1971)

32. B. Beauzamy, *Introduction to Banach Spaces and Their Geometry* (North-Holland, Amsterdam, 1985)

33. P.R. Beesack, E. Hughes, M. Ortel, Rotund complex normed linear spaces. Proc. Am. Math. Soc. **75**(1), 42–44 (1979)

34. S.K. Berberian, *Notes on Spectral Theory* (Van Nostrand, Princeton, NJ, 1966)

35. S.K. Berberian, *Introduction to Hilbert Space* (Chelsea, New York, 1976)

36. S.J. Bernau, F. Smithies, A note on normal operators. Proc. Camb. Philos. Soc. **59**, 727–729 (1963)

37. M. Bernkopf, The development of functions spaces with particular reference to their origins in integral equation theory. Arch. Hist. Exact Sci. **3**, 1–66 (1966)

38. D. Bernoulli, Réflexions et éclaircissements sur les nouvelles vibrations des cordes. Hist. Mém Acad. R. Sci. Lett. Berlin **9**, 147–172 (1753) (published in 1755)

39. S.N. Bernstein, Démonstration du théorème de Weierstrass fondée sur le calcul de probabilités. Commun. Kharkov Math. Soc. **13**, 1–2 (1912)

40. C. Bessaga, A. Pełczyński, *Selected Topics in Infinite-Dimensional Topology.* Monografje Matematyczne, Tome 58 (Państwowe Wydawnictwo Naukowe, Warszawa, 1975)

41. F.W. Bessel, *Über das Dollond'sche Mittagsfernrohr etc.* Astronomische Beobachtungen etc. Bd. 1, Königsberg 1815, [43] II, 24–25

42. F.W. Bessel, Über die Bestimmung des Gesetzes einer priosischen Erscheinung, Astron. Nachr. Bd. 6, Altona 1828, 333–348, [43] II, 364–368

43. F.W. Bessel, *Abhandlungen von Friedrich Wilhelm Bessel I-III*, ed. by R. Engelman (Engelmann, Leipzig, 1875–1876)

44. P. Billingsley, Van der Waerden's continuous nowhere differentiable function. Am. Math. Mon. **89**, 691 (1982)

45. G. Birkhoff, E. Kreyszig, The establishment of functional analysis. Hist. Math. **11**, 258–321 (1984)

46. S. Bochner, Integration von Funktionen, deren Wert die Elemente eines Vektorraumes sind. Fund. Math. **20**, 262–276 (1933)

47. H. Bohman, On approximation of continuous and analytic functions. Ark. Mat **2**, 43–56 (1952)

48. H.F. Bohnenblust, A. Sobczyk, Extensions of functionals on complex linear spaces. Bull. Am. Math. Soc. **44**, 91–93 (1938)

49. P. du Bois-Reymond, Ueber die Fourier'schen Reihen. Nachr. K. Ges. Wiss. Göttingen **21**, 571–582 (1873)

50. P. du Bois Reymond, Versuch einer Classification der willkürlichen Functionen reeller Argumente nach ihren Aenderungen in den kleinsten Intervallen [German]. J. Reine Angew. Math. **79**, 21–37 (1875)

51. P. du Bois-Reymond, Untersuchungen über die Convergenz und Divergenz der Fourierschen Darstellungsformeln (mit drei lithografierten Tafeln). Abh. Math. Phys. Kl. Bayer. Akad. Wiss. **12**, 1–103 (1876)

52. P. du Bois-Reymond, *Die allgemeine Funktionentheorie* (Laapp, Tübingen, 1882); French translation: *Théorie générale des fonctions*, Nice, 1887

53. P. du Bois-Reymond, Ueber das Doppelintegral. J. Math. Pures Appl. **94**, 273–290 (1883)

54. B. Bolzano, *Rein analytischer Beweis des Lehrsatzes, dass zwischen je zwei Werthen, die ein entgegengesetztes Resultat gewähren, wenigstens eine reelle Wurzel der Gleichung liege* (Prag, Haase, 1817). New edition: Ostwald's Klassiker der exakten Wissenschaften, No. 153, Leipzig, 1905. English translation: Purely analytic proof of the theorem the between any two values which give results of opposite sign there lies at least one real root of the equation. Historia Math. **7**, 156–185 (1980); See also in [57].

55. B. Bolzano, *Functionenlehre*, around 1832. Partially published in Königliche böhmische Gesellschaft der Wissenschaften, Prága, 1930; complete publication in [56] 2A10/1, 2000, 23–165; English translation: [57], 429–572.

56. B. Bolzano, *Bernard Bolzano Gesamtausgabe* (Frommann–Holzboog, Stuttgart, 1969)

57. B. Bolzano, *The Mathematical Works of Bernard Bolzano*, translated by S. Russ, (Oxford University Press, Oxford, 2004)

58. E. Borel, Sur quelques points de la théorie des fonctions. Ann. Éc. Norm. Sup. (3) **12**, 9–55 (1895); [62] I, 239–285

59. E. Borel, *Leçons sur la théorie des fonctions* (Gauthier-Villars, Paris, 1898)

60. E. Borel, *Leçons sur les fonctions de variables réelles* (Gauthier-Villars, Paris, 1905)

61. E. Borel (ed.), L. Zoretti, P. Montel, M. Fréchet, *Recherches contemporaines sur la théorie des fonctions: Les ensembles de points, Intégration et dérivation, Développements en séries.* Encyclopédie des sciences mathématiques, II-2 (Gauthier-Villars/Teubner, Paris/Leipzig, 1912) German translation and adaptation by A. Rosenthal: *Neuere Untersuchungen über Funktionen reeller Veränderlichen: Die Punktmengen, Integration und Differentiation, Funktionenfolgen*, Encyklopädie der mathematischen Wissenschaften, II-9 (Teubner, Leipzig, 1924)

62. E. Borel, *Oeuvres* I-IV (CNRS, Paris, 1972)

63. M.W. Botsko, An elementary proof of Lebesgue's differentiation theorem. Am. Math. Mon. **110**, 834–838 (2003)

64. N. Bourbaki, Sur les espaces de Banach. C. R. Acad. Sci. Paris 206, 1701–1704 (1938)

65. H. Brezis, *Functional Analysis, Sobolev Spaces and Partial Differential Equations* (Springer, New York, 2010)

66. H. Brunn, Zur Theorie der Eigebiete. Arch. Math. Phys. (3) **17**, 289–300 (1910)

67. J.R. Buddenhagen, Subsets of a countable set. Am. Math. Mon. **78**, 536–537 (1971)

68. J.C. Burkill, *The Lebesgue Integral* (Cambridge University Press, Cambridge, 1951)

69. G. Cantor, Ueber trigonometrische Reihen. Math. Ann. **4**, 139–143 (1871); [76], 87–91

70. G. Cantor, Über eine Eigenschaft des Inbegriffes aller reellen algebraischen Zahlen. J. Reine Angew. Math. **77**, 258–262 (1874); [76], 115–118

71. G. Cantor, Über unendliche, lineare Punktmannigfaltigkeiten III. Math. Ann. **20**, 113–121 (1882); [76], 149–157

72. G. Cantor, Über unendliche, lineare Punktmannigfaltigkeiten V. Math. Ann. **21**, 545–586 (1883); [76], 165–208

73. G. Cantor, De la puissance des ensembles parfaits de points. Acta Math. **4**, 381–392 (1884)

74. G. Cantor, Über unendliche, lineare Punktmannigfaltigkeiten VI. Math. Ann. **23**, 451–488 (1884); [76], 210–244

75. G. Cantor, Über eine Frage der Mannigfaltigkeitslehre. Jahresber. Deutsch. Math. Verein. **1**, 75–78 (1890–1891); [76], 278–280

76. G. Cantor, *Gesammelte Abhandlungen* (Springer, Berlin, 1932)

77. C. Carathéodory, Constantin *Vorlesungen über reelle Funktionen* [German], 3rd edn. (Chelsea, New York, 1968), x+718 pp. [1st edn. (Teubner, Leipzig, 1918); 2nd edn. (1927); reprinting (Chelsea, New York, 1948)]

78. L. Carleson, On convergence and growth of partial sums of Fourier series. Acta Math. **116**, 135–157 (1966)

79. A.L. Cauchy, Cours d'analyse algébrique (Debure, Paris, 1821); [82] (2) III, 1–476

80. L.-A. Cauchy, *Résumé des leçons données à l'École Polytechnique sur le calcul infinitésimal. Calcul intégral* (1823); [82] (2) IV, 122–261

81. L.-A. Cauchy, Mémoire sur les intégrales définies. Mém. Acad. Sci. Paris **1** (1827); [82] (1) I, 319–506

82. A.L. Cauchy, *Oeuvres*, 2 sorozat, 22 kötet (Gauthier-Villars, Paris, 1882–1905)

83. P.L. Chebyshev [Tchebychef], Sur les questions de minima qui se rattachent à la représentation approximative des fonctions. Mém. Acad. Imp. Sci. St. Pétersb. (6) Sciences math. et phys. **7**, 199–291 (1859); [84] I, 273–378

84. P.L. Chebyshev [Tchebychef], *Oeuvres I-II* (Chelsea, New York, 1962)

85. E.W. Cheney, *Introduction to Approximation Theory* (Chelsea, New York, 1982)

86. P.R. Chernoff, Pointwise convergence of Fourier series. Am. Math. Monthly **87**, 399–400 (1980)

87. P.G. Ciarlet, *Linear and Nonlinear Functional Analysis with Applications* (SIAM, Philadelphia, 2013)

88. A.C. Clairaut, Mémoire sur l'orbite apparente du soleil autour de la Terre, en ayant égard aux perturbations produites par des actions de la Lune et des Planètes principales. Hist. Acad. Sci. Paris, 52–564 (1754, appeared in 1759)

89. J.A. Clarkson, Uniformly convex spaces. Trans. Am. Math. Soc. **40**, 396–414 (1936)

90. J.A. Clarkson, P. Erdős, Approximation by polynomials. Duke Math. J. **10**, 5–11 (1943)

91. R. Courant, D. Hilbert, *Methoden de matematischen Physik I* (Springer, Berlin, 1931)

92. Á. Császár, *Valós analízis I–II [Real Analysis I–II]* (Tankönyvkiadó, Budapest, 1983)

93. P. Daniell, A general form of integral. Ann. Math. **19**, 279–294 (1917/18)

94. G. Dantzig, *Linear Programming and Extensions* (Princeton University Press/RAND Corporation, Princeton/Santa Monica, 1963)

95. G. Darboux, Mémoire sur les fonctions discontinues. Ann. Éc. Norm. Sup. Paris (2) **4**, 57–112 (1875)

96. M.M. Day, The spaces L^p with $0 < p < 1$. Bull. Am. Math. Soc. **46**, 816–823 (1940)

97. M.M. Day, *Normed Linear Spaces* (Springer, Berlin, 1962)

98. A. Denjoy, Calcul de la primitive de la fonction dérivée la plus générale. C. R. Acad. Sci. Paris **154**, 1075–1078 (1912)

99. A. Denjoy, Une extension de l'intégrale de M. Lebesgue. C. R. Acad. Sci. Paris **154**, 859–862 (1912)

100. A. Denjoy, Mémoire sur la totalisation des nombres dérivées non sommables. Ann. Éc. Norm. Sup. Paris (3) **33**, 127–222 (1916)

101. M. de Vries, V. Komornik, Unique expansions of real numbers. Adv. Math. **221**, 390–427 (2009)

102. M. de Vries, V. Komornik, P. Loreti, Topology of univoque bases. Topol. Appl. (2016). doi:10.1016/j.topol.2016.01.023,

103. J. Diestel, *Geometry of Banach Spaces. Selected Topics* (Springer, New York, 1975)

104. J. Diestel, *Sequences and Series in Banach Spaces* (Springer, New York, 1984)

105. J. Dieudonné, Sur le théorème de Hahn–Banach. Revue Sci. **79**, 642–643 (1941)

106. J. Dieudonné, *History of Functional Analysis* (North-Holland, Amsterdam, 1981)

107. U. Dini, *Sopra la serie di Fourier* (Nistri, Pisa, 1872)

108. U. Dini, Sulle funzioni finite continue di variabili reali che non hanno mai derivata. Atti R. Acc. Lincei, (3), **1**, 130–133 (1877); [111] II, 8–11

109. U. Dini, Fondamenti per la teoria delle funzioni di variabili reali (Nistri, Pisa, 1878)
110. U. Dini, *Serie di Fourier e altre rappresentazioni analitiche delle funzioni di una variabile reale* (Nistri, Pisa, 1880); [111] IV, 1–272
111. U. Dini, *Opere I-V* (Edizioni Cremonese, Roma, 1953–1959)
112. G.L. Dirichlet, Sur la convergence des séries trigonométriques qui servent à représenter une fonction arbitraire entre des limites données. J. Reine Angew. Math. **4**, 157–169 (1829); [114] I, 117–132
113. G.L. Dirichlet, Über eine neue Methode zur Bestimmung vielfacher Integrale. Ber. Verh. K. Preuss. Acad. Wiss. Berlin, 18–25 (1839); [114] I, 381–390
114. G.L. Dirichlet, *Werke I-II* (Reimer, Berlin, 1889–1897)
115. J.L. Doob, *The Development of Rigor in Mathematical Probability* (1900–1950), [361], 157–169
116. N. Dunford, Uniformity in linear spaces. Trans. Am. Math. Soc. **44**, 305–356 (1938)
117. N. Dunford, J.T. Schwartz, *Linear Operators I-III* (Wiley, New York, 1957–1971)
118. W.F. Eberlein, Weak compactness in Banach spaces I. Proc. Natl. Acad. Sci. USA **33**, 51–53 (1947)
119. R.E. Edwards, *Functional Analysis. Theory and Applications* (Holt, Rinehart and Winston, New York, 1965)
120. R.E. Edwards, *Fourier Series. A Modern Introduction I-II* (Springer, New York, 1979–1982)
121. D.Th. Egoroff, Sur les suites de fonctions mesurables. C. R. Acad. Sci. Paris **152**, 244–246 (1911)
122. M. Eidelheit, *Zur Theorie der konvexen Mengen in linearen normierten Räumen.* Stud. Math. **6**, 104–111 (1936)
123. H.W. Ellis, D.O. Snow, On $(L^1)^*$ for general measure spaces, Can. Math. Bull. **6**, 211–229 (1963)
124. P. Erdős, P. Turán, On interpolation I. Quadrature- and mean-convergence in the Lagrange interpolation. Ann. Math. **38**, 142–155 (1937); [126] I, 50–51, 97–98, 54–63
125. P. Erdős, P. Vértesi, On the almost everywhere divergence of Lagrange interpolatory polynomials for arbitrary system of nodes. Acta Math. Acad. Sci. Hungar. **36**, 71–89 (1980), **38**, 263 (1981)
126. P. Erdős (ed.), *Collected Papers of Paul Turán I-III* (Akadémiai Kiadó, Budapest, 1990)
127. P. Erdős, I. Joó, V. Komornik, Characterization of the unique expansions $1 = \sum q^{-n_i}$ and related problems. Bull. Soc. Math. France **118**, 377–390 (1990)
128. L. Euler, De summis serierum reciprocarum. Comm. Acad. Sci. Petrop. **7**, 123–134 (1734/1735, appeared in 1740); [132] (1) 14, 73–86
129. L. Euler, *Introductio in analysin infinitorum I* (M.-M. Bousquet, Lausanne, 1748); [132] (1) 8
130. L. Euler, De formulis integralibus duplicatis. Novi Comm. Acad. Sci. Petrop. **14** (1769): I, 1770, 72–103; [132] (1) 17, 289–315
131. L. Euler, Disquisitio ulterior super seriebus secundum multipla cuius dam anguli progredientibus. Nova Acta Acad. Sci. Petrop. **11**, 114–132 (1793, written in 1777, appeared in 1798); [132] (1) 16, Part 1, 333–355
132. L. Euler, *Opera Omnia*, 4 series, 73 volumes (Teubner/Füssli, Leipzig/Zürich, 1911)
133. G. Faber, Über die interpolatorische Darstellung stetiger Funtionen. Jahresber. Deutsch. Math. Verein. **23**, 192–210 (1914)
134. K. Falconer, *Fractal Geometry. Mathematical Foundations and Applications*, 2nd edn. (Wiley, Chicester, 2003)
135. J. Farkas, Über die Theorie der einfachen Ungleichungen. J. Reine Angew. Math. **124**, 1–27 (1902)
136. P. Fatou, Séries trigonométriques et séries de Taylor. Acta Math. **30**, 335–400 (1906)
137. L. Fejér, Sur les fonctions bornées et intégrables. C. R. Acad. Sci. Paris **131**, 984–987 (1900); [143] I, 37–41
138. L. Fejér, Untersuchungen über Fouriersche Reihen. Math. Ann. **58**, 51–69 (1904); [143] I, 142–160

139. L. Fejér, Eine stetige Funktion deren Fourier'sche Reihe divergiert. Rend. Circ. Mat. Palermo
     28, 1, 402–404 (1909); [143] I, 541–543
140. L. Fejér, Beispiele stetiger Funktionen mit divergenter Fourierreihe. J. Reine Angew. Math.
     137, 1–5 (1910); [143] I, 538–541
141. L. Fejér, Lebesguesche Konstanten und divergente Fourierreihen. J. Reine Angew. Math. **138**,
     22–53 (1910); [143] I, 543–572
142. L. Fejér, Über Interpolation. Götting. Nachr. 66–91 (1916); [143] II, 25–48
143. L. Fejér, *Leopold Fejér. Gesammelte Arbeiten I-II*, ed. by Turán Pál (Akadémiai Kiadó,
     Budapest, 1970)
144. G. Fichera, Vito Volterra and the birth of functional analysis, [361], 171–183
145. G.M. Fichtenholz, L.V. Kantorovich [Kantorovitch], Sur les opérations linéaires dans l'espace
     des fonctions bornées. Stud. Math. **5**, 69–98 (1934)
146. E. Fischer, Sur la convergence en moyenne. C. R. Acad. Sci. Paris **144**, 1022–1024, 1148–
     1150 (1907)
147. S.R. Foguel, On a theorem by A.E. Taylor. Proc. Am. Math. Soc. **9**, 325 (1958)
148. J.B.J. Fourier, *Théorie analytique de la chaleur* (Didot, Paris 1822); [149] I
149. J.B.J. Fourier *Oeuvres de Fourier* I-II (Gauthier-Villars, Paris, 1888–1890)
150. I. Fredholm, Sur une nouvelle méthode pour la résolution du problème de Dirichlet. Kong.
     Vetenskaps-Akademiens Förh. Stockholm, 39–46 (1900); [152], 61–68
151. I. Fredholm, Sur une classe d'équations fonctionnelles. Acta Math. **27**, 365–390 (1903);
     [152], 81–106
152. I. Fredholm, *Oeuvres complètes de Ivar Fredholm* (Litos Reprotryck, Malmo, 1955)
153. G. Freud, *Über positive lineare Approximationsfolgen von stetigen reellen Funktionen auf
     kompakten Mengen*. On Approximation Theory, Proceedings of the Conference in Oberwol-
     fach, 1963 (Birkhäuser, Basel, 1964), pp. 233–238
154. M. Fréchet, Sur quelques points du calcul fonctionnel. Rend. Circ. Mat. Palermo **22**, 1–74
     (1906)
155. M. Fréchet, Sur les ensembles de fonctions et les opérations linéaires. C. R. Acad. Sci. Paris
     144, 1414–1416 (1907)
156. M. Fréchet, Sur les opérations linéaires III. Trans. Am. Math. Soc. **8**, 433–446 (1907)
157. M. Fréchet, Les dimensions d'un ensemble abstrait. Math. Ann. **68**, 145–168 (1910)
158. M. Fréchet, Sur l'intégrale d'une fonctionnelle étendue àä un ensemble abstrait. Bull. Soc.
     Math. France **43**, 248–265 (1915)
159. M. Fréchet, L'écart de deux fonctions quelconques. C. R. Acad. Sci. Paris **162**, 154–155
     (1916)
160. M. Fréchet, Sur divers modes de convergence d'une suite de fonctions d'une variable réelle.
     Bull. Calcutta Math. Soc. **11**, 187–206 (1919–1920)
161. M. Fréchet, *Les espaces abstraits* (Gauthier-Villars, Paris, 1928)
162. G. Frobenius, Über lineare Substitutionen und bilineare Formen. J. Reine Angew. Math. **84**,
     1–63 (1878); [163] I, 343–405
163. G. Frobenius, *Gesammelte Abhandlungen I-III* (Springer, Berlin, 1968)
164. G. Fubini, Sugli integrali multipli. Rend. Accad. Lincei Roma **16**, 608–614 (1907)
165. G. Fubini, Sulla derivazione per serie. Rend. Accad. Lincei Roma **24**, 204–206 (1915)
166. I.S. Gál, On sequences of operations in complete vector spaces. Am. Math. Mon. **60**, 527–538
     (1953)
167. B.R. Gelbaum, J.M.H. Olmsted, *Counterexamples in Mathematics* (Holden-Day, San Fran-
     cisco, 1964)
168. B.R. Gelbaum, J.M.H. Olmsted, *Theorems and Counterexamples in Mathematics* (Springer,
     New York, 1990)
169. L. Gillman, M. Jerison, *Rings of Continuous Functions* (D. Van Nostrand Company,
     Princeton, 1960)
170. I.M. Glazman, Ju.I. Ljubic, *Finite-Dimensional Linear Analysis: A Systematic Presentation
     in Problem Form* (Dover, New York, 2006)
171. H.H. Goldstine, Weakly complete Banach spaces. Duke Math. J. **4**, 125–131 (1938)

172. D.B. Goodner, Projections in normed linear spaces. Trans. Am. Math. Soc. **69**, 89–108 (1950)
173. J.P. Gram, *Om Rackkendvilklinger bestemte ved Hjaelp af de mindste Kvadraters Methode*, Copenhagen, 1879. German translation: *Ueber die Entwicklung reeller Funktionen in Reihen mittelst der Methode der kleinsten Quadrate*, J. Reine Angew. Math. **94**, 41–73 (1883)
174. A. Grothendieck, Sur la complétion du dual d'un espace vectoriel localement convexe. C. R. Acad. Sci. Paris **230**, 605–606 (1950)
175. B.L. Gurevich, G.E. Shilov, *Integral, Measure and Derivative: A Unified Approach* (Prentice-Hall, Englewood Cliffs, NJ, 1966)
176. A. Haar, Zur Theorie der orthogonalen Funktionensysteme. Math. Ann. **69**, 331–371 (1910); [179], 47–87
177. A. Haar, Reihenentwicklungen nach Legendreschen Polynomen. Math. Ann. **78**, 121–136 (1918); [179], 124–139
178. A. Haar, A folytonos csoportok elméletéről. Magyar Tud. Akad. Mat. és Természettud. Ért. **49**, 287–306 (1932); [182], 579–599; German translation: *Die Maßbegriffe in der Theorie der kontinuierlichen Gruppen*, Ann. Math. **34**, 147–169 (1933); [182], 600–622
179. A. Haar, *Gesammelte Arbeiten*, ed. by B. Sz.-Nagy (Akadémiai Kiadó, Budapest, 1959)
180. H. Hahn, *Theorie der reellen Funktionen I* (Springer, Berlin, 1921)
181. P. Hahn, Über Folgen linearer Operationen. Monatsh. Math. Physik **32**, 3–88 (1922); [183] I, 173–258
182. P. Hahn, Über linearer Gleichungssysteme in linearer Räumen. J. Reine Angew. Math. **157**, 214–229 (1927); [183] I, 259–274
183. P. Hahn, *Gesammelte Abhandlungen* I-III (Springer, New York, 1995–1997)
184. P.R. Halmos, *Measure Theory* (D. Van Nostrand Co., Princeton, NJ, 1950)
185. P.R. Halmos, *Introduction to Hilbert Space and the Theory of Spectral Multiplicity* (Chelsea, New York, 1957)
186. P.R. Halmos, L *Naive Set Theory* (Van Nostrand, Princeton, NJ, 1960)
187. P.R. Halmos, *A Hilbert Space Problem Book* (Springer, Berlin, 1974)
188. G. Halphén, Sur la série de Fourier. C. R. Acad. Sci. Paris **95**, 1217–1219 (1882)
189. H. Hanche-Olsen, H. Holden, The Kolmogorov-Riesz compactness theorem. Expo. Math. **28**, 385–394 (2010)
190. H. Hankel, Untersuchungen über die unendlich oft oszillierenden und unstetigen Funktionen. Math. Ann. **20**, 63–112 (1882)
191. G. Hardy, J.E. Littlewood, G. Pólya, *Inequalities*, 2nd edn. (Cambridge University Press, Cambridge, 1952)
192. A. Harnack, *Die Elemente der Differential- und Integralrechnung* (Teubner, Lepizig, 1881)
193. A. Harnack, Die allgemeinen Sätze über den Zusammenhang der Functionen einer reellen Variabelen mit ihren Ableitungen II. Math. Ann. **24**, 217–252 (1884)
194. A. Harnack, Ueber den Inhalt von Punktmengen. Math. Ann. **25**, 241–250 (1885)
195. F. Hausdorff, *Grundzüge der Mengenlehre* (Verlag von Veit, Leipzig, 1914); [197] II
196. F. Hausdorff, Dimension und äusseres Mass. Math. Ann. **79**, 1–2, 157–179 (1918); [197] IV
197. F. Hausdorff, *Gesammelte Werke* I-VIII (Springer, Berlin, 2000)
198. T. Hawkins, *Lebesgue's Theory of Integration. Its Origins and Development* (AMS Chelsea Publishing, Providence, 2001)
199. T.L. Heath (ed.), *The Works of Archimedes* (Cambridge University Press, Cambridge, 1897)
200. E. Heine, Die Elemente der Funktionenlehre. J. Reine Angew. Math. **74**, 172–188 (1872)
201. E. Hellinger, O. Toeplitz, Grundlagen für eine Theorie der unendlichen Matrizen. Nachr. Akad. Wiss. Göttingen. Math.-Phys. Kl., 351–355 (1906)
202. E. Hellinger, O. Toeplitz, Grundlagen für eine Theorie der unendlichen Matrizen. Math. Ann. **69**, 289–330 (1910)
203. E. Hellinger, O. Toeplitz, *Integralgleichungen une Gleichungen mit unendlichvielen Unbekannten*. Encyklopädie der Mathematischen Wissenschaften, II C 13 (Teubner, Leipzig, 1927)
204. E. Helly, Über lineare Funktionaloperationen. Sitzber. Kais. Akad. Wiss. Math.-Naturwiss. Kl. Wien **121**, 2, 265–297 (1912)

205. R. Henstock, The efficiency of convergence factors for functions of a continuous real variable. J. Lond. Math. Soc. **30**, 273–286 (1955)
206. R. Henstock, Definitions of Riemann type of the variational integrals. Proc. Lond. Math. Soc. (3) **11**, 402–418 (1961)
207. E. Hewitt, K. Stromberg, *Real and Abstract Analysis* (Springer, Berlin, 1965)
208. D. Hilbert, Grundzüge einer allgemeinen Theorie der Integralgleichungen I. Nachr. Akad. Wiss. Göttingen. Math.-Phys. Kl., 49–91 (1904)
209. D. Hilbert, Grundzüge einer allgemeinen Theorie der Integralgleichungen IV. Nachr. Akad. Wiss. Göttingen. Math.-Phys. Kl., 157–227 (1906)
210. T.H. Hildebrandt, Necessary and sufficient conditions for the interchange of limit and summation in the case of sequences of infinite series of a certain type. Ann. Math. (2) **14**, 81–83 (1912–1913)
211. T.H. Hildebrandt, On uniform limitedness of sets of functional operations. Bull. Am. Math. Soc. 29, 309–315 (1923)
212. T.H. Hildebrandt, Über vollstetige, lineare Transformationen. Acta. Math. **51**, 311–318 (1928)
213. T.H. Hildebrandt, On bounded functional operations. Trans. Am. Math. Soc. **36**, 868–875 (1934)
214. E.W. Hobson, On some fundamental properties of Lebesgue integrals in a two-dimensional domain. Proc. Lond. Math. Soc. (2) **8**, 22–39 (1910)
215. H. Hochstadt, Eduard Helly, father of the Hahn-Banach theorem. Math. Intell. **2**(3), 123–125 (1979)
216. R.B. Holmes, *Geometric Functional Analysis and Its Applications* (Springer, Berlin, 1975)
217. O. Hölder, Ueber einen Mittelwerthsatz. Götting. Nachr. 38–47 (1889)
218. L. Hörmander, *Linear Differential Operators* (Springer, Berlin, 1963)
219. L. Hörmander, *The Analysis of Linear Partial Differential Operators I* (Springer, Berlin, 1983)
220. R.A. Hunt, On the convergence of Fourier series orthogonal expansions and their continuous analogues, in *Proceedings of the Conference at Edwardsville 1967* (Southern Illinois University Press, Carbondale, IL, 1968), pp. 237–255
221. D. Jackson, *Über die Genauigkeit der Annäherung stetiger Funktionen durch ganze rationale Funktionen gegebenen Grades und trigonometrische Summen gegebener Ordnung*, Dissertation, Göttingen, (1911)
222. D. Jackson, On approximation by trigonometric sums and polynomials. Trans. Am. Math. Soc. **13**, 491–515 (1912)
223. D. Jackson, *The Theory of Approximation*, vol. 11 (American Mathematical Society, Colloquium Publications, Providence, RI, 1930)
224. C.G.J. Jacobi, De determinantibus functionalibus. J. Reine Angew. Math. **22**, 319–359 (1841); [225] III, 393–438
225. C.G.J. Jacobi, *Gesammelte Werke I-VIII* (G. Reimer, Berlin, 1881–1891)
226. R.C. James, Characterizations of reflexivity. Stud. Math. **23**, 205–216 (1964)
227. V. Jarník, *Bolzano and the Foundations of Mathematical Analysis* (Society of Czechoslovak Mathematicians and Physicists, Prague, 1981)
228. I. Joó, V. Komornik, On the equiconvergence of expansions by Riesz bases formed by eigenfunctions of the Schrödinger operator. Acta Sci. Math. (Szeged) **46**, 357–375 (1983)
229. C. Jordan, Sur la série de Fourier. C. R. Acad. Sci. Paris **92**, 228–230 (1881); [232] IV, 393–395
230. C. Jordan, *Cours d'analyse de l'École Polytechnique I-III* (Gauthier-Villars, Paris, 1883)
231. C. Jordan, Remarques sur les intégrales définies. J. Math. (4) **8**, 69–99 (1892); [232] IV, 427–457
232. C. Jordan, *Oeuvres I-IV* (Gauthier-Villars, Paris, 1961–1964)
233. P. Jordan, J. von Neumann, On inner products in linear, metric spaces. Ann. Math. (2) **36**(3), 719–723 (1935)
234. M.I. Kadec, On strong and weak convergence (in Russian). Dokl. Akad. Nauk SSSR **122**, 13–16 (1958)

235. M.I. Kadec, On the connection between weak and strong convergence (in Ukrainian). Dopovidi. Akad. Ukraïn RSR **9**, 949–952 (1959)
236. M.I. Kadec, Spaces isomorphic to a locally uniformly convex space (in Russian). Izv. Vysš. Učebn. Zaved. Matematika **13**(6), 51–57 (1959)
237. J.-P. Kahane, *Fourier Series*; see J.-P. Kahane, P.-G. Lemarié-Rieusset, *Fourier Series and Wavelets* (Gordon and Breach, New York, 1995)
238. J.-P. Kahane, Y. Katznelson, Sur les ensembles de divergence des séries trigonométriques. Stud. Math. **26**, 305–306 (1966)
239. S. Kakutani, Weak topology and regularity of Banach spaces. Proc. Imp. Acad. Tokyo **15**, 169–173 (1939)
240. S. Kakutani, Concrete representations of abstract (M)-spaces. (A characterization of the space of continuous functions). Ann. Math. (2) **42**, 994–1024 (1941)
241. N.J. Kalton, An F-space with trivial dual where the Krein–Milman theorem holds. Isr. J. Math. **36**, 41–50 (1980)
242. N.J. Kalton, N. Peck, A re-examination of the Roberts example of a compact convex set without extreme points. Math. Ann. **253**, 89–101 (1980)
243. L.V. Kantorovich, G.P. Akilov, *Functional Analysis*, 2nd edn. (Pergamon Press, Oxford, 1982)
244. W. Karush, Minima of Functions of Several Variables with Inequalities as Side Constraints. M.Sc. Dissertation. University of Chicago, Chicago (1939)
245. Y. Katznelson, *An Introduction to Harmonic Analysis* (Dover, New York, 1976)
246. J. Kelley, Banach spaces with the extension property. Trans. Am. Math. Soc. **72**, 323–326 (1952)
247. J. Kelley, *General Topology* (van Nostrand, New York, 1954)
248. J. Kindler, A simple proof of the Daniell-Stone representation theorem. Am. Math. Mon. **90**, 396–397 (1983)
249. A.A. Kirillov, A.D. Gvisiani, *Theorems and Problems in Functional Analysis* (Springer, New York, 1982)
250. V.L. Klee, Jr., Convex sets in linear spaces I-II. Duke Math. J. **18**, 443–466, 875–883 (1951)
251. A.N. Kolmogorov, Über Kompaktheit der Funktionenmengen bei der Konvergenz im Mittel. Nachr. Ges. Wiss. Göttingen **9**, 60–63 (1931); English translation: *On the compactness of sets of functions in the case of convergence in the mean*, in V.M. Tikhomirov (ed.), Selected Works of A.N. Kolmogorov, vol. I (Kluwer, Dordrecht, 1991), pp. 147–150
252. A.N. Kolmogorov, *Grundbegriffe der Wahrscheinlichkeitsrechnung* (Springer, Berlin, 1933); English translation: *Foundations of the Theory of Probability* (Chelsea, New York, 1956)
253. A. Kolmogorov, Zur Normierbarkeit eines allgemeinen topologischen Raumes. Stud. Math. **5**, 29–33 (1934)
254. A.N. Kolmogorov, S.V. Fomin, *Elements of the Theory of Functions and Functional Analysis* (Dover, New York, 1999)
255. V. Komornik, An equiconvergence theorem for the Schrödinger operator. Acta Math. Hungar. **44**, 101–114 (1984)
256. V. Komornik, Sur l'équiconvergence des séries orthogonales et biorthogonales correspondant aux fonctions propres des opérateurs différentiels linéaires. C. R. Acad. Sci. Paris Sér. I Math. **299**, 217–219 (1984)
257. V. Komornik, On the equiconvergence of eigenfunction expansions associated with ordinary linear differential operators. Acta Math. Hungar. **47**, 261–280 (1986)
258. V. Komornik, A simple proof of Farkas' lemma. Am. Math. Mon. **105**, 949–950 (1998)
259. V. Komornik, D. Kong, W. Li, *Hausdorff dimension of univoque sets and Devil's staircase*, arxiv: 1503.00475v1 [math. NT]
260. V. Komornik, P. Loreti, On the topological structure of univoque sets. J. Number Theory **122**, 157–183 (2007)
261. V. Komornik, M. Yamamoto, On the determination of point sources. Inverse Prob. **18**, 319–329 (2002)
262. V. Komornik, M. Yamamoto, Estimation of point sources and applications to inverse problems. Inverse Prob. **21**, 2051–2070 (2005)

263. P.P. Korovkin, On the convergence of positive linear operators in the space of continuous functions (in Russian). Dokl. Akad. Nauk. SSSR **90**, 961–964 (1953)

264. P.P. Korovkin, *Linear Operators and Approximation Theory*. Russian Monographs and Texts on Advanced Mathematics and Physics, vol. III (Gordon and Breach Publishers, Inc/Hindustan Publishing Corp., New York/Delhi, 1960)

265. C.A. Kottman, Subsets of the unit ball that are separated by more than one. Stud. Math. **53**, 15–27 (1975)

266. G. Köthe, *Topological Vector Spaces I-II* (Springer, Berlin, 1969, 1979)

267. M.A. Krasnoselskii, Ya.B. Rutitskii, *Convex Functions and Orlicz Spaces* (P. Noordhoff Ltd., Groningen, 1961)

268. M.G. Krein, S.G. Krein, On an inner characteristic of the set of all continuous functions defined on a bicompact Hausdorff space. Dokl. Akad. Nauk. SSSR **27**, 429–430 (1940)

269. M. Krein, D. Milman, On extreme points of regularly convex sets. Stud. Math. **9**, 133–138 (1940)

270. P. Krée, *Intégration et théorie de la mesure. Une approche géométrique* (Ellipses, Paris, 1997)

271. H.W. Kuhn, A.W. Tucker, Nonlinear programming, in *Proceedings of 2nd Berkeley Symposium*. (University of California Press, Berkeley, 1951), pp. 481–492

272. K. Kuratowski, La propriété de Baire dans les espaces métriques. Fund. Math. **16**, 390–394 (1930)

273. K. Kuratowski, Quelques problèmes concernant les espaces métriques non-séparables. Fund. Math. **25**, 534–545 (1935)

274. J. Kurzweil, Generalized ordinary differential equations and continuous dependence on a parameter (in Russian). Czechoslov. Math. J. **7**(82), 418–449 (1957)

275. J. Kürschák, *Analízis és analitikus geometria [Analysis and Analytic Geometry]* (Budapest, 1920).

276. H.E. Lacey, The Hamel dimension of any infinite-dimensional separable Banach space is *c*. Am. Math. Mon. **80**, 298 (1973)

277. M. Laczkovich, Equidecomposability and discrepancy; a solution of Tarski's circle-squaring problem. J. Reine Angew. Math. **404**, 77–117 (1990)

278. M. Laczkovich, *Conjecture and Proof* (Typotex, Budapest, 1998)

279. J.L. Lagrange, Solution de différents problèmes de calcul intégral. Miscellanea Taurinensia **III**, (1762–1765); [281] I, 471–668

280. J.L. Lagrange, Sur l'attraction des sphéroïdes elliptiques. Nouv. Mém. Acad. Royale Berlin (1773); [281] III, 619–649

281. J.L. Lagrange, *Oeuvres I-XIV* (Gauthier-Villars, Paris, 1867–1882)

282. E. Landau, Über einen Konvergenzsatz. Nachr. Akad. Wiss. Göttingen. Math.-Phys. Kl. IIa, 25–27 (1907); [284] III, 273–275

283. E. Landau, Über die Approximation einer stetigen Funktion durch eine ganze rationale Funktion. Rend. Circ. Mat. Palermo 25, 337–345 (1908); [284] III, 402–410

284. E. Landau, *Collected Works I-VIII* (Thales, Essen, 1986)

285. R. Larsen, *Functional Analysis. An Introduction* (Marcel Dekker, New York, 1973)

286. H. Lebesgue, Sur l'approximation des fonctions. Bull. Sci. Math. **22**, 10 p. (1898); [298] III, 11–20

287. H. Lebesgue, Sur une généralisation de l'intégrale définie. C. R. Acad. Sci. Paris **132**, 1025–1027 (1901); [298] I, 197–199

288. H. Lebesgue, Intégrale, longueur, aire. Ann. Mat. Pura Appl. (3) **7**, 231–359 (1902); [298] I, 201–331

289. H. Lebesgue, Sur les séries trigonométriques. Ann. Éc. Norm. (3) **20**, 453–485 (1903); [298] III, 27–59

290. H. Lebesgue, *Leçons sur l'intégration et la recherche des fonctions primitives* (Gauthier-Villars, Paris, 1904); [298] II, 11–154

291. H. Lebesgue, Sur la divergence et la convergence non-uniforme des séries de Fourier. C. R. Acad. Sci. Paris **141**, 875–877 (1905)

292. H. Lebesgue, *Leçons sur les séries trigonométriques* (Gauthier-Villars, Paris, 1906)

293. H. Lebesgue, Sur les fonctions dérivées. Atti Accad. Lincei Rend. **15**, 3–8 (1906); [298] II, 159–164

294. H. Lebesgue, Sur la méthode de M. Goursat pour la résolution de l'équation de Fredholm. Bull. Soc. Math. France **36**, 3–19 (1908); [298] III, 239–254

295. H. Lebesgue, Sur l'intégration des fonctions discontinues. Ann. Éc. Norm. (3) **27**, 361–450 (1910); [298] II, 185–274

296. H. Lebesgue, Notice sur les travaux scientifiques de M. Henri Lebesgue (Impr. E. Privat, Toulouse, 1922); [298] I, 97–175

297. H. Lebesgue, *Leçons sur l'intégration et la recherche des fonctions primitives*, 2nd edn. (Paris, Gauthier-Villars, 1928)

298. H. Lebesgue, *Oeuvres scientifiques I-V* (Université de Genève, Genève, 1972–1973)

299. J. Le Roux, Sur les intégrales des équations linéaires aux dérivées partielles du 2e ordre à 2 variables indépendantes. Ann. Éc. Norm. (3) **12**, 227–316 (1895)

300. B. Levi, Sul principio di Dirichlet. Rend. Circ. Mat. Palermo **22**, 293–360 (1906)

301. B. Levi, Sopra l'integrazione delle serie. Rend. Instituto Lombardo Sci. Lett. (2) **39**, 775–780 (1906)

302. J.W. Lewin, A truly elementary approach to the bounded convergence theorem. Am. Math. Mon. **93**(5), 395–397 (1986)

303. J. Lindenstrauss, L. Tzafriri, *Classical Banach Spaces II : Function Spaces* (Springer, Berlin, 1979)

304. J.-L. Lions, *Quelques méthodes de résolution des problèmes aux limites non linéaires* (Dunod–Gauthier-Villars, Paris, 1969)

305. J.-L. Lions, E. Magenes, *Problèmes aux limites non homogènes et applications I-III* (Dunod, Paris, 1968–1970)

306. J. Liouville, Troisième mémoire sur le développement des fonctions ou parties de fonctions en séries dont les divers termes sont assoujettis à satisfaire à une même équation différentielle du second ordre, contenant un paramètre variable. J. Math. Pures Appl. **2**, 418–437 (1837)

307. J.S. Lipiński, Une simple démonstration du théorème sur la dérivée d'une fonction de sauts. Colloq. Math. **8**, 251–255 (1961)

308. R. Lipschitz, De explicatione per series trigonometricas instituenda functionum unius variabilis arbitrariarum, etc. J. Reine Angew. Math. **63**(2), 296–308 (1864)

309. L.A. Ljusternik, W.I. Sobolev, *Elemente der Funtionalanalysis* (Akademie-Verlag, Berlin, 1979)

310. S. Lozinski, On the convergence and summability of Fourier series and interpolation processes. Mat. Sb. N. S. **14**(56), 175–268 (1944)

311. S. Lozinski, On a class of linear operators (in Russian). Dokl. Akad. Nauk SSSR **61**, 193–196 (1948)

312. H. Löwig, Komplexe euklidische Räume von beliebiger endlicher oder unendlicher Dimensionszahl. Acta Sci. Math. Szeged **7**, 1–33 (1934)

313. N. Lusin, Sur la convergence des séries trigonométriques de Fourier. C. R. Acad. Sci. Paris **156**, 1655–1658 (1913)

314. J. Marcinkiewicz, Quelques remarques sur l'interpolation. Acta Sci. Math. (Szeged) **8**, 127–130 (1937)

315. A. Markov, On mean values and exterior densities. Mat. Sb. N. S. **4**(46), 165–190 (1938)

316. R.D. Mauldin (ed.), *The Scottish Book: Mathematics from the Scottish Café* (Birkhäuser, Boston, 1981)

317. S. Mazur, Über konvexe Mengen in linearen normierten Räumen. Stud. Math. **4**, 70–84 (1933)

318. S. Mazur, On the generalized limit of bounded sequences. Colloq. Math. **2**, 173–175 (1951)

319. J. McCarthy, An everywhere continuous nowhere differentiable function. Am. Math. Mon. **60**, 709 (1953)

320. E.J. McShane, Linear functionals on certain Banach spaces. Proc. Am. Math. Soc. **1**, 402–408 (1950)

321. R.E. Megginson, *An Introduction to Banach Space Theory* (Springer, New York, 1998)

322. D.P. Milman, On some criteria for the regularity of spaces of the type (B). C. R. (Doklady) Acad. Sci. URSS (N. S.) **20**, 243–246 (1938)

323. H. Minkowski, *Geometrie der Zahlen I* (Teubner, Leipzig, 1896)

324. H. Minkowski, *Geometrie der Zahlen* (Teubner, Leipzig, 1910)

325. H. Minkowski, *Theorie der konvexen Körper, insbesondere Begründung ihres Oberflächenbegriffs*, [326] II, 131–229

326. H. Minkowski, *Gesammelte Abhandlungen I-II* (Teubner, Leipzig, 1911)

327. A.F. Monna, *Functional Analysis in Historical Perspective* (Oosthoek, Utrecht, 1973)

328. F.J. Murray, On complementary manifolds and projections in spaces L_p and l_p. Trans. Am. Math. Soc. **41**, 138–152 (1937)

329. Ch.H. Müntz, *Über den Approximationssatz von Weierstrass*. Mathematische Abhandlungen II.A. Schwarz gewidmet, (Springer, Berlin, 1914), pp. 303–312

330. L. Nachbin, A theorem of Hahn-Banach type for linear transformations. Trans. Am. Math. Soc. **68**, 28–46 (1950)

331. L. Narici, E. Beckenstein, *Topological Vector Spaces* (Marcel Dekker, New York, 1985)

332. I.P. Natanson, *Theory of functions of a real variable I-II* (Frederick Ungar Publishing, New York, 1955, 1961)

333. I.P. Natanson, *Constructive Function Theory I-III* (Ungar, New York, 1964)

334. J. von Neumann, Mathematische Begründung der Quantenmechanik. Nachr. Gesell. Wiss. Göttingen. Math.-Phys. Kl., 1–57 (1927); [340] I, 151–207

335. J. von Neumann, Zur allgemeinen Theorie des Masses. Fund. Math. **13**, 73–116 (1929); [340] I, 599–643

336. J. von Neumann, Zur Algebra der Funktionaloperationen und Theorie der Normalen Operatoren. Math. Ann. **102**, 370–427 (1929–1930); [340] II, 86–143

337. J. von Neumann, *Mathematische Grundlagen der Quantenmechanik* (Springer, Berlin, 1932)

338. J. von Neumann, On complete topological spaces. Trans. Am. Math. Soc. **37**, 1–20 (1935); [340] II, 508–527

339. J. von Neumann, On rings of operators III. Ann. Math. **41**, 94–161 (1940); [340] III, 161–228

340. J. von Neumann, *Collected works I-VI* (Pergamon Press, Oxford, 1972–1979).

341. M.A. Neumark, *Normierte Algebren* (VEB Deutscher Verlag der Wissenschaften, Berlin, 1959)

342. O. Nikodým, Sur une généralisation des intégrales de M. Radon. Fund. Math. **15**, 131–179 (1930)

343. O. Nikodým, Sur le principe du minimum dans le problème de Dirichlet. Ann. Soc. Polon. Math. **10**, 120–121 (1931)

344. O. Nikodým, Sur le principe du minimum. Mathematica (Cluj) **9**, 110–128 (1935)

345. W.P. Novinger, Mean convergence in L^p spaces. Proc. Am. Math. Soc. **34**(2), 627–628 (1972)

346. V.F. Nikolaev, On the question of approximation of continuous functions by means of polynomials (in Russian). Dokl. Akad. Nauk SSSR **61**, 201–204 (1948)

347. W. Orlicz, Über eine gewisse Klasse von Räumen vom Typus *B*. Bull. Int. Acad. Polon. Sci. A (8/9), 207–220 (1932)

348. W. Orlicz, Über Räume (L^M), Bull. Int. Acad. Polon. Sci. A, 93–107 (1936)

349. W. Orlicz, *Linear Functional Analysis*, Series in Real Analysis, vol. 4 (World Scientific Publishing, River Edge, NJ, 1992)

350. W. Osgood, Non uniform convergence and the integration of series term by term. Am. J. Math. **19**, 155–190 (1897)

351. J.C. Oxtoby, *Measure and Category* (Springer, New York, 1971)

352. M.-A. Parseval, Mémoire sur les séries et sur l'intégration complète d'une équation aux différences partielles linéaires du second ordre, à coefficients constants. Mém. prés. par divers savants, Acad. des Sciences, Paris, (1) **1**, 638–648 (1806)

353. G. Peano, *Applicazioni geometriche del calcolo infinitesimale* (Bocca, Torino, 1887)

354. G. Peano, Sur une courbe qui remplit toute une aire. Math. Ann. **36**, 157–160 (1890)

355. R.R. Phelps, Uniqueness of Hahn-Banach extensions and unique best approximation. Trans. Am. Math. Soc. **95**, 238–255 (1960)

356. O. Perron, Über den Integralbegriff. Sitzber. Heidelberg Akad. Wiss., Math. Naturw. Klasse Abt. A **16**, 1–16 (1914)
357. B.J. Pettis, A note on regular Banach spaces. Bull. Am. Math. Soc. **44**, 420–428 (1938)
358. B.J. Pettis, A proof that every uniformly convex space is reflexive. Duke Math. J. **5**, 249–253 (1939)
359. R.R. Phelps, *Lectures on Choquet's Theorem* (Van Nostrand, New York, 1966)
360. J.-P. Pier, *Intégration et mesure 1900–1950*; [361], 517–564
361. J.-P. Pier (ed.), *Development of Mathematics 1900–1950* (Birkhäuser, Basel, 1994)
362. J.-P. Pier (ed.), *Development of Mathematics 1950–2000* (Birkhäuser, Basel, 2000)
363. H. Poincaré, *Science and Hypothesis* (The Walter Scott Publishing, New York, 1905)
364. L.S. Pontryagin, *Topological Groups*, 3rd edn., Selected Works, vol. 2 (Gordon & Breach Science, New York, 1986)
365. A. Pringsheim, *Grundlagen der allgemeinen Funktionenlehre*, Encyklopädie der mathematischen Wissenschaften, II-1 (Teubner, Leipzig, 1899); French translation and adaptation: A. Pringsheim, J. Molk, *Principes fondamentaux de la théorie des fonctions*, Encyclopédie des sciences mathématiques, II-1, (Gauthier-Villars/Teubner, Paris/Leipzig, 1909)
366. J. Radon, Theorie und Anwendungen der absolut additiven Mengenfunktionen. Sitsber. Akad. Wiss. Wien **122**, Abt. II a, 1295–1438 (1913)
367. M. Reed, B. Simon, *Methods of Modern Mathematical Physics I-IV* (Academic Press, New York, 1972–1979)
368. F. Rellich, Spektraltheorie in nichtseparabeln Räumen. Math. Ann. **110**, 342–356 (1935)
369. I. Richards, On the Fourier inversion theorem for R^1. Proc. Amer. Math. Soc. **19**(1), 145 (1968)
370. B. Riemann, *Grundlagen für eine allgemeine Theorie der Functionen einer veränderlichen complexen Grösse*, Inaugural dissertation, Göttingen, 1851; [372], 3–45; French translation: *Principes fondamentaux pour une théorie générale des fonctions d'une grandeur variable complexe*, Dissertation inaugurale de Riemann, Göttingen, 1851; [372], 1–56
371. B. Riemann, *Ueber die Darstellberkeit einer Function durch eine trigonometrische Reihe*, Habilitationsschrift, 1854, Abhandlungen der Königlichen Gesellschaft der Wissenschaften zu Göttingen 13 (1867); [372], 213–251. French translation: *Sur la possibilité de représenter une fonction par une série trigonométrique*, Bull. des Sciences Math. et Astron. (1) 5 (1873); [372], 225–272.
372. B. Riemann, *Werke* (Teubner, Leipzig, 1876); French translation: *Oeuvres mathématiques de Riemann* (Gauthier-Villars, Paris, 1898). English translation: *Collected Works of Bernhard Riemann* (Dover Publications, New York, 1953)
373. F. Riesz, Sur les systèmes orthogonaux de fonctions. C. R. Acad. Sci. Paris **144**, 615–619 (1907); [392] I, 378–381
374. F. Riesz, Sur les systèmes orthogonaux de fonctions et l'équation de Fredholm. C. R. Acad. Sci. Paris **144**, 734–736 (1907); [392] I, 382–385
375. F. Riesz, Sur une espèce de géométrie analytique des systèmes de fonctions sommables. C. R. Acad. Sci. Paris **144**, 1409–1411 (1907); [392] I, 386–388
376. F. Riesz, Über orthogonale Funktionensysteme. Göttinger Nachr. 116–122 (1907); [392] I, 389–395
377. F. Riesz, Sur les suites de fonctions mesurables. C. R. Acad. Sci. Paris **148**, 1303–1305 (1909); [392] I, 396–397, 405–406
378. F. Riesz, Sur les opérations fonctionnelles linéaires à une. C. R. Acad. Sci. Paris **149**, 974–977 (1909); [392] I, 400–402
379. F. Riesz, Sur certains systèmes d'équations fonctionnelles et l'approximation des fonctions continues. C. R. Acad. Sci. Paris **150**, 674–677 (1910); [392] I, 403–404, 398–399
380. F. Riesz, Untersuchungen über Systeme integrierbar Funktionen. Math. Ann. **69**, 449–497 (1910); [392] I, 441–489
381. F. Riesz, Integrálható függvények sorozatai [Sequences of integrable functions]. Matematikai és Physikai Lapok **19**, 165–182, 228–243 (1910); [392] I, 407–440
382. F. Riesz, *Les systèmes d'équations linéaires à une infinité d'inconnues* (Gauthier-Villars, Paris, 1913); [392] II, 829–1016

383. F. Riesz, Lineáris függvényegyenletekről [On linear functional equations]. Math. Term. Tud. Ért. **35**, 544–579 (1917); [392] II, 1017–1052; German translation: Über lineare Funktionalgleichungen. Acta Math. **41**, 71–98 (1918); [392] II, 1053–1080

384. F. Riesz, Su alcune disuguglianze. Boll. Unione Mat. Ital. **7**, 77–79 (1928); [392] I, 519–521

385. F. Riesz, Sur la convergence en moyenne I-II. Acta Sci. Math. (Szeged) **4**, 58–64, 182–185 (1928–1929); [392] I, 512–518, 522–525

386. F. Riesz, A monoton függvények differenciálhatóságáról [On the differentiability of monotone functions]. Mat. Fiz. Lapok **38**, 125–131 (1931); [392] I, 243–249

387. F. Riesz, Sur l'existence de la dérivée des fonctions monotones et sur quelques problèmes qui s'y rattachent. Acta Sci. Math. (Szeged) **5**, 208–221 (1930–32); [392] I, 250–263

388. M. Riesz, Sur les ensembles compacts de fonctions sommables. Acta Sci. Math. Szeged **6**, 136–142 (1933)

389. F. Riesz, Zur Theorie der Hilbertschen Raumes. Acta Sci. Math., 34–38 (1934–1935); [392] II, 1150–1154

390. F. Riesz, L'évolution de la notion d'intégrale depuis Lebesgue. Ann. Inst. Fourier **1**, 29–42 (1949); [392] I, 327–340

391. F. Riesz, Nullahalmazok és szerepük az analízisben. Az I. Magyar Mat. Kongr. Közl., 204–214 (1952); [392] I, 353–362; French translation: *Les ensembles de mesure nulle et leur rôle dans l'analyse*, Az I. Magyar Mat. Kongr. Közl., English translation: *Proceedings of the First Hungarian Mathematical Congress*, 214–224 (1952); [392] I, 363–372

392. F. Riesz, *Oeuvres complètes I-II*, ed. by Á. Császár (Akadémiai Kiadó, Budapest, 1960)

393. F. Riesz, B. Sz.-Nagy, Über Kontraktionen des Hilbertschen Raumes [German]. Acta Univ. Szeged. Sect. Sci. Math. **10**, 202–205 (1943)

394. F. Riesz, B. Sz.-Nagy, *Leçons d'analyse fonctionnelle* (Akadémiai Kiadó, Budapest, 1952); English translation: *Functional Analysis,* (Dover, New York, 1990)

395. J. Roberts, *Pathological compact convex sets in the spaces L_p, $0 < p < 1$*, The Altgeld Book 1975/76, University of Illinois

396. J. Roberts, A compact convex set without extreme points. Stud. Math. **60**, 255–266 (1977)

397. A.P. Robertson, W. Robertson, *Topological Vector Spaces* (Cambridge University Press, 1973)

398. R.T. Rockafellar, *Conves Analysis* (Princeton University Press, New Jersey, 1970)

399. L.J. Rogers, An extension of a certain theorem in inequalities. Messenger Math. **17**, 145–150 (1888)

400. S. Rolewicz, *Metric Linear Spaces* (Państwowe Wydawnictwo Naukowe, Varsovie, 1972)

401. L.A. Rubel, Differentiability of monotonic functions. Colloq. Math. **10**, 277–279 (1963)

402. W. Rudin, *Fourier Analysis on Groups* (Wiley, New York, 1962)

403. W. Rudin, *Principles of Mathematical Analysis*, 3rd edn. (McGraw-Hill, New York, 1976)

404. W. Rudin, Well-distributed measurable sets. Am. Math. Mon. **90**, 41–42 (1983)

405. W. Rudin, *Real and Complex Analysis* (McGraw-Hill, New York, 1987)

406. W. Rudin, *Functional Analysis* (McGraw-Hill, New York, 1991)

407. S. Russ, Bolzano's analytic programme. Math. Intell. **14**(3), 45–53 (1992)

408. S. Saks, On some functionals. Trans. Am. Math. Soc. **35**(2), 549–556 and **35**(4), 965–970 (1933)

409. S. Saks, *Theory of The Integral* (Hafner, New York, 1937)

410. S. Saks, Integration in abstract metric spaces. Duke Math. J. **4**, 408–411 (1938)

411. H. Schaefer, *Topological vector spaces,* Second edition with M.P. Wolff, 1st edn. (Springer, Berlin, 1966), 2nd edn. (1999)

412. J. Schauder, Zur Theorie stetiger Abbildungen in Funktionenräumen. Math. Z. **26**, 47–65 (1927)

413. J. Schauder, Über die Umkehrung linearer, stetiger Funktionaloperationen. Stud. Math. **2**, 1–6 (1930)

414. J. Schauder, Über lineare, vollstetige Funktionaloperationen. Stud. Math. **2**, 183–196 (1930)

415. E. Schmidt, Entwicklung willkürlicher Funktionen nach Systemen vorgeschriebener. Math. Ann. **63**, 433–476 (1907)

416. E. Schmidt, Über die Auflösung linearer Gleichungen mit unendlich vielen Unbekannten. Rend. Circ. Mat. Palermo **25**, 53–77 (1908)

417. I.J. Schoenberg, On the Peano curve of Lebesgue. Bull. Am. Math. Soc. **44**(8), 519 (1938)

418. I. Schur, Über lineare Transformationen in der Theorie der unendlichen Reihen. J. Reine Angew. Math. **151**, 79–111 (1920)

419. J.T. Schwartz, A note on the space L_p^*. Proc. Am. Math. Soc. **2**, 270–275 (1951)

420. L. Schwartz, *Théorie des distributions* (Hermann, Paris, 1966)

421. Z. Semadeni, *Banach Spaces of Continuous Functions I* (Pánstwowe Wydawnictvo Naukowe, Warszawa, 1971)

422. A. Shidfar, K. Sabetfakhiri, On the continuity of Van der Waerden's function in the Hölder sense. Am. Math. Mon. **93**, 375–376 (1986)

423. H.J.S. Smith, On the integration of discontinuous functions. Proc. Lond. Math. Soc. **6**, 140–153 (1875)

424. V.L. Šmulian, Linear topological spaces and their connection with the Banach spaces. Doklady Akad. Nauk SSSR (N. S.) **22**, 471–473 (1939)

425. V.L. Šmulian, Über lineare topologische Räume. Mat. Sb. N. S. (7) **49**, 425–448 (1940)

426. S. Sobolev, Cauchy problem in functional spaces. Dokl. Akad. Nauk SSSR **3**, 291–294 (1935) (in Russian)

427. S. Sobolev, Méthode nouvelle à résoudre le problème de Cauchy pour les équations linéaires hyperboliques normales. Mat. Sb. **1**(43), 39–72 (1936)

428. S.L. Sobolev, *Some Applications of Functional Analysis in Mathematical Physics* (American Mathematical Society, Providence RI, 1991)

429. R. Solovay, A model of set theory where every set of reals is Lebesgue measurable. Ann. Math. **92**, 1–56 (1970)

430. G.A. Soukhomlinov, Über Fortsetzung von linearen Funktionalen in linearen komplexen Räumen und linearen Quaternionräumen (in Russian, with a German abstract). Mat. Sb. N. S. (3) **4**, 355–358 (1938)

431. L.A. Steen, Highlights in the history of spectral theory, Am. Math. Mon. **80**, 359–381 (1973)

432. H. Steinhaus, Additive und stetige Funktionaloperationen. Math. Z. **5**, 186–221 (1919)

433. V.A. Steklov, Sur la théorie de fermeture des systèmes de fonctions orthogonales dépendant d'un nombre quelconque de variables. Petersb. Denkschr. (8) **30**, 4, 1–86 (1911)

434. V.A. Steklov, Théorème de fermeture pour les polynômes de Tchebychev–Laguerre. Izv. Ross. Akad. Nauk Ser. Mat. (6) **10**, 633–642 (1916)

435. T.J. Stieltjes, Recherches sur les fractions continues. Ann. Toulouse **8**, 1–122 (1894), **9**, 1–47 (1895); [436] II, 402–566

436. T.J. Stieltjes, *Oeuvres complètes I-II* (Springer, Berlin, 1993)

437. O. Stolz, Ueber einen zu einer unendlichen Punktmenge gehörigen Grenzwerth. Math. Ann. **23**, 152–156 (1884)

438. O. Stolz, Die gleichmässige Convergenz von Functionen mehrerer Veränderlichen zu den dadurch sich ergebenden Grenzwerthen, dass einige derselben constanten Werthen sich nähern. Math. Ann. **26**, 83–96 (1886)

439. M.H. Stone, *Linear Transformations in Hilbert Space* (Amer. Math. Soc., New York, 1932)

440. M.H. Stone, Application of the theory of Boolean rings to general topology, Trans. Am. Math. Soc. **41**, 325–481 (1937)

441. M.H. Stone, The generalized Weierstrass approximation theorem. Math. Mag. **11**, 167–184, 237–254 (1947/48)

442. R.S. Strichartz, How to make wavelets. Am. Math. Mon. **100**, 539–556 (1993)

443. P.G. Szabó, *A matematikus Riesz testvérek. Válogatás Riesz Frigyes és Riesz Marcel levelezéséből [The Mathematician Riesz Brothers. Selected letters from the correspondence between Frederic and Marcel Riesz]* (Magyar Tudománytörténeti Intézet, Budapest, 2010)

444. P.G. Szabó, *Kiváló tisztelettel: Fejér Lipót és a Riesz testvérek levelezése magyar matematikusokkal [Respectfully: Correspondence of Leopold Fejér and the Riesz Brothers with Hungarian Mathematicians]* (Magyar Tudománytörténeti Intézet, Budapest, 2011)

445. O. Szász, Über die Approximation stetiger Funktionen durch lineare Aggregate von Potenzen. Math. Ann. **77**, 482–496 (1915–1916)

446. G. Szegő, *Orthogonal Polynomials* (American Mathematical Society, Providence, 1975)

447. B. Sz.-Nagy, *Spektraldarstellung linearer Transformationen des Hilbertschen Raumes* (Springer, Berlin, 1942)

448. B. Sz.-Nagy, *Introduction to Real Functions and Orthogonal Expansions* (Oxford University Press, New York, 1965)

449. T. Takagi, A simple example of a continuous function without derivative. Proc. Phys. Math. Japan **1**, 176–177 (1903)

450. A.E. Taylor, The extension of linear functionals. Duke Math. J. **5**, 538–547 (1939)

451. S.A. Telyakovsky, *Collection of Problems on the Theory of Real Functions* (in Russian) (Nauka, Moszkva, 1980)

452. K.J. Thomae, Ueber bestimmte integrale. Z. Math. Phys. **23**, 67–68 (1878)

453. H. Tietze, Über Funktionen, die auf einer abgeschlossenen Menge stetig sind. J. Reine Angew. Math. **145**, 9–14 (1910)

454. A. Tychonoff, Ein Fixpunktsatz. Math. Ann. **111**, 767–776 (1935)

455. A. Toepler, Bemerkenswerte Eigenschaften der periodischen Reihen. Wiener Akad. Anz. **13**, 205–209 (1876)

456. O. Toeplitz, Das algebrischen Analogen zu einem Satze von Fejér. Math. Z. **2**, 187–197 (1918)

457. L. Tonelli, Sull'integrazione per parti. Atti Accad. Naz. Lincei (5), **18**, 246–253 (1909)

458. V. Trénoguine, B. Pissarevski, T. Soboléva, *Problèmes et exercices d'analyse fonctionnelle* (Mir, Moscou, 1987)

459. N.-K. Tsing, Infinite-dimensional Banach spaces must have uncountable basis—an elementary proof. Am. Math. Mon. **91**(8), 505–506 (1984)

460. J.W. Tukey, Some notes on the separation of convex sets. Port. Math. **3**, 95–102 (1942)

461. P. Urysohn, Über die Mächtigkeit der zusammenhängenden Mengen. Math. Ann. **94**, 262–295 (1925)

462. S. Vajda, *Theory of Games and Linear Programming* (Methuen, London, 1956)

463. Ch.-J. de la Vallée-Poussin, Sur l'approximation des fonctions d'une variable réelle et de leurs dérivées par des polynômes et des suites limitées de Fourier. Bull. Acad. R.. Cl. Sci. 3 (1908), 193–254

464. Ch.-J. de la Vallée-Poussin, Réduction des intégrales doubles de Lebesgue: application à la définition des fonctions analytiques. Bull. Acad. Sci. Brux, 768–798 (1910)

465. Ch.-J. de la Vallée-Poussin, Sur l'intégrale de Lebesgue. Trans. Am. Math. Soc. **16**, 435–501 (1915)

466. G. Vitali, *Sul problema della misura dei gruppi di punti di una retta* (Gamberini e Parmeggiani, Bologna, 1905)

467. N.Ya. Vilenkin, *Stories About Sets* (Academic Press, New York, 1968)

468. N.Ya. Vilenkin, *In Search of Infinity* (Birkhäuser, Boston, 1995)

469. C. Visser, Note on lincar operators. Proc. Acad. Amst. **40**, 270–272 (1937)

470. G. Vitali, Sulle funzioni integrali. Atti Accad. Sci. Torino **40**, 1021–1034 (1905)

471. G. Vitali, Sull'integrazione per serie. Rend. Circ. Mat. Palermo **23**, 137–155 (1907)

472. V. Volterra, Sulla inversione degli integrali definiti I-IV. Atti Accad. Torino **31**, 311–323, 400–408, 557–567, 693–708 (1896); [476] II, 216–254

473. V. Volterra, Sulla inversione degli integrali definiti. Atti R. Accad. Rend. Lincei. Cl. Sci. Fis. Mat. Nat. (5) **5**, 177–185 (1896); [476] II, 255–262

474. V. Volterra, Sulla inversione degli integrali multipli. Atti R. Accad. Rend. Lincei. Cl. Sci. Fis. Mat. Nat. (5) **5**, 289–300 (1896); [476] II, 263–275

475. V. Volterra, Sopra alcuni questioni di inversione di integrali definiti. Annali di Mat. (2) **25**, 139–178 (1897); [476] II, 279–313

476. V. Volterra, *Opere matematiche. Memorie e Note I-V* (Accademia Nazionale dei Lincei, Roma, 1954–1962)

477. B.L. van der Waerden, Ein einfaches Beispiel einer nichtdifferenzierbaren stetiges Funktion. Math. Z. **32**, 474–475 (1930)

478. S. Wagon, *The Banach-Tarski Paradox* (Cambridge University Press, Cambridge, 1985)

479. J.V. Wehausen, Transformations in linear topological spaces. Duke Math. J. **4**, 157–169 (1938)

480. K. Weierstrass, *Differential Rechnung*. Vorlesung an dem Königlichen Gewerbeinstitute, manuscript of 1861, Math. Bibl., Humboldt Universität, Berlin

481. K. Weierstrass, *Über continuirliche Functionen eines reellen Arguments, die für keinen Werth des letzteren einen bestimmten Differentialquotienten besitzen*. Gelesen in der Königlich. Akademie der Wissenschaften, 18. Juli 1872; [484] II, 71–74

482. K. Weierstrass, *Theorie der analytischen Funktionen*, Vorlesung an der Univ. Berlin, manuscript of 1874, Math. Bibl., Humboldt Universität, Berlin

483. K. Weierstrsass, Über die analytische Darstellbarkeit sogenannter willkürlicher Funktionen reeller Argumente, Erste Mitteilung. Sitzungsberichte Akad. Berlin, 633–639 (1885); [484] III, 1–37. French translation: *Sur la possibilité d'une représentation analytique des fonctions dites arbitraires d'une variable réelle*, J. Math. Pures Appl. 2 (1886), 105–138

484. K. Weierstrsass, *Mathematische Werke* vol. I-VI, (Mayer & Müller, Berlin, 1894–1915), vol. VII, (Georg Olms Verlagsbuchhandlung, Hildesheim, 1927)

485. A. Weil, *L'intégration dans les groupes topologiques et ses applications* (Hermann, Paris, 1940)

486. R. Whitley, An Elementary Proof of the Eberlein–Šmulian Theorem. Math. Ann. 172, 116–118 (1967)

487. N. Wiener, Limit in terms of continuous transformation. Bull. Soc. Math. France **50**, 119–134 (1922)

488. K. Yosida, *Functional Analysis* (Springer, Berlin, 1980)

489. W.H. Young, On classes of summable functions and their Fourier series. Proc. R. Soc. (A) **87**, 225–229 (1912)

490. W.H. Young, The progress of mathematical analysis in the 20th century. Proc. Lond. Math. Soc. (2) 24 (1926), 421–434

491. L. Zajícek, An elementary proof of the one-dimensional density theorem. Am. Math. Mon. **86**, 297–298 (1979)

492. M. Zorn, A remark on a method in transfinite algebra. Bull. Am. Math. Soc. **41**, 667–670 (1935)

493. A. Zygmund, *Trigonometric Series* (Cambridge University Press, London, 1959)

Subject Index

$\cup^*$, 212
$\overset{*}{\rightharpoonup}$, 136
$\rightharpoonup$, 30, 79

a.e., 156
absolutely continuous
    function , 198
    measure, 235
    signed measure, 239
accumulation point
    of a net, xv
    of a sequence, xvi
adjoint operator, 35, 99
affine hyperplane, 15, 16, 57
almost everywhere, 156
anti-discrete topology, xiii
automorphism, 103
axiom of choice, 62

$\mathcal{B}(K)$ space, 76
$\mathcal{B}(K, X)$ space, 76
Baire
    measure, 289
    set, 289
Baire's lemma, 32
balanced set, 58
ball, xvi, 120
Banach
    algebra, 43
    space, 55, 76
Bernstein polynomials, 282
Bessel
    equality, 25

inequality, 25
bidual space, 79
Bochner integral, 305
Borel set, 195
boundary, xiii
bounded
    function, xvi
    set, xvi, 122
broken line, xix

c_0 is not a dual space, 140
c_0 space, 70, 77
C_0 function class, 171
C_1 function class, 174
C_2 function class, 176
$C(K)$ space, 77
$C(K, X)$ space, 77
$C_b(K)$ space, 77
$C_b(K, X)$ space, 77
$C_c(I)$ space, 312
$C_{2\pi}$ space, 263
$C_b^k(I, Y)$ space, 77
$C_b^k(U, Y)$ space, 78
Cantor function, 199, 204
Cantor's
    diagonal method, 34, 90
    ternary set, 154, 209, 254, 303
Cauchy sequence, xvii
Cauchy–Schwarz inequality, xix, 4, 46
change of variable in integrals, 246
characteristic function, 173
Chebyshev's characterization of closest
        polynomials, 301
$C([0, 1])$ is not a dual space, 259

© Springer-Verlag London 2016
V. Komornik, *Lectures on Functional Analysis and the Lebesgue Integral*,
Universitext, DOI 10.1007/978-1-4471-6811-9

Name Index

Alabau-Boussouira, 367
Alaoglu, 139
Alexander the Great, 341
Archimedes, 149
Arzelà, 181, 268
Ascoli, 61, 154, 268, 301
Austin, 161

Baire, 32, 149, 178, 209, 289
Banach, 1, 32, 65, 67, 76, 79, 81, 96, 99, 139,
     140, 192, 276, 291
Bauer, 117
Benner, 117
Bernoulli, Daniel, 270
Bernstein, 282
Besenyei, 367
Bessel, 25
Bochner, 211, 305
Bohman, 281
Bohnenblust, 112
du Bois-Reymond, 154, 229, 270, 271
Bolzano, 29
Bolzano–Weierstrass, 6
Borel, 149, 156, 195, 212, 282
Botsko, 161
Bourbaki, 140
Brunn, 61
Buddenhagen, 369

Cantor, 34, 36, 90, 149, 151, 152, 199, 212
Carathéodory, 252
Carleson, 271, 315, 316
Cauchy, 4, 46, 149, 229, 275, 276

Chebyshev, 301
Chernoff, 301
Clairaut, 26
Clarkson, 265, 323, 329
Császár, 160
Czách, 251

Darboux, 204
Day, 350
Denjoy, 197, 204
Dieudonné, 1, 21
Dini, 161, 198, 204, 271, 292, 301
Dirac, 213
Dirichlet, 149, 169, 224, 271, 272, 303
Dunford, 332

Eberlein, 140
Egorov, 361
Eidelheit, 61
Ellis, 336
Erdős, 154, 265, 288
Euclid, 341
Euler, 26, 28, 224

Faber, 288
Farkas, 133
Fatou, 183
Fejér, 265, 271, 275, 276, 303
Fichtenholz, 67
Fischer, 149, 184, 306
Foguel, 65
Fourier, 26, 149, 270, 301, 320

© Springer-Verlag London 2016
V. Komornik, *Lectures on Functional Analysis and the Lebesgue Integral*,
Universitext, DOI 10.1007/978-1-4471-6811-9

Printed in the United States
By Bookmasters